Diversity and change in food wellbeing

Diversity and change in food wellbeing

Cases from Southeast Asia and Nepal

edited by:
A. Niehof
H.N. Gartaula
M. Quetulio-Navarra

Wageningen Academic
Publishers

ISBN: 978-90-8686-316-7
e-ISBN: 978-90-8686-864-3
EAN: 9789086863167
e-EAN: 9789086868643
DOI: 10.3920/978-90-8686-864-3

First published, 2018

Wageningen Academic Publishers
P.O. Box 220
6700 AE Wageningen
the Netherlands
www.WageningenAcademic.com
copyright@WageningenAcademic.com

This publication has been made possible by the financial and technical support of the Neys-Van Hoogstraten Foundation (NHF) in The Hague, the Netherlands

NHF

Neys - van Hoogstraten Foundation

Contents

Preface 13

1. Introduction 15
H.N. Gartaula, A. Niehof and M. Quetulio Navarra
1.1 Background and context 15
1.2 The theoretical framework 16
1.3 Definition and discussion of key concepts 17
1.4 Layout and chapters 20
References 23

2. Changing value of food and perception of food security in the context of male out-migration in Nepal 27
H.N. Gartaula and A. Niehof
Abstract 27
2.1 Introduction 27
2.2 Labour migration and food security in the literature 29
2.3 Research area and methodology 31
2.4 Results 32
2.5 Discussion and conclusion 37
Acknowledgement 39
References 39

3. Indigenous knowledge and household food security: the role of root and tuber crops among indigenous peoples in the Northern Philippines 43
B.T. Gayao, D.T. Meldoz and G.S. Backian
Abstract 43
3.1 Introduction 43
3.2 Conceptual framework 46
3.3 Research area and methodology 47
3.4 Results and discussion 52
3.5 Conclusions and recommendations 66
Acknowledgement 67
References 67

4. Gathering practices and actual use of wild edible mushrooms among ethnic groups in the Cordilleras, Philippines 71
D. Corazon M. Licyayo
Abstract 71
4.1 Background of the study 71
4.2 Conceptual framework of the study 73
4.3 Research area and methodology 74
4.4 Results and discussion 75
4.5 Conclusion 84
Acknowledgment 85
References 85

5. Household food security, food care, child nutrition, and wellbeing of ethnic Tengger families in East Java, Indonesia 87
D. Hastuti
Abstract 87
5.1 Introduction 87
5.2 Conceptual framework 89
5.3 Research methodology 91
5.4 Results and discussion 93
5.5 Conclusion and recommendations 110
References 112

6. The role of social capital and institutions in food security and wellbeing of children under five for resettled households in Central Java, Indonesia 115
M. Quetulio-Navarra, E. Frunt and A. Niehof
Abstract 115
6.1 Introduction 115
6.2 Food security and under-fives through the lens of social capital and institutions 117
6.3 Study design and methodology 119
6.4 Results 124
6.5 Conclusions and discussion 132
References 135

7. What explains stunting among children living in a rice surplus area in Central Java, Indonesia? 137
R.C. Purwestri, Z. Barati, N.N. Wirawan, I. Fahmi, J. Lauvai and V. Scherbaum
Abstract 137
7.1 Introduction 137
7.2 Methodology 140
7.3 Results 142
7.4 Discussion and conclusion 145
Acknowledgement 148
References 148

8. Gender, intra-household food allocation and social change in two Himalayan communities in Nepal 153
D.S. Madjdian
Abstract 153
8.1 Introduction 153
8.2 Conceptual framework 155
8.3 Research area and methodology 158
8.4 Results 161
8.5 Conclusion and discussion 168
Acknowledgements 172
References 172

9. Parents' characteristics, food habits and physical activity of overweight schoolchildren in Bogor City, Indonesia 177
K.R. Ekawidyani, I. Karimah, B. Setiawan and A. Khomsan
Abstract 177
9.1 Introduction 177
9.2 Conceptual framework 179
9.3 Research area and methodology 181
9.4 Results 182
9.5 Conclusion and discussion 190
Acknowledgements 191
References 192

10. Climate change and nutritional status of preschool children living in Infanta, Quezon, and Los Baños, Laguna, Philippines 195
M.T.M. Talavera, M.R.L. Pangan, J.O. Colladilla, J.N. Sarol, I.A. Agdeppa and F.M.E. Lorenzo
Abstract 195
10.1 Introduction 195
10.2 Conceptual framework 197
10.3 Research area and methodology 199
10.4 Results 201
10.5 Discussion and conclusion 218
Acknowledgements 222
References 222

11. Does backyard vegetable gardening enhance perceived household food security and dietary diversity? A case study in the Philippines 225
J.B. Dorado, R.V. Viajar, G.P. Azaña, G.S. Caraig and M.V. Capanzana
Abstract 225
11.1 Introduction 225
11.2 Conceptual framework of the study 226
11.3 Methodology 227
11.4 Results and discussion 229
11.5 Implementation of the BVGP and the role of the program implementers 237
11.6 Conclusion and recommendations 239
Acknowledgment 240
References 241

12. Developing homestead food production in Muslim Mindanao, Philippines: does mode of implementation matter? 243
M.F.D. Reario, P.J. Pacheco, E. Lafuente and A. Talukder
Abstract 243
12.1 Introduction 243
12.2 Conceptual framework 246
12.3 Methods 248
12.4 Results 249
12.5 Discussion and conclusion 264
Acknowledgement 266
References 266

13. How non-targeted households benefited from implementing a homestead food production program in Son La, Vietnam 267
K. Shaw, P.K. Ngoc, H. Lan, G. Mundy and Z. Talukder
Abstract 267
13.1 Introduction 267
13.2 Methodology 271
13.3 Results 271
13.4 Discussion and conclusions 275
Acknowledgement 277
References 277

14. Conclusion: diversity and dimensions of food wellbeing 279
A. Niehof, H. Gartaula and M. Quetulio-Navarra
The kaleidoscope of food wellbeing 279
What do the studies tell us? 279
Food wellbeing: dimensions and constraints 284
References 289

About the authors 293

Neys - van Hoogstraten Foundation 299

Preface

This book presents the results of a series of studies conducted in the last seven years in Indonesia, the Philippines, Vietnam and Nepal with financial and technical support of the Neys-Van Hoogstraten Foundation (NHF) in The Hague, the Netherlands. Since its inception, the foundation has supported over two hundred of academic and applied research projects on the social, economic and cultural aspects of food and nutrition. The founder, Dr Karel Neijs, bequeathed his estate to a foundation to support research about household food security and family nutrition in Indonesia and other (mainly) Southeast Asian countries. He named the foundation after his parents. Dr Karel Neijs died in 1992. More information about the life of Dr Neijs and his motivation to establish the foundation can be found in in the booklet by the late Dr Adel den Hartog, entitled *Dr Karel Neys 1920-1992 and the Neys-van Hoogstraten Foundation* (The Hague, Neys-Van Hoogstraten Foundation). The Neys-Van Hoogstraten Foundation became operational in 1994.

The present book is a sequel to the book that was published in 2010, entitled *Food, Diversity, Vulnerability and Social Change: Research Findings from Insular Southeast Asia* (A. Niehof, Wageningen Academic Publishers), which covered results of NHF-supported research prior to 2010. Both books are a tribute to the late Dr Karel Neijs. For their research, the authors of the present book all received financial and technical support from the Neys-Van Hoogstraten Foundation. Although the studies presented are quite diverse in terms of research questions, methods and study area, together they form a kaleidoscope of the various aspects that constitute the book's focal theme of people's food wellbeing. The studies are based on research conducted according to academic standards of rigour and accountability. In 1946, Dr Neijs was appointed Assistant to the Rector Magnificus of the Universitas Indonesia in Jakarta. We think that he would have been pleased with the relevance and quality of the studies presented in this book.

Anke Niehof
Hom Nath Gartaula
Melissa Quetulio-Navarra
Wageningen, January 2018

1. Introduction

H.N. Gartaula[1], A. Niehof[2] and M. Quetulio Navarra[3]*

*[1]International Program Coordinator, Manitoba Council for International Cooperation (MCIC), Winnipeg, Canada; [2]Professor Emeritus, Wageningen University, Wageningen School of Social Sciences (WASS), the Netherlands; [3]Presidential Commission for the Urban Poor, Quezon City, Metro Manila, the Philippines; *hom.gartaula@gmail.com*

1.1 Background and context

Universally, food security is considered a basic requirement for the functioning of the human body and a fundamental human right. However, over 12% of the people in the world go to bed hungry in every single night. Even though the incidence of malnutrition has significantly declined in the past 20 years, over 98% undernourished people live in the developing countries (FAO, 2017). The current Sustainable Development Goals (SDGs) set the stage to address global hunger, malnutrition, and poverty through two main goals (SDGs 1 and 2) and many other goals that provide complementary support to achieve food and nutrition security. However, the world in which the goals have to be met is changing in many ways. In many parts of the world, including in East Asia – the regional focus of the research presented in this book – increasing numbers of people are living in urban areas. Ruel *et al.* (2017: 24) note that worldwide 'one in three stunted children now lives in an urban area' and that 'rapid increases in overweight and obesity have been concentrated in urban areas.' In East and South Asia, the volume of international labour migration has substantially increased, which has mixed effects on the food environment of migrant families (Zezza *et al.*, 2011). At the same time, food crops and the livelihoods of small farmers are affected by climate change. In East Asia, rice has been gaining prominence as the preferred staple, at the expense of traditional staples (maize, root and tuber crops) that could be more resilient to climate change variables such as erratic rainfall. The increasing cultural preference for rice (cf. Niehof, 2010) is bolstered by pro-rice policies of the governments in the region (Chen *et al.*, 2017). These dynamics are all, in one way or another, reflected in the studies presented in the book.

Since the concept of food security entered academic and policy discussion in the early second half of the last century, there has been an ongoing discussion among researchers and policy makers about its conceptualisation and measurement. Food security is perhaps the most studied concept in the field of international development studies, as indicated by the emergence of over 200 definitions and 450 indicators of food security (Mechlem, 2004). There have been a number of shifts in the theoretical and methodological orientation in the study of the concept. Maxwell (2001: 14) reports three major shifts: (1) from the global and the national to the household and the individual; (2) from a 'food first' perspective to

DOI 10.3920/978-90-8686-864-3_1, © Wageningen Academic Publishers 2018

a livelihood perspective; and (3) from objective indicators to subjective perception. The above-mentioned paradigm shifts are well articulated in this book.

In the early 1970s, food security was defined as physical food availability, but more recent conceptualisations of food security include the cultural and relational dimensions of food provision and consumption. These conceptualisations also expand the notion of food as a means of human wellbeing (Gartaula *et al.*, 2017; Patel *et al.*, 2015). This book aims at exploring the dimensions of food wellbeing. The concept encompasses the complex realities of food and nutrition security on the one hand and food sovereignty on the other, which metaphorically can be seen as the two sides of the same coin. A food wellbeing approach allows for capturing the multifaceted nature of food and nutrition security by applying a holistic and interdisciplinary perspective, based on the natural as well as the social sciences. The natural sciences enable the quantitative analysis of food production and consumption, nutrition and climate change. The social sciences contribute insights on attitudes to and perceptions of food in the contemporary changing world, indigenous knowledge and traditional food crops, social capital and institutions, the moral household economy, gender inequalities, and social wellbeing. The authors of the chapters have a variety of disciplinary backgrounds and often work in a multidisciplinary environment.

1.2 The theoretical framework

The book uses food wellbeing as the overarching concept, because there can be no overall wellbeing without food wellbeing and the degree of food wellbeing determines overall wellbeing to a large extent. According to Gartaula *et al.* (2017), the concept of food wellbeing synthesizes the insights from food security, food sovereignty and social wellbeing approaches. Through such a synthesis, a more profound understanding may emerge of the complex nature of food security in the context of social and environmental changes and of people's responses to those changes.

Our conceptualisation of food wellbeing is pictured in the Flip-Coin Model of Food Wellbeing below (Figure 1.1). The model shows food wellbeing as having two key components that are visualised as two sides of the same coin: (1) food and nutrition security; and (2) food sovereignty.

Accomplishing food and nutrition security and food sovereignty requires the use of resources by actors (individuals, households and communities), who need to have access to and control over these resources to enable them to exercise their agency. Food and nutrition security and food sovereignty are three dimensional, having (1) an objective dimension, (2) a subjective dimension, and (3) a relational dimension. The first relates to issues of production and consumption, such as food availability, access, sufficiency and quality, food utilisation capacity, and climatic variations. The subjective dimension is about people's own views, perceptions and experiences regarding these issues. The relational dimension addresses the

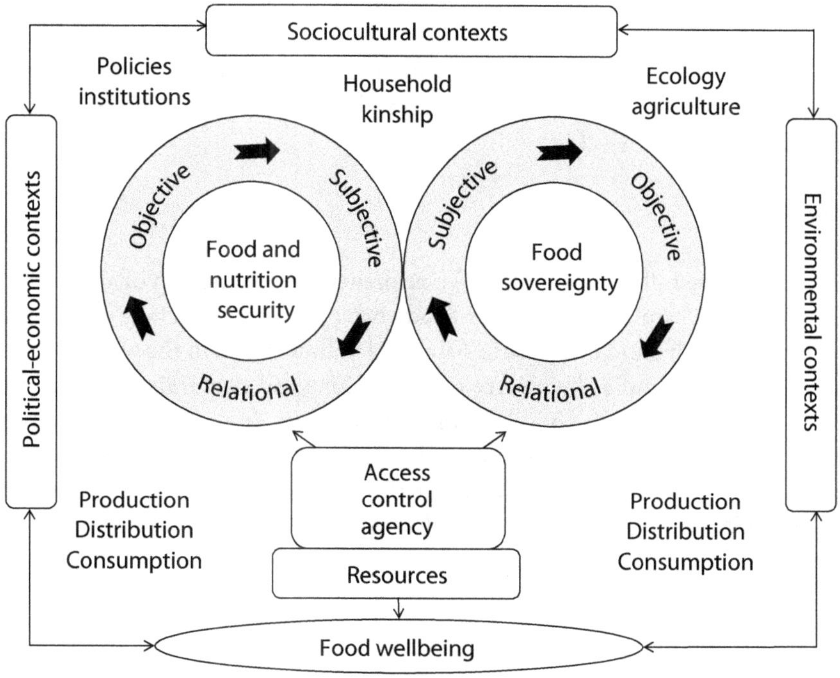

Figure 1.1. The flip-coin model of food wellbeing.

social relations and inequalities ensuing from gender, class, ethnicity, cultural frames and institutional structures that influence food and nutrition outcomes.

The food wellbeing approach draws attention to relational concerns connected to food access, with particular attention for the influence of gender and social differences on decision-making and choices regarding food. This is about people's interactions with others, their agency, and their relations with the state, social institutions, rules and norms, which can all dictate access to food (Gartaula *et al.*, 2017; McGregor, 2006; Noack and Pouw, 2015). The processes relating to (1) food and nutrition security and (2) food sovereignty are reflected in the food production, procurement, distribution, consumption and utilisation. These processes interface with the immediate contexts of household, kinship systems, policies and institutions, local ecologies and farming systems, which, in turn, are embedded in the socio-cultural, political-economic, and environmental meta contexts.

1.3 Definition and discussion of key concepts

In the chapters of this book, the authors use concepts in the areas of food and nutrition security and food sovereignty that are contained in the model. These are discussed below.

1.3.1 Wellbeing

Wellbeing can be defined as 'a state of being with others, where human needs are met, where one can act meaningfully to pursue one's goals, and where one enjoys a satisfactory quality of life' (McGregor, 2008: 1).

1.3.2 Food security

For food security, we follow FAO's (2013) comprehensive description of all people having access to sufficient, safe and nutritious food that meets their dietary needs and food preferences for an active and healthy life. Safe food is understood in the sense of containing no harmful chemicals and as being free of microbiological contamination (relevance of hygiene) that may obstruct the bodily uptake of nutrients. In actual cases, however, food preferences may be at odds with dietary needs (e.g. food taboos for pregnant women, fast foods). We see food security as anchored at the household level, since households are resource-managing, family-based groups that are geared towards meeting the primary and daily needs of their members (Niehof, 2011).

1.3.3 Nutrition security

Nutrition security is a sub-set of food security and relates to food consumption and utilisation. Food security at household level does not necessarily imply that all individual household members always get the nutrients they need, also depending on individual health status. The nutrition security of minors in particular is mediated by (usually the mother's) food care practices.

1.3.4 Gender and other inequalities

Gender and age are important variables along which inequality in households, communities and societies is aligned, also with regard to meeting food needs. People's dietary needs differ according to sex and age, the dietary needs of children under five years old and the importance of meeting these in view of the children's healthy and balanced growth being a case in point.

Gender is a crosscutting variable throughout the model. It plays a role in access to and control over resources, for example. Since resources are the means by which agency is exercised (Kabeer, 2005), women's agency may be curtailed by lack of control over resources and by gender-biased policies and power structures. Culturally underpinned discriminatory practices in the intra-household distribution of food may favour men over women and boys over girls. In a large cross-country survey, the data on South Asia revealed 'a clear pattern of male advantage reflected in anthropometric data on children' (DeRose *et al.*, 2000: 526).

Universally, women play a dominant role in household-based food and nutrition practices (Allen and Sachs, 2012). These practices are listed in Table 1.1 (cf. Niehof, 2016).

The division of labour in almost all of these practices is gendered, as is access to the resources needed to perform the practices. Above we already alluded to gender inequalities in household food distribution and consumption. The studies in the book highlight the various practices involved.

1.3.5 Change

Change is another crosscutting theme in our model, comprising both social and environmental (climate) changes. Social change both affects and is induced by the practices and relationships comprised by the model. Life style changes and urbanisation are examples of sociocultural change. Climate change influences food crop production and the availability of wild foods. Changes in the political-economic environment (policies, prices and institutions) can have an impact not only on food and nutrition security, but also on food sovereignty.

1.3.6 Food sovereignty

Yaro (2004) distinguishes three main approaches to food security: the food availability approach, the livelihood, entitlement approach, and the food sovereignty approach. Although the livelihood and entitlement approach is a dominant approach in this book, in our view the food sovereignty approach is inextricably linked to it. In the model it is the flipside of the food wellbeing coin. Based on a human rights perspective, the food sovereignty approach highlights the role and right of people to define their own food and agricultural system to achieve self-sufficiency and sustainable food security. Though

Table 1.1. Key practices in household production of food and nutrition security.

Food security practices	Nutrition security practices
Food procurement: • Food crop production • Buying, borrowing and exchanging food, receiving food gifts • Gathering of wild foods	Selective food procurement practices aimed at enhancing food safety
Food processing and storage	Food processing, storage and preparation
Changing the composition of the food-sharing unit	Intra-household food distribution
	Food consumption and hygiene, including supervision and food care of minors

originally defined as the right to food self-reliance of nations, the approach was extended to all actors involved in the food chain, including family farms and consumers (Agarwal, 2014). It is especially this extended coverage that plays a key role in this book. The application of the food sovereignty concept to consumers and their rights to and control over the foods they prefer, is relevant in view of the proliferation of urban life styles in Southeast Asia and the increasing availability of affordable, and sometimes aggressively marketed, fast foods.

The food sovereignty approach is fraught with a number of complexities and contradictions, such as the lack of conviction and legitimacy of policy-making on food security (Godek, 2015) and the investment challenge for small-farmer based ecological agricultural development that goes against mainstream politics (Thiemann, 2015). The gender blindness of the concept – as if food preferences, rights and practices are not affected by gender differences and inequalities – has been criticised by Agarwal (2014) and Park *et al.* (2015). For these reasons, Alonso-Fradejas *et al.* (2015) call for more empirical research that can contextualise the challenges for food sovereignty posed by the changing global contexts. This book contributes results of empirical research to address some of these challenges and complexities.

1.3.7 Objective, subjective and relational

By objective, we refer to the use of objective indicators and measurements of the variables involved. By subjective, we refer to people's own culturally underpinned emic models and views on farming styles, food security, nutritious foods and wellbeing. The relational dimension highlights food wellbeing as a social construction based on social relations, positions and normative frameworks. In the different chapters, the different perspectives are differently emphasised, require different research methods, and may also be contrasted.

1.4 Layout and chapters

The book contains 14 chapters. This first chapter provides a general introduction to the book. Chapter 2 (Gartaula and Niehof, 2018) presents the first empirical study, a case from Nepal. The study addresses the issue of the impacts of male labour out-migration on perceived household food security, division of labour in agriculture, and on values associated with land and agricultural work. On the short term, the balance seems to be positive, but on the long term the changed local dynamics of agriculture and food provision may pose a challenge to food sovereignty at the national level.

The Chapters 3 and 4 report on wild and domesticated indigenous crops and the knowledge associated with their collection, cultivation and consumption in the Cordilleras in the northern Philippines. Chapter 3 (Gayao *et al.*, 2018) addresses the role of root and tuber crops in the food system of the indigenous people in the region. The many different species of root and tuber crops known are documented and the authors show how their farming inculcates the virtues of industriousness and self-reliance, leading to food sovereignty. The authors also investigated the transfer of knowledge on root and tuber crops and found the younger generation to be less knowledgeable than the older people. Chapter 4 (Licyayo, 2018) reports the abundance and diversity of wild edible fungi and shows how this has kindled a strong local tradition of mushroom gathering. Men proved to primarily responsible for gathering edible mushrooms in the wild, women for preparing the mushrooms for family consumption and for selling mushrooms. Regarding the latter, the contribution to the household income from mushrooms sales is currently very little.

In Chapter 5, Hastuti (2018) profiles family life of Tengger people, an ethnic minority group in East Java, Indonesia in terms of their household's food security, food caring practices, children's nutrition and family wellbeing. She used qualitative and quantitative methods of data collection, in this way covering subjective feelings about and objective indicators of food and nutrition security. The assumption at the start of the research that among the Tengger people food and nutrition security would be problematic because of their position as an ethnic minority living in a relatively remote area, is not entirely confirmed. Regarding food and nutrition, to a certain extent the Tengger households and children are doing better than expected. However, according to objective measurements there still are considerable proportions of food insecure households, stunted children, and mothers whose nutritional status is inadequate.

In Chapter 6, Quetulio Navarra *et al.* (2018) put another vulnerable group in the limelight, that of households in Central Java, Indonesia, who in 2001 had to be resettled after landslides had destroyed their houses in the mountains. The researchers use the lens of the role of social capital and institutions in investigating the dynamics of household food security and wellbeing of children under five years old. Applying a longitudinal perspective, they analyse the growth of institutions and social capital in the resettlement site since displacement, and they pay attention to cultural mechanisms of social protection as well. As in the study presented in the foregoing chapter, also in this case the situation is more positive than could be anticipated based on the literature. In the study presented in Chapter 7 (Purwestri *et al.*, 2018), the nutritional status of children under five is the main focus. However, in this case the children do not live in a poorly endowed resettlement site but in the rice surplus area of Demak district in Central Java. All children in the research are from farmer families. Quite contrary to expectation, household farm income and expenditures do not make a difference for the nutritional status of these children. What proved to be determining factors in the odds of stunting among children were diet quality and snacking habits. Hence, living in a rice surplus area does not prevent children from being undernourished.

In Chapter 8, Madjdian (2018) takes us to Humla district in Nepal. Using a gender perspective the author unravels intra-house food allocation and relates it to women's vulnerability and nutritional insecurity. Ethnicity and family system are other explanatory variables, which – to some extent – crosscut the impact of gender. The role of these variables is investigated by comparing two different ethnic communities, Hindu and Buddhist, in the same area. The Hindu women, especially the Dalit women, were found to be more vulnerable and nutrition insecure. They also seemed to be less able to benefit less from social change than the Buddhist women were.

In Chapter 9 (Ekawidyani *et al.*, 2018), the urban context of social change proved to have a considerable impact on the nutritional status of children. The middle-class schoolchildren in the city of Bogor, Indonesia, are exposed to an environment where snacks and fast foods are widely available and, with pocket money, accessible. They often have higher educated and working mothers, who have little time for quality food care. In addition, technological advancement has made children prefer watching television and playing virtual games to playing outdoors and physical exercise. All these factors are contributing to children being overweight and obese. Not social change but climate change is the focus of the study by *Talavera et al.* (2018) presented in Chapter 10. Data on climate variables were collected in the Philippines for two areas, Infanta and Los Baños, with a different climate type. In each location, a survey was done and anthropometric measurements were taken to assess household characteristics and the children's nutritional status. From the relationship between the two data sets, the authors infer that especially high rainfall affects the nutritional status of children through higher incidence of illness. Increased rainfall and flooding induced by climate change could thus increase stunting and underweight in children.

The Chapters 11 (Dorado *et al.*, 2018), 12 (Reario *et al.*, 2018) and 13 (Shaw *et al.*, 2018) all address the effects of food and nutrition interventions, the first two in the Philippines, and the last in Vietnam. The study presented in Chapter 11 looks at the effects of the Backyard Vegetable Gardening Program on perceived household food security and dietary diversity of households and schoolchildren. The researchers conducted interviews among households with and without a backyard garden and among program implementers. In spite of the positive attitude to vegetables and gardening of the respondents, regardless whether they had a garden of their own or not, the program proved to have no positive impact on dietary diversity. The authors explore the reasons for this and formulate recommendations. In Chapter 12, the intervention is the homestead food production (HFP) model developed by Helen Keller International. This intervention worked well, in terms of strengthening household income, dietary diversity and women's empowerment. The mode of implementation did not make a significant difference; the effects of the intervention were positive, whether managed by the local government or the NGO. Chapter 13 also presents a case of the implementation of the (enhanced) HFP model, this time in Son La province, Vietnam. The question addressed in this study is whether households not included as project households in the intervention, would also experience project benefits. This so-called spillover effect could indeed be observed, albeit for some activities more than for others. Apparently, such an intervention enhances food sovereignty at the level of the community and not just at that of the individual participating households. The authors conclude that the finding that the program coverage went far beyond the direct beneficiaries should inform future monitoring and evaluation.

References

Agarwal, B., 2014. food sovereignty, food security and democratic choice: critical contradictions, difficult conciliations. *The Journal of Peasant Studies* 41(6): 1247-1268.

Allan, P. and Sachs, C., 2012. Women and food chains: the gendered politics of food. In: William-Forson, P. and Counihan, C. (eds) *Taking food public: redefining foodways in a changing world.* Routledge, New York, NY, USA, pp. 23-40.

Alonso-Fradejas, A. Borras, J.S.M., Holmes, T., Holt-Giménez, E. and Robbins, M.J., 2015. Food sovereignty: convergence and contradictions, conditions and challenges. *Third World Quarterly* 36(3): 431-448.

Chen, K., Timmer, P. and Dawe, D., 2017. East Asia. In: *Global food policy report 2017.* International Food Policy Research Institute (IFPRI), Washington, DC, USA, pp. 75-78.

Corazon M. Licyayo, D., 2018. Gathering practices and actual use of wild edible mushrooms among ethnic groups in the Cordilleras, Philippines. Chapter 4. In: Niehof, A., Gartaula, H.N. and Quetulio-Navarra, M. (eds) *Diversity and change in food wellbeing: cases from Southeast Asia and Nepal.* Wageningen Academic Publishers, Wageningen, the Netherlands, pp. 71-86.

DeRose, L.F., Das, M. and Millman, S.R., 2000. Does female disadvantage mean lower access to food? *Population and Development Review* 26(3): 517-547.

Dorado, J.B., Viajar, R.V., Azaña, G.P., Caraig G.S. and Capanzana, M.V., 2018. Does backyard vegetable gardening enhance perceived household food security and dietary diversity? A case study in the Philippines. Chapter 11. In: Niehof, A., Gartaula, H.N. and Quetulio-Navarra, M. (eds) *Diversity and change in food wellbeing: cases from Southeast Asia and Nepal.* Wageningen Academic Publishers, Wageningen, the Netherlands, pp. 225-242.

Ekawidyani, K.R., Karimah, I., Setiawan, B. and Khomsan, A., 2018. Parents' characteristics, food habits and physical activity of overweight schoolchildren in Bogor City, Indonesia. Chapter 9. In: Niehof, A., Gartaula, H.N. and Quetulio-Navarra, M. (eds) *Diversity and change in food wellbeing: cases from Southeast Asia and Nepal.* Wageningen Academic Publishers, Wageningen, the Netherlands, pp. 177-194.

FAO, 2013. *The state of food security in the world.* Food and Agricultural Organization, Rome, Italy.

FAO, 2017. *The state of food insecurity in the world: building resilience for peace and food security.* Food and Agricultural Organization, Rome, Italy.

Gartaula, H.N. and Niehof, A., 2018. Changing value of food and perception of food security in the context of male out-migration in Nepal. Chapter 2. In: Niehof, A., Gartaula, H.N. and Quetulio-Navarra, M. (eds) *Diversity and change in food wellbeing: cases from Southeast Asia and Nepal.* Wageningen Academic Publishers, Wageningen, the Netherlands, pp. 27-42.

Gartaula, H., Patel, K., Johnson, D., Devkota, R., Khadka, K. and Chaudhary, P. 2017. From food security to food wellbeing: examining food security through the lens of food wellbeing in Nepal's rapidly changing agrarian landscape. *Agriculture and Human Values* 34(3): 573-589.

Gayao, B.T., Meldoz, D.T. and Backian, G.S., 2018. Indigenous knowledge and household food security: the role of root and tuber crops among indigenous peoples in the Northern Philippines. Chapter 3. In: Niehof, A., Gartaula, H.N. and Quetulio-Navarra, M. (eds) *Diversity and change in food wellbeing: cases from Southeast Asia and Nepal.* Wageningen Academic Publishers, Wageningen, the Netherlands, pp. 43-70.

Godek, W., 2015. Challenges for food sovereignty policy making: the case of Nicaragua's Law 693. *Third World Quarterly* 36(3): 526-543.

Hastuti, D., 2018. Household food security, food care, child nutrition, and wellbeing of ethnic Tengger families in East Java, Indonesia. Chapter 5. In: Niehof, A., Gartaula, H.N. and Quetulio-Navarra, M. (eds) *Diversity and change in food wellbeing: cases from Southeast Asia and Nepal.* Wageningen Academic Publishers, Wageningen, the Netherlands, pp. 87-113.

Kabeer, N., 2005. Gender equality and women's empowerment: a critical analysis of the Third Millennium Development Goal. *Gender and Development* 13(1): 13-24.

Madjdian, D.S., 2018. Gender, intra-household food allocation and social change in two Himalayan communities in Nepal. Chapter 8. In: Niehof, A., Gartaula, H.N. and Quetulio-Navarra, M. (eds) *Diversity and change in food wellbeing: cases from Southeast Asia and Nepal.* Wageningen Academic Publishers, Wageningen, the Netherlands, pp. 153-175.

Maxwell, S., 2001. The evolution of thinking about food security. In: Devereux, S. and Maxwell, S. (eds) *Food Security in Sub-Saharan Africa.* ITDG Publishing, Warwickshire, UK, pp. 13-27.

McGregor, J.A., 2006. *Researching wellbeing: from concepts to methodology.* Working Paper 20. Bath: University of Bath, ESRC Research Group on Wellbeing in Developing Countries. Available at: www.welldev.org.uk.

McGregor, J.A., 2008. *Wellbeing, poverty and conflict.* Briefing Paper 1/08. Bath: University of Bath, ESRC Research Group on Wellbeing in Developing Countries. Available at: www.welldev.org.uk.

Mechlem, K., 2004. Food security and the right to food in the discourse of the United Nations. *European Law Journal* 10(5): 631-648.

Niehof, A., 2010. *Food, diversity, vulnerability and social change: research findings from insular southeast Asia.* Wageningen Academic Publishers, Wageningen, the Netherlands.

Niehof, A., 2011. Conceptualizing the household as an object of study. *International Journal of Consumer Studies* 35(5): 488-497.

Niehof, A., 2016. Food and nutrition security as gendered social practice. *Apstract* 10(2-3): 59-67.

Noack, A.-L. and Pouw, N.R.M., 2015. A blind spot in food and nutrition security: where culture and social change shape the local food plate. *Agriculture and Human Values* 32(2): 169-182.

Park, C.M.Y., White, B. and Julia, 2015. We are not all the same: taking gender seriously in food sovereignty discourse. *Third World Quarterly* 36(3): 584-599.

Patel, K., Gartaula, H. Johnson, D. and Karthikeyan, M., 2015. The interplay between household food security and wellbeing among small-scale farmers in the context of rapid agrarian change in India. *Agriculture and Food Security* 4: 1-16.

Purwestri, R.C., Barati, Z., Wirawan, N.N., Fahmi, I., Lauvai J. and Scherbaum, V., 2018. What explains stunting among children living in a rice surplus area in Central Java, Indonesia? Chapter 7. In: Niehof, A., Gartaula, H.N. and Quetulio-Navarra, M. (eds) *Diversity and change in food wellbeing: cases from Southeast Asia and Nepal.* Wageningen Academic Publishers, Wageningen, the Netherlands, pp. 137-152.

Quetulio-Navarra, M., Frunt, E. and Niehof, A., 2018. The role of social capital and institutions in food security and wellbeing of children under five for resettled households in Central Java, Indonesia. Chapter 6. In: Niehof, A., Gartaula, H.N. and Quetulio-Navarra, M. (eds) *Diversity and change in food wellbeing: cases from Southeast Asia and Nepal.* Wageningen Academic Publishers, Wageningen, the Netherlands, pp. 115-136.

Reario, M.F.D., Pacheco, P.J., Lafuente, E. and Talukder, A., 2018. Developing homestead food production in Muslim Mindanao, Philippines: Does mode of implementation matter? Chapter 12. In: Niehof, A., Gartaula, H.N. and Quetulio-Navarra, M. (eds) *Diversity and change in food wellbeing: cases from Southeast Asia and Nepal.* Wageningen Academic Publishers, Wageningen, the Netherlands, pp. 243-266.

Ruel, M., Garrett, J. and Yosef, S., 2017. Food security and nutrition: growing cities, new challenges. In: *Global food policy report 2017.* International Food Policy Research Institute (IFPRI), Washington, DC, USA, pp. 24-34.

Shaw, K., Ngoc, P.K., Lan, H., Mundy, G. and Talukder, Z., 2018. How non-targeted households benefited from implementing a homestead food production program in Son La, Vietnam. Chapter 13. In: Niehof, A., Gartaula, H.N. and Quetulio-Navarra, M. (eds) *Diversity and change in food wellbeing: cases from Southeast Asia and Nepal.* Wageningen Academic Publishers, Wageningen, the Netherlands, pp. 267-278.

Talavera, M.T.M., Pangan, M.R.L., Colladilla, J.O., Sarol, J.N., Agdeppa, I.A. and Lorenzo, F.M.E., 2018. Climate change and nutritional status of preschool children living in Infanta, Quezon, and Los Baños, Laguna, Philippines. Chapter 10. In: Niehof, A., Gartaula, H.N. and Quetulio-Navarra, M. (eds) *Diversity and change in food wellbeing: cases from Southeast Asia and Nepal.* Wageningen Academic Publishers, Wageningen, the Netherlands, pp. 195-224.

Thiemann, L., 2015. Operationalizing food sovereignty through an investment lens: how agro-ecology is putting 'big push theory' back on the table. *Third World Quarterly* 36(3): 544-562.

Yaro, J.A., 2004. Theorizing food insecurity: building a livelihood vulnerability framework for researching food insecurity. *Norwegian Journal of Geography* 58(1): 23-37.

Zezza, A., Carletto, C., Davis, B. and Winters, P., 2011. Assessing the impact of migration on food and nutrition security. *Food Policy* 36(1): 1-6.

2. Changing value of food and perception of food security in the context of male out-migration in Nepal

H.N. Gartaula[1] and A. Niehof[2]*

*[1]International Program Coordinator, Manitoba Council for International Cooperation (MCIC), Winnipeg, Canada; [2]Professor Emeritus, Wageningen University, Wageningen School of Social Sciences (WASS), the Netherlands; *hom.gartaula@gmail.com*

Abstract

International development practitioners argue that migration improves household food security in the origin areas, by providing access to capital for investing in agriculture or purchasing food. However, these debates have occurred without paying enough attention to the values that justify food production and consumption in the areas of origin. This paper questions the assumption that a shift from an agricultural-based economy to an economy based on remittances increases the ability of households to secure access to food in the face of rapid economic and cultural change. Based on the results of fieldwork conducted in Nepal, the paper argues that male out-migration adds to the workload of the women left behind, reduces women's and men's subjective attachment to agriculture in the area of origin and changes the values associated with land and agriculture. This makes the sector vulnerable and poses a threat to longer-term food security.

Keywords: food security, wellbeing, rural out-migration, Nepal

2.1 Introduction

In the past few decades, Nepal has undergone a number of major social, political and cultural changes. In the 1990s, the country became a constitutional monarchy and entered into an open, liberal economy (Hachhethu, 2000; Lecomte-Tilouine, 2009). In 2006, another political transformation resulted in the country becoming a federal republican state. These changes not only improved people's access to information and knowledge about the outside world, but also increased their expectations that the state would do something about the country's prevalent food and social security problems. Even though the changes increased people's socio-political awareness, the problems of unemployment, poverty, and livelihood insecurity were not adequately addressed. These realities of everyday life in Nepal led people to look for opportunities outside the country. Furthermore, the decade-long Maoist insurgency (1996-2006) impelled many rural youths to leave their homes to escape joining the Maoists' militant force. At the same time, the 'oil boom' in the Gulf States increased the demand for international labourers (Russell, 1992). This all resulted in many contract labourers working abroad, for whom migration has now become part and parcel of their lives (see also Lecomte-Tilouine, 2009).

DOI 10.3920/978-90-8686-864-3_2, © Wageningen Academic Publishers 2018

At present, out of the country's total population of 26.5 million about two million people live outside the country and one in every four households has at least one migrant member (CBS, 2011). Recent World Bank data suggest that remittances contributed by these migrants in the year 2016 accounted for about 30% of the national gross domestic product (World Bank, 2017). From the country's only international airport (Tribhuvan Airport), almost 1,500 young Nepalese leave the country every day in search of better wage employment mainly in the Gulf countries, Malaysia and other Asian countries. During the period 1994-2012, the country recorded a 100 times increase in the number of migrants, not including the almost equal number of irregular and unofficial (illegal) migrants to India and other countries (DOFE, 2014).

This trend of increasing labour out-migration is endorsed by Nepal's policy environment, in which foreign employment is envisaged as a safe, manageable and decent option to solve the country's unemployment problem. To regulate and streamline the process and to protect the rights of the workers, the Foreign Employment Act 2007 and the Foreign Employment Regulations 2007 were put in place (GON, 2007). Migration is promoted through the Foreign Employment Promotion Board, chaired by the Minister of Labour and Employment and established in the same year (FEPB, 2007). The establishment in 2010 of the Department of Foreign Employment further attests to the government's commitment to promote international labour migration and to make this sector important for people's livelihoods and – through the generated remittances – for the country's economy.

The policies and developments described above raise questions about the impact of migration on food security as a consequence of changes in agriculture and livelihood practices in the migrant-sending areas. These questions were the subject of the fieldwork conducted in Nepal during 2009-2010 by the first author. In existing scholarship on the migration-food security nexus the focus is very much on its objective aspects, such as the effects of migration on access to food and on nutrition (Nguyen and Winters, 2011; Zezza *et al.*, 2011). The results presented here, however, highlight the changing and subjective meanings of food security in a migrant-sending area.

Thus, this paper investigates how rural out-migration exerts a pressure on agriculture and food production in the migrant-sending areas of Nepal, and how that results in changing patterns of access to food and food consumption, as well as changes in people's perception of the role of agriculture in food security. These changes could render the agricultural sector in such areas more vulnerable, less productive, and unattractive, which would result in a long-term negative impact on food security. We see out-migration is not just a coping strategy of the rural households involved, but as part of rural development processes. These processes can only be sustainable if the linkages between migration, poverty alleviation, agriculture, and food security are understood and taken into account in policies and programmes.

2.2 Labour migration and food security in the literature

Labour migration is an important livelihood strategy for many rural households in the developing world, where remittances are considered to contribute positively to the wellbeing of the households at origin (Toyota *et al.*, 2007). Recent scholarship in migration studies has paid significant attention to the development impact of remittances. The contemporary literature tends to focus on how migration improves household income and finances the acquisition of capital assets in the area of origin (Adams and Page, 2005; Adger *et al.*, 2002; Berthélemy *et al.*, 2009; Gupta *et al.*, 2009). Less emphasis has been placed on understanding the empirical relationship between migration and food security, perhaps owing to the fact that food security is generally assumed to be an obvious element of the positive development impacts of remittances (Crush, 2013). Moreover, the current body of work tends to conceptualize food security primarily in terms of material access to food (FAO, 2013; Vargas-Lundius *et al.*, 2008; Zezza *et al.*, 2011), thereby largely ignoring the socio-cultural, relational, and political nature of food access and consumption. We would like to argue that these elements of food security have important subjective dimensions, which include people's perceptions and subjective experience of food security, their attitudes toward agricultural land, and other cultural and social factors that affect people's food choices. While out-migration may increase purchasing power and, thus, contribute positively to physical access, the multi-facetted nature of food security necessitates that studies also engage with these subjective aspects. So, if migration and development policies are to be attentive to the longer-term sustainability of food security in migrant-sending areas, people's own emic conceptions of food security must be integrated into the analysis of the social and cultural dynamics of local contexts (Scoones, 2009).

The concept of food security was first introduced to the global debate at the World Food Conference organized by the United Nations General Assembly in 1974. Since then, an intensive debate has evolved surrounding the conceptualization and measurement of food security. In the 1970s, food security was defined by fluctuations in food availability, in relation to the stability of food grain prices (FAO, 1974). It was realized that adequacy of world food supplies was needed at all times to sustain a steady expansion of food consumption, as well as to offset fluctuations in grain prices (FAO, 2003). However, scholars like (Drèze and Sen (1989), Sen (1981) and Watts and Bohle (1993) emphasized a multi-dimensional understanding of food insecurity and poverty. They argued that unequal access to and distribution of food – due to a lack of economic resources and capabilities of individuals – are key aspects of food security. This new paradigm led to a distinction between the ability of the state to ensure a constant supply of food available at the national level and the capability of individuals or households to access available food. In this discourse, the focus of food security studies has also shifted from regional or national levels to household-level access to food (Niehof, 2010). The World Food Summit 1996 reassessed the concept and provided a new definition of food security as existing when all people, at all times, have physical and economic access to sufficient, safe and nutritious food to meet their dietary needs and food preferences for an active and healthy life (FAO, 1996).

In an historical overview of the food security literature, Yaro (2004) identified three types of then existing food security approaches: the food availability approach, the livelihood and entitlement approach, and the food sovereignty approach. While the 1996 definition has remained a pillar in food security studies, scholars continued to reinterpret and redefine the concept. This gave rise to extended definitions that include access to nutritionally adequate food at all times and procured in conformity with human aspirations, wellbeing and dignity (FAO, 2013; Yaro, 2004). A closer look at all these definitions reveals two interrelated but distinct lines in the conceptualization of food security: first, food security as output or process, and, second, food security as an objective or a subjective condition. This analytical distinction is useful to classify food security scholarship. For example, the food availability approach focuses on output and objective aspects, whereas the entitlement, livelihood and food sovereignty approaches emphasize process and objective aspects (Drèze and Sen, 1989; Maxwell and Frankenberger, 1992; Patel, 2009; Scoones, 1998). However, the existing literature is rather silent on the subjective aspects of food security, which are the focus of this paper. In other words, rather than just investigating *what* people get to eat and *how* they get it, we are interested in the perceptions and notions behind people's food choices in relation to the context in which they make those choices. To investigate this, we have applied the social wellbeing approach as a theoretical framework (McGregor *et al.*, 2007).

The social wellbeing approach embraces both the objective and subjective aspects of wellbeing, including food security as a component of wellbeing (Gartaula *et al.*, 2016). It also recognizes how these are shaped by the livelihood systems to which people's food security is anchored. Since wellbeing depends on what resources people have, whether the available resources can fulfil the requirements for living, including access to food, and how people perceive the quality of life they achieve, it measures people's capabilities and choices in relation to food and means of acquiring it (McGregor *et al.*, 2007; Noack and Pouw, 2015). The three dimensions of wellbeing are accessibility, adequacy and perception. These are interconnected and are reproduced in interaction with the wider structures of family, community and society (McGregor, 2006; Newton, 2007). Therefore, a social wellbeing approach offers a comprehensive focus that integrates material resources, social relationships, people's psychological states, and the subjective meanings attached to these (White and Ellison, 2006). Applying this approach to food security allows for an emphasis on the status and symbolic value assigned to different food systems and lifestyles and what these mean for safeguarding food security. It broadens the perspective of existing food security research by seeing resources through an understanding of a person's agency and the context specific values and meanings of resources. For example, 'education' can be seen as qualifying the stock of labour of a household in attaining food security, but could equally be seen as a social or cultural resource to enhance social status. Likewise, 'land' is a natural resource but can also be an economic, commercial, social or cultural resource, depending on its valuation by individuals, households and communities (White, 2009). A social wellbeing approach challenges orthodox measurements of food security by ownership and use of types of capital and assets only. Instead, it recognizes that lived experience, perception and context are important for understanding how social change affects people's lives and their ability to be food secure (White and Ellison, 2006).

2.3 Research area and methodology

The fieldwork was carried out in the VDC (Village Development Committee) of Maharani Jhoda, Jhapa district, eastern Nepal. Jhapa district was selected because of its dynamic history of in- and out-migration and its location in the *terai* region, considered the granary of the country. Among the three ecological zones of Nepal (high mountains, mid hills and the *terai*), the *terai* is an extension of the flooding plain of the river Ganges in India. The *terai* is very suitable for lowland rice-based agriculture. Hence, food production in Jhapa district is an important issue because changes in food production here have an impact on the overall food production of the country and may thus affect the country's food sufficiency in the end.

Maharani Jhoda VDC is located 56 kilometres west of the district headquarters and 550 kilometres east of the country's capital of Kathmandu. The available historical sources indicate that the settlement dates from the years 1912-13. In-migration increased when the government opened up the *terai* for settlement after the eradication of malaria in the late 1950s. While in-migration continued, out-migration began in the mid-1970s and has been increasing ever since. However, whereas in-migration took the form of permanent family migration, out-migration is temporary, individual and (mostly) international labour migration (Gartaula and Niehof, 2013).

Wetland rain-fed rice-based farming is the dominant cropping system in the study area. There is no surface irrigation system, but over 50% of the households own motor pumps to draw water from underground boreholes installed in their farms. The water from boreholes is needed mainly for spring season rice (April-June) and winter season crops like wheat, maize, mustard, potato, and green vegetables. Summer (June-August) is the rainy season and is the main season for rice cultivation, during which natural streams or small irrigation channels dug by the farmers can meet the water demand. There is not much agricultural mechanization in the area; most agricultural activities such as hoeing, seeding, transplanting seedlings, and harvesting are done manually.

The research combined qualitative and quantitative methods of data collection. Qualitative methods included key informant interviews, focus group discussions, in-depth interviews, and participant observation. Twenty-six persons, comprising older and younger people, wives of migrant workers, returned migrants, local political leaders, and early settlers, were interviewed or participated in the focus groups. Quantitative data were collected in a household survey among 277 households using stratified random sampling (for details see Gartaula, 2011).

2.4 Results

All households in the research area are engaged in both farm and non-farm economic activities within and outside the village, including foreign employment, which appears to be the main sources of cash income. Among the 277 households surveyed, 49% had migrant members and 40% had received remittances in the previous year. The labour migrants were predominantly male (87%), married (82%), young (on average 29 years old), and unskilled (60%), working in both formal and informal sectors at destination. Apart from in-country rural-urban migration (27%), major destinations were Qatar (21%), United Arab Emirates (14%), Malaysia (12%), and Saudi Arabia (10%).

The results show that the process of labour out-migration is linked with cultural change and food security dynamics in the research area by the ways in which people from different social groups perceive food security. The first sub-section describes the impact of out-migration on agriculture, since male out-migration reduces the labour available for agricultural production. The second sub-section describes how these changed agricultural labour dynamics are affecting food access and perceptions of food security. Household remittances can be used for immediate consumption or for long-term productive investment. Either use will entail changes in access to food and perception of food security. In the research area, expenditures are mainly consumptive, such as on children's education, health, housing, and consumer durables. The level of investment for the longer term, such as in agriculture, businesses development, or other economic activities, is low. We also found that the use of remittances affects social norms, values, and social organisation within and outside the households (Gartaula *et al.,* 2012).

2.4.1 Migration, labour dynamics and agriculture

The nature of migration in the research area proved to be temporary and cyclic. An individual lives and works abroad for two to three years on average and then returns home for three to six months, depending on the type of contract at destination. The cycle may continue for many years, sometimes up to the worker's retirement age. Such cycles give rise to two simultaneous phenomena: the loss of labour force at home and the increase of household income from financial remittances. Shortage of household labour could in theory be solved by hiring labour, but in practice it is not as simple that. When a large section of the able-bodied population leaves, there is a labour deficit at village level. This leads to the remaining labour becoming hard to get when needed or too expensive to hire.

Among the 134 households that responded to the labour issue, 70% said they hired agricultural labour in the previous year, 16% exchanged labour with neighbours, and the remaining 14% had no labour constraint. Those who hired labour said that the wage rate is increasing every year and that it is difficult to get labour in time. Due to the excessive male labour out-migration, more responsibility for agriculture has accrued to elderly men and women, in the case of women resulting in what has been referred to as the feminisation of agriculture (Radel *et al.*, 2012; Zuo, 2004). For the women, the additional agricultural work comes on top of their domestic chores. Opting for exchanging labour would also pose an extra burden because, due to the differences between men's and women's work, the labour of men is seen as worth twice that of women. Feminisation of agriculture could have the positive effect of empowering women. However, the author of a study on the effects of the feminisation of agriculture caused by male labour migration in China concludes that the results 'do not support the argument that men's migration and labour feminization in agriculture provide the space for women to plead for more equal gender relations with men. The unequal gender relations are re-affirmed and re-produced through this process' (Meng, 2014: 163). In our case, wives of migrants were more engaged in groups and agricultural cooperatives than women with a husband at home (Gartaula *et al.*, 2012), which could be seen as a positive spin off.

For households with a labour constraint there are in fact two options. The first one is to hire scarce and expensive labour, the second to engage in a sharecropping arrangement or to rent out the land. Among the 21 households who rented-out land or gave their land out for sharecropping in the previous year, 13 were migrant households. The following two statements illustrate why the respondents ended up with the second option:

> Yes, I have given all the land out for sharecropping. There is no one at home to work on the farm. When my daughter-in-law was here, we cultivated land on our own. We hired male labourers for ploughing, but we were handling the land ourselves. Since she joined her husband in Kathmandu three years ago, I remained alone at home so I gave the land to others for cultivation. However, this is not my choice. [...] I get half of the harvest from my sharecropper. (JKL, 55, migrant's mother)

> Since the main able-bodied male member of the household is absent, we as women and children [herself, her mother-in-law, and two children, a boy and a girl] are having hard time to work on the farm. The decision of renting-out land was actually not mine though. As we were doing agriculture with difficulty, he [husband] asked me to rent out the land. We have given our land out for sharecropping and we get half of the produce. I could handle the land but finding labour is hard. We have money to hire, but we cannot find the labourers on time. By the time they are available, the season would be over. Running after this profitless business is not good for us, as he is making good money out there. On top of it, I have my own local liquor business, which is easy and more profitable than agriculture. (TDK, 38, migrant's wife)

Another side of the story is that for wives of migrants the absence of a husband has provided them with an opportunity to open up to the outside world and interact with other people. They engage in activities that they would rarely engage in when the husband was present, such as going to the market for groceries or other household purposes, going to the bank, joining groups and cooperatives, and so on. This has increased their self-confidence and self-esteem and has enhanced their role in household decision-making (Gartaula *et al.*, 2012). Moreover, the remittances they receive have decreased their sense of dependence on agriculture as a livelihood source. Some of the women indicated that upon return of their husband, they plan to move out of agriculture:

> After his final return, I do not think we will be involved in agriculture. I am fed up with this. Probably we do businesses (small local trade) or, as I have taken a beautician training, I may open a beauty parlour. Even though we are not sure what we are going to do, we have agreed that we are not going to continue farming. It is mainly because agriculture requires lots of labour and it is difficult to work. In return, it does not yield a good income. (RBT, 28, migrant's wife)

The decision of young migrant families not go back again into agriculture upon the return of the migrant, is becoming increasingly common. This has created a conflict between their generation and that of their parents. While the older generation still values agriculture and wants to put an effort into it, the younger generation sees its future outside agriculture. To support their plans many of them have started to buy residential plots in the nearby towns. Out of 20 migrant households considered for in-depth interviews, 12 bought residential plots in nearby towns, while only three households bought agricultural land. The remaining five households did not buy any land. A key informant who used to be a migrant worker but now works as a land broker said: 'People's attitude nowadays once they return from abroad is that they want to buy a plot in town, put up a shop and live an easy life. If the father asks to buy or add agricultural land they reply: 'who is going to plough it'?' (BTR, 56, former migrant). These findings corroborate those of studies carried out in Mexico (Durand *et al.*, 1996) and China (De Brauw and Giles, 2008) where remittances were mainly invested in housing.

2.4.2 Changing access to food and perception of food security

Most research participants reported that the food options and availabilities to compose their food basket have been changing over time. Their daily menu shows an increase of green vegetables and meat. Asked about it, they said it was possible because of the added household income from foreign employment. However, there is also a tendency to eat more junk food. Children used to be happy with homemade snacks and food such as beaten rice and popcorn, but nowadays they do not like these anymore and want noodles, cookies and other items available in the market instead. The following narrative from one of the interviewed women supports this assertion:

> When my kids were young I used to give them fried rice to take to school, but nowadays they do not like to eat this at school. They prefer to have money so that they can buy the items they like to eat. [...] There are many changes in our menu too. In the previous time, it was hard for us to buy meat, but now there are a number of meat shops in the market and we can easily buy meat, vegetables and fruits anytime we want. We used to eat *gundruk* [dried leafy vegetables mainly made of mustard leaves, preserved vegetable during the lean season, also considered a delicacy of the poor], but now we cook many different varieties. (ANK, 36, migrant's wife)

These changes in food consumption also change the meaning of having land and practicing agriculture for food security. To understand what people think about land and agriculture, a focus group discussion was conducted with younger women who were wives of current migrant workers discussing their perceptions about land, agriculture and food security. Following up on the discussion about agriculture as not being a preferred occupation anymore and using remittances for something else than investing in agriculture, the following questions were discussed:

- Why are people here getting less interested in agriculture?
- What do you think is the importance of agriculture for people's access to healthy food?
- What role does the phenomenon of labour out-migration play in this situation?

Excerpts from the discussion are presented below. The figures in parentheses are the respective ages of the participants.

> SBT (28): It is possible to be successful with agriculture, but the labour cost is very high, production is not good, and there is lack of an appropriate market for our produce. Thus, labour migrants appear to be more successful than farmers are, and people are discouraged from farming.
>
> MKN (29): The returned migrants often buy plots at Gauradaha or Damak [nearby towns] and build houses over there. They are moving to those areas.
>
> NBK (35): Most of the returned migrants are found to be more successful than the people who do agriculture here.
>
> SBT (28): The returned migrants have saved good amounts of money so that they can start any business, but it is hard for farmers to save and use the money for other purposes. Farming is just for getting food, but not for money. We cannot get outputs from agriculture that are worth the hard work.
>
> MKN (29): The returned migrants are happy, as they have moved to the cities. They build houses anyhow in the market areas after return.

> NBK (35): Actually, agriculture operates at a loss here. There is labour scarcity as the main person goes out. A woman does not plough the land and the elderly cannot do it, so they have to hire expensive male labour (NPR 350-500 per day) for ploughing and other labourers (NPR 150-200 per day) for other purposes. Take the example of my household: we have one hectare of land, my husband sends about NPR 10,000 per month. We have three kids. I am always busy taking care of them. We have to hire labourers to work for at least NPR 200 a day. In total, we invest about 10-15 thousand rupees, while we get only 5-7 quintals of rice, equivalent to hardly about 10,000 rupees. You can imagine how profitable this business is. It is then very tough to manage a household. Those with a high salary may have become successful, but with a normal salary abroad, it is all the same.
>
> SBT (28): That is why agriculture is for food, not for income. The returned migrants do not want to work on the farm. It is interesting that they can work hard abroad but not here.
>
> MKN (29): They feel ashamed to work on the farm after return. They think that people would laugh at them as if they were unsuccessful abroad so that they started farming again.

The excerpts show that the meaning and value of land and agriculture are changing. People have diversified their livelihoods and are opting for a lifestyle that is not just based on agriculture. The implied meaning of the word 'successful' seems to be having earned enough money to be able to leave agriculture. However, the older generation still values agriculture and agricultural land. In their view, only agriculture provides a secure living, more than any other means of livelihood. One of the older respondents said:

> I can tell you, there is nothing bigger than agriculture. If you have land, you can never go hungry. But the present generation considers agriculture a dirty job. At home, they become lazy and shameful of their work, but outside they work hard in whatever job they get. This is not a good mentality, in my opinion. (RKP, 72, father of a migrant)

A migrant worker expressed the opposite view:

> I would never think of continuing agriculture after my final return. I want to have land but do not like to work in all that dirt and mud. I have bought a piece of land at Kakadbhitta [a town near the Indian border, about 60 km northeast of the research area] where I will move later on to start a business. (DPB, 29, migrant worker)

It seems that the younger generation does not entirely disregard the importance of land and wants to keep it, but – at the same time – they do not want to work on it. The older generation rather finds agriculture a good occupation and a reliable basis for food security.

In a study carried out elsewhere in Nepal, Sharma (2008) reports that going to India or other countries opens up possibilities of being modern and developed, thus demonstrating a 'modern' conception of manhood. In this conception migration becomes a pathway to experience the outside world while – at the same time – money is sent home for the upkeep of the family. This is quite different from the notion of migration as propelled by economic necessity or as the result of exploitation in the ideological space of development and modernity (Mills, 1997; Sharma, 2008). Thus, our case shows how labour out-migration induces social and cultural change. It also reveals a shift from an agriculture-based economy to an economy that also includes income from remittances and other non-agricultural sources.

2.5 Discussion and conclusion

Out-migration of a family member not only changes a migrant household's economic situation, but also the social norms, values and social organization in the households and communities at origin. Some of the changes that occur in rural Nepal in general and in the research area in particular relate to the value attached to land and agriculture and the way the link with food security is perceived. Additionally, there we found indications of an increased awareness of the importance of a good diet by including vegetables, fruits and meat in the daily meals. This is accomplished in different ways. First, although migrant households tend to rent out agricultural land, many are involved in kitchen gardening and produce fresh vegetables for home consumption. Second, migrant households can use the additional income from remittances for purchasing meat and fruit at the market. Eating fruits and meat also conveys social status, as these items are much more expensive than cereals and vegetables. Third, and related to the second point, in the past few decades there has been a rapid expansion of market infrastructure both within and outside the research area, which has broadened the range of food and non-food items available.

The findings of this study do not provide unqualified support for the assumptions of many international development practitioners that migration from rural areas has a positive impact on food security in the area of origin. It is not that migration cannot contribute positively – it can – but the relationship is far more complex than it appears at first glance. While migration may bolster immediate access to food at the household level through remittances, as we shall argue below the associated value changes could threaten food security in the longer term.

The findings show that the process of labour out-migration is closely linked with cultural change and food security dynamics in the area of origin. These changes manifest themselves predominantly in a preference for purchased food over food from own agricultural production and in the way households restructure their livelihoods in order to satisfy that preference. In areas where livelihoods have historically been agriculture-based, migration fundamentally changes the values that underpin food procurement from agriculture. In the migration context, considerations of convenience, new markers of social status and modernity aspirations are playing an important role in people's livelihood and food choices. The results suggest that a 'culture of migration' gives rise to a shift away from subsistence agriculture. This changes the way food security is achieved at the household level, that is by buying food rather than by own food production. Such a development may threaten Nepal's food sovereignty in the end, as the *terai* area where so many migrants come from is the country's granary.

The decision to migrate is born largely from a need for cash income and the desire to improve family wellbeing. However, migration and the associated labour loss in the migrant-sending communities render subsistence agriculture less viable as a means to procure food. This, in turn, increases households' reliance on bought food, necessitating those households to generate additional income. If migration is the best (or the only) opportunity to generate such income – and in many cases it is – then a culture of migration will inevitability decrease agricultural production. Households may find themselves with an increased need for cash and improved wellbeing, but with limited and unsecure means to meet this need. Consequently, migration-driven dependency on bought food at the expense of own food production could be unsustainable and eventually lead to stress in food systems at macro level. In general, those we spoke to viewed migration as positive, acceptable and even inevitable. However, its inevitability is not a given. The opportunity to migrate is subject to change when the demands in the area of destination or the regulatory environment undergo significant changes.

Given the agrarian structure of rural areas like the research area, a decrease in the quantity of locally produced food will adversely affect non-migrant households and the community as a whole. Then, subsistence agriculture can no longer provide a safety net for communities in Nepal (cf. Halbrendt *et al.*, 2014). Migration is engendering a change in the role that subsistence agriculture always played in rural livelihoods and lifestyles, with migrants and their families unwilling to hold on to the traditional ways of life. There is a problem of incompatibility, to be resolved only if agriculture is transformed in such a way that (1) it becomes more profitable and less 'dirty' and (2) it can also be part of a 'modern' younger-generation migrant-family lifestyle. Lifestyle encompasses practices, perceptions, attitudes and values. It is not just what people do but rather why they are doing it the way they do and how they would like to do it.

The above applies to both migration and practicing agriculture. The responsibility of the government is to steer these processes for the wellbeing of its citizens. However, whereas for migration a regulatory system has been put in place to protect the rights of migrant workers, agricultural work seems to have been more or less left to its own devices. In our view, the situation as pictured in this paper contains a threat to longer-term food security, which can only be averted by a concerted effort to make agriculture more attractive. If at the start of the migration era it was anticipated that remittances would be profitably invested in the agricultural sector, the results of this research can dismiss this as an illusion.

In the development literature it is argued that although migrants are de-linking themselves from land and farming, the higher incomes associated with migration are linked with improved access to food among migrant households (Zezza *et al.*, 2011). Accordingly, migration would not have any negative impacts on household food security. The problem with this argument is that it lacks a temporal dimension – it holds all factors constant when making assumptions about the future state of a given household's food security status. This point of view is also rather optimistic. Not only may the opportunity to migrate change, but the 'culture of migration' is going to affect future patterns of agricultural production as well. By framing the migration-food security relationship in a one-dimensional way, policy makers risk overlooking the potential longer-term insecurities in the areas of origin.

The conclusion is that only by conceptualising the relationship between out-migration and food security as dynamic and multi-dimensional – comprising practices, resources and people's perceptions on these – the incompatibilities we have shown to exist, can be resolved. Policies should be geared towards sustainable lifestyles, in areas like the *terai* based on sustainable and profitable agriculture.

Acknowledgement

The authors acknowledge the generous support of the Neys-van Hoogstraten Foundation, the Netherlands, for the fieldwork of the first author in Nepal. We are also indebted to the peer-reviewers Dr Derek Johnson (University of Manitoba, Canada) and Dr Kishor Atreya (Asian Network for Sustainable Agriculture and Bioresources, Nepal) for their constructive comments that improved the quality of this paper.

References

Adams, R.H. and Page, J., 2005. Do international migration and remittances reduce poverty in developing countries? *World Development* 33(10): 1645-1669.

Adger, W.N., Kelly, P.M., Winkels, A., Huy, L.Q. and Locke, C., 2002. Migration, remittances, livelihood trajectories, and social resilience. *AMBIO: A Journal of the Human Environment* 31(4): 358-366.

Berthélemy, Jean-C., Beuran, M. and Maurel, M., 2009. Aid and migration: substitutes or complements? *World Development* 37(10): 1589-1599.

CBS, 2011. *National population census 2011: major highlights*. Central Bureau of Statistics, National Planning Commission, Government of Nepal, Kathmandu, Nepal. Available at: http://cbs.gov.np/sectoral_statistics/population/major_highlights.

Crush, J., 2013. Linking food security, migration and development. *International Migration* 51(5): 61-75.

De Brauw, A. and Giles, J., 2008. *Migrant labour markets and the welfare of rural households in the developing world: evidence from China.* The World Bank, Washington DC, USA.

DOFE, 2014. Labour Migration for Employment: A Status Report for Nepal 2013/14. Kathmandu: Department of Foreign Employment, Ministry of Labour and Transport Management, Government of Nepal. Available at: http://tinyurl.com/y9xq6wqk.

Drèze, J. and Sen, A.K., 1989. *Hunger and public action.* Clarendon Press, Oxford, UK.

Durand, J., Kandel, W., Parrado, E.A. and Massey, D.S., 1996. International migration and development in Mexican communities. *Demography* 33(2): 249-264.

FAO, 1974. *The state of world food and agriculture situation.* Available at: www.fao.org/docrep/017/f3350e/f3350e.pdf.

FAO, 1996. *World food summit. Rome declaration on world food security.* Available at: www.fao.org/docrep/003/W3613E/W3613E00.HTM.

FAO, 2003. *Trade reforms and food security: conceptualizing the linkages.* Available at: www.fao.org/docrep/005/y4671e/y4671e00.htm.

FAO, 2013. *The state of food insecurity in the world 2013: the multiple dimensions of food security.* Available at: www.fao.org/docrep/018/i3434e/i3434e00.htm.

FEPB, 2007. Introduction of the board. Available at: http://fepb.gov.np/content.php?id=1

Gartaula, H.N., 2011. *The other side of migration in rural Nepal: sociocultural transformation and the women left behind.* PhD Thesis Wageningen University, Wageningen, the Netherlands. Available at: http://edepot.wur.nl/185275.

Gartaula, H.N. and Niehof, A., 2013. Migration to and from the Nepal terai: shifting movements and motives. *The South Asianist* 2(2): 28-50.

Gartaula, H.N., Patel, K., Johnson, D., Devkota, R., Khadka, K. and Chaudhary, P., 2016. From food security to food wellbeing: examining food security through the lens of food wellbeing in Nepal's rapidly changing agrarian landscape. *Agriculture and Human Values* 34: 573-589.

Gartaula, H.N., Visser, L. and Niehof, A., 2012. Sociocultural dispositions and wellbeing of the women left behind: a case of migrant households in Nepal. *Social Indicators Research* 108(3): 401-420.

GON, 2007. *Foreign employment act, 2064 (2007).* Government of Nepal, Kathmandu, Nepal.

Gupta, S., Pattillo, C.A. and Wagh, S., 2009. Effect of remittances on poverty and financial development in Sub-Saharan Africa. *World Development* 37(1): 104-115.

Hachhethu, K., 2000. Nepali politics: political parties, political crisis and problem of governance. In: Kumar, D. (ed.) *Domestic conflicts and crisis of governability in Nepal.* Centre for Nepal and Asian Studies, Trubhuvan University, Kathmandu, Nepal, pp. 1-19.

Halbrendt, J., Kimura, A.H., Gray, S.A., Radovich, T., Reed, B. and Tamang, B.B., 2014. Implications of conservation agriculture for men's and women's workloads among marginalized farmers in the central middle hills of Nepal. *Mountain Research and Development* 34(3): 214-222.

Lecomte-Tilouine, M., 2009. *Hindu kingship, ethnic revival, and Maoist rebellion in Nepal.* Oxford University Press, New Delhi, India.

Maxwell, S. and Frankenberger, T.R., 1992. *Household food security: concepts, indicators and measurements.* UNICEF and IFAD, New York and Rome, USA and Italy.

McGregor, A.J., 2006. *Researching wellbeing: from concepts to methodology*. In: Gough, I. and McGregor, A.J. (eds) Wellbeing in developing countries: from theory to research. Cambridge University Press, Cambridge, UK, pp. 316-350.

McGregor, J.A., McKay, A. and Velazco, J., 2007. Needs and resources in the investigation of well-being in developing countries: illustrative evidence from Bangladesh and Peru. *Journal of Economic Methodology* 14(1): 107-131.

Meng, X., 2014. *Feminization of agricultural production in rural China: a sociological analysis*. PhD Thesis Wageningen University, Wageningen, the Netherlands. Available at: http://edepot.wur.nl/287465.

Mills, M.B., 1997. Contesting the margins of modernity: women, migration, and consumption in Thailand. *American Ethnologist* 24(1): 37-61.

Newton, J., 2007. *Structures, regimes and wellbeing*. WeD Working Paper 30, University of Bath, Bath, UK.

Nguyen, M.C. and Winters, P., 2011. The impact of migration on food consumption patterns: the case of Vietnam. [Assessing the Impact of Migration on Food and Nutrition Security]. *Food Policy* 36(1): 71-87.

Niehof, A., 2010. *Food, diversity, vulnerability and social change: research findings from insular southeast Asia*. Wageningen Academic Publishers, Wageningen, the Netherlands.

Noack, A.-L. and Pouw, N.R.M., 2015. A blind spot in food and nutrition security: where culture and social change shape the local food plate. *Agriculture and Human Values* 32(2): 169-182.

Patel, R., 2009. Food sovereignty: what does food sovereignty look like? *The Journal of Peasant Studies* 36(3): 663-706.

Radel, C., Schmook, B., McEvoy, J., Méndez, C. and Petrzelka, P., 2012. Labour migration and gendered agricultural relations: the feminization of agriculture in the Ejidal sector of Calakmul, Mexico. *Journal of Agrarian Change* 12(1): 98-119.

Russell, S.S., 1992. International migration and political turmoil in the Middle East. *Population and Development Review* 18(4): 719-727.

Scoones, I., 1998. *Sustainable rural livelihoods: a framework for analysis*. Institute of Development Studies, Brighton, Sussex, UK.

Scoones, I., 2009. Livelihoods perspectives and rural development. *The Journal of Peasant Studies* 36(1): 171-196.

Sen, A., 1981. *Poverty and famines: an essay on entitlement and deprivation*. Clarendon Press, Oxford, UK.

Sharma, J.R., 2008. Practices of male labor migration from the hills of Nepal to India in development discourses: which pathology? *Gender, Technology and Development* 12(3): 303-323.

Toyota, M., Yeoh, B.S.A. and Nguyen, L., 2007. Bringing 'the left behind' back into view in Asia: a framework for understanding the 'migration-left behind nexus'. *Population, Space and Place* 13(3): 157-161.

Vargas-Lundius, R., Lanly, G., Villarreal, M. and Osorio, M., 2008. *International migration, remittances and rural development*. Available at: www.ifad.org/pub/remittances/migration.pdf.

Watts, M.J. and Bohle, H.G., 1993. Hunger, famine and the space of vulnerability. *GeoJournal* 30(2): 117-125.

White, S., 2009. *Bringing wellbeing into development practice.* Working Papers: WeD Working Paper 09. University of Bath, Bath, UK.

White, S. and Ellison, M., 2006. *Wellbeing, livelihoods and resources in social practice.* Available at: http://tinyurl.com/y9ypq9ej.

World Bank., 2017. *Personal remittances, received (% of GDP).* Available at: http://data.worldbank.org/indicator/bx.trf.pwkr.dt.gd.zs

Yaro, J.A., 2004. Theorizing food insecurity: building a livelihood vulnerability framework for researching food insecurity. *Norsk Geografisk Tidsskrift – Norwegian Journal of Geography* 58(1): 23-37.

Zezza, A., Carletto, C., Davis, B. and Winters, P., 2011. Assessing the impact of migration on food and nutrition security. *Food Policy* 36(1): 1-6.

Zuo, J., 2004. Feminization of agriculture, relational exchange, and perceived fairness in China: a case in Guangxi province. *Rural Sociology* 69(4): 510-531.

3. Indigenous knowledge and household food security: the role of root and tuber crops among indigenous peoples in the Northern Philippines

B.T. Gayao, D.T. Meldoz and G.S. Backian*

*Northern Philippines Root Crops Research and Training Center, Benguet State University, La Trinidad, Benguet, 2601 Philippines; *btgayao@yahoo.com*

Abstract

The indigenous peoples of Northern Philippines have always considered roots and tubers as staple food and vegetable. With the changing lifestyles of the younger generation and the lack of documentation, indigenous knowledge on roots and tubers is at risk of disappearing. This study aimed at mapping the knowledge on roots and tubers and assessing the role of roots and tubers in household food security among 15 ethno-linguistic indigenous groups in the Philippines. As part of the study, secondary data analysis, key informant interviews, focus-group discussion, field visits, administered tests and validation follow-ups were conducted. More than 20 roots and tubers consisting of 10 cultivated species and more than 10 species of wild roots and tubers were identified. The first include cassava, sweet potato, greater yam, taro, tannia, lesser yam, arrowroot, potato, yacon and yam bean, the second nami, elephant yam, giant taro, canna, wild yam bean and wild species of taro, greater yam and lesser yam. Among the indigenous people, the roots and tubers are not only used for meeting household food needs, but are also traded, used for medicinal purposes and as gifts and offerings. In the eyes of the people, root and tuber crops farming inculcates the virtues of industriousness and self-reliance. Unfortunately, only those born before the 1980's retain traditional knowledge pertaining to the diversity and variety of roots and tubers, cropping practices, and utilisation. The younger generation already exhibited loss in both conceptual knowledge and practical skills regarding roots and tubers.

Keywords: socio-economic importance, retention of traditional knowledge

3.1 Introduction

Root and tuber crops have been identified as food crops to feed the world in coming decades (Scott *et al.*, 2000). Many of the developing world's poorest and most food insecure households will continuously value these crops for their contribution food availability, nutrition, and cash income. Roots and tubers also serve as staple food when there are for rice shortages. In the past 25 years, the Philippines have been importing 10% of its rice requirement annually because of shortages in domestic rice production, and in 2008 became the world's largest importer of rice (IRRI, 2008). Increasing food crops production for cash also contributed to this situation. These trends have repressed indigenous knowledge

DOI 10.3920/978-90-8686-864-3_3,

on the production and utilisation of traditional staple foods such as roots and tubers, especially among mountain dwellers, indigenous peoples or ethnoscapes. At the same time, documentation of indigenous knowledge on roots and tubers is needed for further research and development, to use their potential for improving the livelihoods, diets and nutrition of the indigenous peoples and rural households.

Basic to research and development is the availability of a database. At present, there is limited documentation of indigenous knowledge on roots and tubers practices in the Philippines. Roots and tubers is a collective term for plants that are grown for their modified, thickened root or stem which generally develops underground. Special terms for specific types are root crops for modified roots, tuber crops and corm crops for modified stems (Bareja, 2010). A rhizome is a modified stem. Balaki and Solimen (1991) and Verdonk (1991) have mentioned root and tuber crops, particularly sweet potato, as survival crops and for weathering the ill effects of typhoons and food crises among marginal swidden farmers who do not own paddies in the rice terraces of Ifugao. Dayo *et al.* (1998) described the year-round cultivation of sweet potato (*Ipomoea batatas*), ubi (*Dioscorea alata*) and tugui (*Dioscorea esculenta*) as subsistence crop in the typhoon prone Batanes island province and the consumption of wild yams by El Niño drought- affected upland farmers, and by fishermen and farm labourers while awaiting harvest of the wet season rice crop. Such studies have contributed to the image of roots and tubers as starchy staples (energy source) only for resource-poor families and during crises, even though the crops contain varied quantities of protein, essential vitamins and minerals (Horton, 1988).

Indigenous knowledge and practice is the knowledge or practice that people in a given community have developed over time and continuously developed to adapt to local culture and environment (NAARAP, 2009). Indigenous knowledge can be lost naturally as practices are modified or when knowledge is left unused for a long time.

The establishment of the root crops centre in the Northern Philippines was based on the knowledge that sweet potato and other roots and tubers are staple food of the indigenous people in the hinterlands of the Caraballo and the Gran Cordillera mountain ranges. However, the focus in root crops research and development has shifted towards commercial uses to meet industry demands for starch, feed and energy, and for the export market. For example, white potato varieties are now developed for farmers to plant to meet the demands of potato chips and fries manufacturers, importers and fast food chains. In the case of cassava, high-yielding varieties needed by the feeds industry and alcohol distillers are prioritised. In the Cordillera Region, however, indigenous people still use roots and tubers as primary foods, and whatever is not eaten, such as peelings, is given as feed to pigs, chicken, ducks, rabbit and even fish (Gayao *et al.*, 2013-2014). But more support is needed to improve this kind of indigenous food system. The food sovereignty activists are fighting for the people's right to decide what they eat and what they produce (Coulibaly, 2011). Food sovereignty gained momentum because of the 2008 global food price crises and calls for fundamental change in food and agricultural policy so that risks and benefits of food systems are shared equitably. It is about the right of people to healthy and culturally appropriate food produced through ecologically sound and sustainable methods, and their right to define own food and agriculture systems (Shawki, 2012).

Roots and tubers are traditionally planted as mixed crops in shifting cultivation. Some crop species like yam, tannia and varieties of sweet potato are left to grow as wild crops when the swidden farm is fallowed or reforested. Mixed crops grown in subsistence agriculture such as roots, tubers, legumes, cereals, vegetables, spices and fruits as practiced by the Aeta, Mangyan and Bugkalot indigenous people in the Philippines (Gayao *et al.*, 2013-2014) contribute to biodiversity and food diversity.

At three levels, those of ecosystems, species, and genetic diversity within species, biodiversity is important (Sunderland, 2011). Biodiversity provides a safety net during times of food insecurity, as the consumption of sweet potato as major staple food in the Cordillera highlands during World War II shows (Solimen *et al.*, 1998). Other cases are the consumption of wild yam *kamangeg* (*Dioscorea* spp.) during seasonal food gaps by fishermen and farm labourers in the coastal village of Ilocos Sur and by the population of Benguet during the aftermath of the 1990 earthquake (Sano *et al.*, 1991), and the consumption of wild yams (*Dioscorea hispida*) by upland farmers in Mindanao due to crop losses resulting from the El Niño drought.

The contribution of roots and tubers to the energy supply in different populations varies per country. The relative importance of these crops is evident from their annual global production of approximately 836 million tonnes. Asia is the main producer followed by Africa, Europe, and America. Asian and African regions produce 43 and 33%, respectively, of the global production of roots and tubers. Cassava, potato, and sweet potato make up 90% of the global production of root and tuber crops (Chandrasekara and Kumar, 2016). Starchy root and tuber crops are second only in importance to cereals as global sources of carbohydrates. They provide a substantial part of the world's food supply and are an important source of animal feed and processed products for human consumption and industrial use. In Asian countries, some edible tubers are used as traditional medicine (Chandrasekara and Kumar, 2016). The extraction and processing of starch from agricultural commodities, notably from cassava, potato and sweet potato, is one of the most important rural agro-industries in Asia (Fuglie *et al.*, 2006).

The Chennai Declaration states: '*Biodiversity is the raw material for food and health security, as well as for the biotechnology industry, and it must be conserved to ensure that it can continue in this function so that farming systems become climate resilient*' (Sunderland, 2011). The declaration reads as a response to the intensive commercialisation and crop specialisation that limit the crops and varieties planted by farmers, the abuse of external chemical farm inputs, and the indiscriminate conversion of forestlands to commercial farms. While these developments also brought about economic prosperity, the negative impacts include those on food security, livelihoods and culture, especially among indigenous people (Davies, 1998; Agada and Igokwe, 2016; Hens, 2006), on ecological systems, and on human health as evidenced in many studies cited by Gadgil *et al.* (1993), Sunderland (2011), and Shawki (2012). According to Gollin (2008), biodiversity is an essential requirement for food security.

Studies have also shown that the diversity in local food species and indigenous practices of mixed cropping correlate positively with dietary quality (Koppmair *et al.*, 2017; Kuhnlein, 2007). Hence, it is important to improve markets for farm produce and technologies that increase farm productivity (Sibhatu *et al.*, 2015). The advent of the Green Revolution that increased rice productivity in particular, also resulted in the erosion of indigenous knowledge on the production of other subsistence crops like roots and tubers, vegetables, legumes and fruits.

Indigenous practices and traditional knowledge may not always be the answer (Briggs, 2005), but instead of re-inventing the wheel, development workers, researchers and scientists could use indigenous knowledge or combine it with modern technology for feasible solutions (Thiagarajan, 2008). Indigenous knowledge is valuable given that large numbers of the world's population are still dependent on agriculture and pastoralism, forestry and fisheries, and on products crafted from natural resources such as housing for their survival and livelihoods. Even in textile and pharmaceutical industries traditional knowledge is applied (Kothari, 2007). Indigenous knowledge systems are slowly disappearing due to the changing needs and interests of indigenous peoples (Camacho *et al.*, 2016), but given their potential for sustainable development and the threat of climate change (Bruyere *et al.*, 2016), governments and other stakeholders should take these seriously. Experiences in China attest to the complementarity of indigenous and foreign innovation efforts (Fu and Soete, 2011). Information on indigenous knowledge, skills and practices remains indispensable to food sovereignty, food availability, and livelihood and entitlement – the three major approaches in addressing the four pillars of food security, namely: availability, access, utilisation and stability (Gartaula *et al.*, 2017).

3.2 Conceptual framework

Supporters of indigenous peoples claim that these peoples' knowledge and farming practices have ensured food security by providing households with a diverse diet consisting of cereals, root crops, legumes, and a variety of vegetables and fruits. Niehof (2010) mentions that in Southeast Asia the younger generation no longer recognizes some traditional foods from roots and tubers, and that the knowledge about these foods is declining. The lack of documentation and dissemination of indigenous knowledge is coupled with the changes in environment, culture, food and farming systems and policies, and in political and economic structures. These trends could result in the erosion of indigenous knowledge, loss of root and tuber crop diversity and in the labelling of roots and tubers as an inferior and poor man's food. Consequently, the potential development of root and tuber crops as sources of food and cash income is also compromised.

This research aimed at answering the following questions: (1) What roots and tubers are known planted or gathered among the indigenous peoples (IPs) in Northern Philippines?; (2) What is the socio-economic significance of these roots and tubers among IPs compared to non-IP farmers?; (3) What is the retention of indigenous knowledge on the diversity and utilization of these roots and tubers among IP groups?; (4) How is indigenous knowledge on roots and tubers gained and transferred?; (5) Will it be prudent to promote roots and tubers as a substitute or complementary staple food, a specialty food, a security food crop, a source of cash income, or shall these crops be confined only to museums or gene banks? The linkages discussed and addressed by the questions are pictured in Figure 3.1.

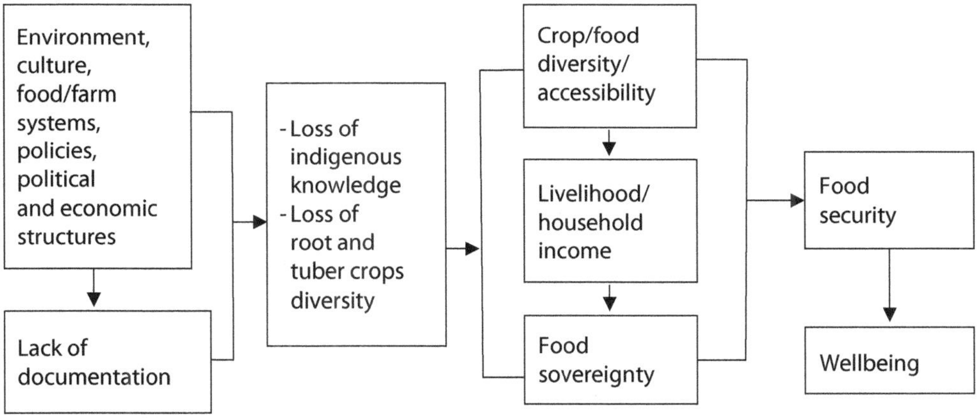

Figure 3.1. Conceptual framework on the consequences of lack of documentation in indigenous knowledge of root and tuber crops.

3.3 Research area and methodology

3.3.1 Location and duration

Fifteen ethnoscapes (ethno-linguistic groups) composed of 13 IPs and two majority ethnoscapes, all known root crops farmers and consumers in the Northern Philippines, were included in the first phase of documentation. These included the indigenous inhabitants of the Cordillera mountain ranges, collectively called *Igorot* (comprising the *Ibaloi, Kankana-ey, Applai, Biga-Kalinga, Iyattuka-Ifugao, Isneg, Tingguian* and *Kalanguya*); those that settled in the mountains bordering the Ilocos plains like the *Bago*; those that roam the Sierra Madre mountain ranges like the *Bugkalot*; IPs in the hinterlands of Central Luzon and Mindoro like the *Aeta* and *Mangyan*; and the native inhabitants of Batanes, the *Ivatan*. However, in the second phase of the return case studies, only the *Ibaloi* and *Bago* IPs and the *Kapampangan/Ilocano* majority ethnoscapes were included. In the third phase, the traditional root crop knowledge (TRK) test was conducted only among the *Ibaloi, Bago, Tingguian* and *Iyattuka*.

The research was conducted from February 2012 to December 2014. Preparations for the implementation of the research started in February 2012. Key informant interview workshops, follow-up interviews, writing of the research report per IP group, and the TRK test were conducted from April 2012 to October 2014. The writing of papers for presentations as well as composing information-education materials went beyond 2014.

3.3.2 Data gathering procedure and methods

The steps in carrying out the research are shown in Figure 3.2.

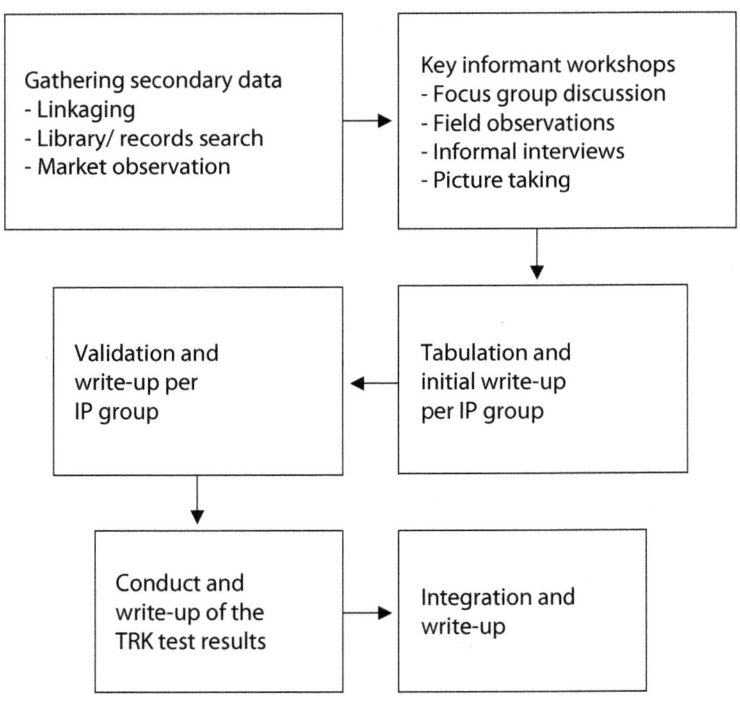

Figure 3.2. Steps in conducting the research.

Selection of key informants and respondents

In Phase 1 key informants and respondents were selected based on familiarity in growing and consuming root crops, age (15 to 70 years) and duration of residence in the community. Based on the completeness of the data gathered in Phase 1, the *Ibaloi* of Taloy Sur in Tuba, Benguet; the *Bago* of Banga and Malikliko in Sugpon, Ilocos Sur and Sudipen, La Union, and the *Kapampangan/Ilocano* in Central Luzon were selected for the Phase 2 cost and return analysis.

Four of the 13 IPs, namely: the *Ibaloi, Bago, Masadiit Tingguians* of Labaan in Bucloc, Abra; and the *Iyattuka* of Amduntog in Asipulo, Ifugao, were selected for the Phase 3 traditional knowledge aptitude test. Thirty test subjects per IP group took the aptitude test. They included adolescents, young adults, the middle-aged and the elderly, and were later on were grouped into three age-groups – 15-35 years old, 36-56 years old and 57-77 years old – and equally divided into male and female subjects (Table 3.1). The respondents were purposively identified with possible replacement by those who were available at the time of the test administration.

Data collection methods

Interviews and focus group discussion were done during the workshop starting with story telling about the village- the ethnic groups or indigenous peoples residing in the village, the crops planted, food eaten and others. This was followed-up with the naming of root crops growing in the village accompanied with showing pictures of roots and tubers, if needed

Table 3.1. Number of traditional knowledge test subjects per IP group and age group.

IP group	Male	Female	Total	%
Ibaloi	16	22	38	24.1
Bago	20	22	42	26.6
Tingguian	19	17	36	22.8
Iyattuka	19	23	42	26.6
Total	74	84	158	100.1
Age group				
15-35 years	29	23	52	32.9
36-56 years	24	29	53	33.5
57-77 years	21	32	53	33.5
Total	74	84	158	
%	46.8	53.2		99.9

to aid recall of what and where planted or gathered. These were listed in cue cards or large sheets of paper as guide for recalling local names of roots and tubers and other information shared and as basis for the assessment of the extent of production (Box 3.1).

Box 3.1. Categorization in assessing extent of production and use as perceived by the key informants (modified, based on understanding extent and distribution of diversity, from Sthapit *et al.*, 2001 and IPGRI, 2002).

a. Large area – many growers (abundant)
b. Large area – few growers (abundant)
c. Small area – many growers/users (occasional)
d. Small area – few growers/users (rare; need of conservation)

Sometimes, the researchers also shared their knowledge on the local names of roots and tubers grown, eaten and special use by other indigenous peoples to liven-up the workshop discussions and to steer the informal interviews as per guide questions. This was followed with a field visit and community walk for observations and taking pictures. Narrative description and photos gave detailed information on the root and tuber crops. Tabulated information, cost and return analysis, and initial write-up were done after the interviews and presented to the collaborating local researchers and some key informants for clarifications, corrections and additional information.

In the last phase of the research, to measure retention or loss of knowledge, the Vitality Index of Traditional Environmental Knowledge (VITEK) methodology was used, which comprises the following features: the listing or documentation of traditional knowledge, defining the conceptual knowledge (know-what) and the practical skills (know-how) knowledge (Zent, 2010a). Based on this methodology, the traditional knowledge aptitude (TRK) test questionnaire was prepared and then administered to a sample of selected subjects. The tests were administered individually with the assistance of the local coordinators who clarified and/or translated the questions especially for those who were less literate or had problems with understanding the questions. Checking and scoring of the completed questionnaires were done *en banc* at each site by the project researchers and local partners.

Analysis

Analysis on roots and tubers diversity and use was largely descriptive, looking at household resource management and use under limited conditions. Indigenous knowledge was described separately as a series per IP group (cf. Gayao, *et al.*, 2013-2014). Included in these series were the results on diversity and use of roots and tubers; the diversity of local varieties for sweet potato, greater yam, taro, cassava, lesser yam and tannia; the land use and cropping systems; the seed system; the crop and management practices; and utilisation.

Socio-economic significance was determined from the traditional utilisation practices and from the cost-benefit analysis. The latter was simplified by converting economic values to monetary values to re-construct the financial costs and benefits statements following the cost accounting process based on prevailing farm-gate prices and unit costs during the field work. Non-use value and existence value of the crop and opportunity costs were not included.

To test differences and relationships in scores among IPs and as influenced by age groups, gender and source of knowledge, the Pearson's chi-square and Pearson's R were used. For the intergroup comparison, the calculations of VITEK statistics were separately done and analysed per IP case that is the intergenerational rate of retention (RG) and the annual rate of change (CA). Formulas for calculation are:

RG indicates the rate of retention between any successive pair of age groups, computed as follows:

$$RG_t = g_t / g_r$$

where g_t = mean score of the target age group (younger age group), and g_r = mean score of the reference age group.

RG_t of the oldest age group is set at 1 based on the logic that no information about the aptitude level of the preceding generation(s) is available and therefore we cannot assume that any difference or changes have occurred in prior periods.

CA expresses the average rate and direction of change per year reflected by the target age group, computed as follows:

$$CA_t = \frac{RC_t - 1}{yg_t}$$

where yg_t is the length in years of the target age group interval.

3.4 Results and discussion

3.4.1 Diversity of roots and tubers

There were more than twenty roots and tubers identified by the IPs in Northern Philippines. These consist of five root crops that have enlarged or modified roots, 12 tuber crops that have modified underground stem; and six corm crops that also have a modified underground stem.

Belonging to the root crops are cassava *(Manihot esculenta)*, sweet potato *(Ipomoea batatas)*, yambean *(Pachyrhizus* spp.*)*, yacon *(Smallanthus sanchifolius),* and wild yam bean (Table 3.2). Cassava and sweet potato are planted by the 13 IP's selected in this study.

Yambean is more of a cash crop among the majority *Kapampangan* and *Ilocano* and in very limited area among the *Applai-Kankana-ey*. Only the Ibaloi and Kalanguya plant Yacon. *Kahoy* is a more popular local name of cassava among the IPs than *katimoro, moro, dutung, balinghoy, balangoy, padpadli, samal, hanglay and kalasan.* Local names of sweet potato are more confusing; the *Kalanguya, Iyattuka* and *Applai-Kankana-ey* call it *ubih*, whereas the *Ibaloi, Kankana-ey* and *Applai* refer to it as *dokto, lokto* or *tugi. Yacon* and *singkamas* for yam bean is the same for all IPs. Local names reflect the IPs' knowledge on the diversity of roots and tubers grown, gathered or eaten.

Table 3.2. Root crops known, grown and used by the indigenous peoples in Northern Philippines.

Roots crops	Scientific name[1]	Local IP names
Cassava	*Manihot esculenta*	*Katimoro, Kahoy, Kamoteng Dutung, Moro, Balinghoy, Padpadli, Kayo-kayo, Samal, Hanglay, Balangoy, Kaka-iw, Molo, Kalasan*
Sweet potato	*Ipomoea batata*	*Dokto, Tugui, Palay, Kamoteng Gapang, Kamochi, Wakay, Kantila, Baynaw, Kamote, Gaselang, Bella, Ubbi, Lokto, Katila, Ubih, Tugi*
Yam bean	*Pachyrhizus* spp.	*Singkamas*
Wild yambean	*Pachyrhizus* spp.	*Badang Singkamas, Sagsag-ot, Ket-el, Battog, Kitkitli, Singkamas ti Bantay*
Yacon	*Smallanthus sanchifolius*	*Yacon*

[1] Cf. Scott *et al.* (2000).

The tuber crops shown in Table 3.3 are greater yam *(D. alata)*, lesser yam *(D. esculenta)*, arrowroot *(Maranta arundinacea)*, potato *(Solanum tuberosum)*, canna *(Cannaceae)*, aerial yam *(Dioscorea bulbifera)*, nami *(D. hispida)*, elephant yam *(Amorphopallus campanulatus)*, wild lesser yam *(Dioscorea* spp.*)*, and three wild species of greater yam.

Greater yam and lesser yam were the tuber crops planted by the 13 IPs. Arrowroot was not among tubers planted by the *Ivatan, Bugkalot* and *Iyattuka*. Only the *Ibaloi* and *Kankana-ey* grow potato. Arrowroot is sometimes a volunteer crop (plants not deliberately planted by farmers, crops that have tendency to re-grow in fallowed farms), whereas canna and aerial yam just like the wild yambean were domesticated. Once planted, arrowroot grows perennially and voluntarily. Some arrowroot varieties were also grown for ornamental purposes. The arrowroot is sometimes interchanged with canna such that it is also called *awwing* or *galumaca* by the *Kalanguya* and *Applai-Kankana-ey*, or *sugod, sago, araro, ahoki, biyas, wikan, luslussi* and *kurita* among the other IPs. Canna is similar to the *kulintas or gaddang* of the *Ibaloi*, *galumaca* of the *Kankana-ey* and *Isneg*, *bonte-ek* of the *Kalinga* and *litogak* or *sag-ot* of the *Tingguian* and *awwing* of the *Applai*. Wild roots and tubers such as nami, elephant yam and unknown species of *Dioscorea* that were once harvested because of

Table 3.3. Tuber crops known, grown and used by the indigenous peoples in Northern Philippines.

Tuber crops	Scientific name[1]	Local IP names
Greater Yam	*Dioscorea alata*	*Uve, Ubi, Liwet, Uvi, Ongo, Ulang, Guhhudan*
Lesser yam	*Dioscorea esculenta*	*Balugan, Tugtugi, Limeng, Dukay, Lufto, Buwang, Lokto, Ontoy, Tugi, Tugui, Lukto, Pukupok*
Arrowroot	*Maranta arundinacea*	*Sugod, Sago, Araro, Ahoki, Biyas, Wikan,Awing, Luslussi, Kurita, Galumaca*
Aerial yam	*Dioscorea bulbifera*	*Kafu-ngaw, Batata, Ilos*
Potato	*Solanum tuberosum*	*Pafas, Patatas, Papas*
Canna	*Cannaceae*	*Kulintas, Gadang, Galumaca, Bonte-ek, Litogak, Sag-ot, Awwing*
Nami	*Dioscorea hispida*	*Kalot, Ka-ut, Karot, Ligaw na Liwet, Mintakey*
Wild greater yam (3 species)	*Dioscorea* spp.	*(Kasey, Lima-lima, Kanapan), Kamangeg, (Durian, Dimurian, Labit, Diyan, Kabwang, Amakey, Borot, Lugiman, Atap Ubi, Dalakit, Gallod, Iyog-iyogan, Ipoy)*
Wild lesser yam	*Dioscorea* spp.	*Boga, Atap Ontoy, Anayed*
Elephant yam	*Amorphopallus campanulatus*	*Tigi, Boton Lakay, Bageng, Pokpoklit*

[1] Cf. Scott, *et al.* (2000).

food and feed scarcity – like during World War II and at the time of the Japanese occupation in the Philippines – were rarely gathered and less known by all IPs. Examples are the aerial yam as *kafu-ngaw, batata* and *ilos,* the *kamangeg, ka-sey* or *lima-lima, durian* or *dimurian, labit, ipoy* or *dalakit, diyan, amakey, borot, lugiman, kanapan, gallod, anayed, atap ontoy* and *boga.* Many of these wild yams hardly exist nowadays, and some were already lost just like the purple-fleshed yams of the *Mangyans.* Nami which is commonly referred to as *kalot, karot, ka-ut, ligaw na liwet* and *mintakey,* requires thorough processing thus is no longer or very seldom gathered from the wild. The *tigi* plants that were profusely growing in the *Tingguian* village are now considered weeds in the farm, but one kind with smooth brownish and white spotted stalk is used as vegetable.

Table 3.4 presents the corm crops, i.e. taro *(Colocasia esculenta),* tannia *(Xanthosoma sagittifolium* and *Xanthosoma violaceum),* giant taro *(Alocasia macrorrhiza)* and three wild taro species. Taro is familiar among the indigenous people in northern Philippines and is used as staple food, specialty vegetable dish, animal feed and now for some as a cash crop. All the 13 indigenous peoples surveyed in Northern Philippines have their own local names for taro: namely, *gabi, aba, ava, abah, pising* and *pihing, gandus, gamey, buyon, putlo, ateng, sudi, lidoy, lapa, loko* and *rabok.* Tannia of which the corm is often mistaken for taro is called *galyang, galiang, galyan, bila, bilah, biga, amasyan, lusya, bulawan* and *singapor.* The giant taro is also called *bila* by the *Kalinga. Galyang* is a cultivated crop but to some IPs, it is a volunteer or a wild plant. Wild taro looks similar to the cultivated varieties and was identified as *lapa, bileng, lubingan, kalapon, wild abah, sed-ing/hed-ing* and *tayagan.* Other wild species of taro grouped in two kinds: those called *bitajon, bitayen, bilagot,* and *bulagot,* and those called *pikaw, pi-aw, pihkaw* and *lanipog.*

Table 3.4. Corm crops known, grown and used by the indigenous peoples in Northern Philippines.

Corm crops	Scientific name[1]	Local IP names
Taro	*Colocasia esculenta*	*Ava, Aba, Gamey, Boyon, Gandus, Sudi, Ateng, Gabi, Lidoy, Putlo, Pihing, Loko, Rabok, Pising*
Tannia	*Xanthosoma sagittifolium (white); Xanthosoma violaceum (purple)*	*Galyang, bila, Galyan, Biga, Canaka, Tanaka, Lusya, Singapor, Bulawan, Galiang Lidoy, Galjang, Amasyan, Bilah,*
Wild taro (3 species)	*Colocasia* spp.	*Pikaw, Bitajon, Lapa, lanipog, Ileng, Pi-aw, Bilagot*
Giant taro	*Alocasia macrorrhiza*	*Bila*

[1] Cf. Scott *et al.* (2000).

Cassava, sweet potato, potato and yam are among the major food crops in developing countries in addition to rice, wheat and corn (Scott *et al.*, 2000). Taro, yambean, tannia, lesser yam, arrowroot, potato and yacon are the root and tuber crops in the CG System (FAO, 2017).

3.4.2 Habitat

According to the IP informants, ten roots and tuber crops were deliberately planted, three were volunteer plants, three were domesticated wild plants and more than six were wild roots and tubers (Table 3.5). Planted crops are found in swidden farms, rained farms, irrigated paddies, home gardens and even riverbanks. Cassava, sweet potato, greater yam, lesser yam, taro and tannia are the roots and tubers planted by all the 13 IPs in this study, while the *Kapampangan* and the *Ilocano* majority commercially plant sweet potato, cassava, yam

Table 3.5. Habitat category, use and extent of production of roots and tubers known by the indigenous peoples in Northern Philippines.

Roots and tubers	Habitat category				Extent of production and use
	Planted crop	Volunteer plant	Domesticated wild plant	Wild plant	
Cassava	X				small to large
Sweet potato	X				small to large
Greater yam	X				small to large
Taro	X				small to large
Potato	X				small to large
Tannia	X	X		X	small and many
Lesser yam	X				small and many
Yambean	X				small and few
Yacon	X	X			small and few
Arrowroot	X	X			small and few
Aerial yam			X		few
Wild yam bean			X		few
Canna			X		few
Wild taro (3 species)			X	X	few
Wild yam (3 species)				X	few
Nami				X	few
Elephant yam				X	few
Wild lesser yam				X	few
Giant taro				X	few

bean and taro. Arrowroot is not planted among the *Ivatan, Bugkalot* and *Iyattuka*. Potato is grown only by the *Ibalois* and *Kankana-ey*. Only the *Ibaloi* and *Kalanguya* plant yacon. Arrowroot, lesser yam and tannia are sometimes volunteer crops, whereas canna, aerial yam and wild species of taro like *bitajon* were domesticated. Volunteer plants are found in old and new farms, pathways and waterways, backyards and riverbanks. Wild plants are found in the forest, old swidden farms and along undisturbed lands near farms, pathways and waterways. Sometimes wild plants are maintained in swidden farms or transferred to backyard gardens. Roots and tubers that thrive in varied habitats and environments in terms of elevation, climate and vegetation (items 6 to 15) contribute to biodiversity, which is an important for food security and dietary diversity (Sunderland, 2011; Gollin, 2008; Chandrasekara and Kumar, 2016).

Also shown in Table 3.5 is the assessment of the extent of production and use of roots and tubers. When the roots and tubers are grown for sale as well as for food, e.g. cassava, sweet potato, greater yam, taro and potato, the extent of production and use as perceived by the IP informants ranged from small to large and many users. When the roots and tubers are mainly for household use as food or feed, the extent of production ranged from a small area to many growers and household users. Small areas of yacon, yambean, and arrowroot are grown or consumed by few groups, though large areas of yam bean are grown in Central Luzon. Cultivation of yacon, arrowroot and the consumption of other roots and tubers has become infrequent nowadays because of the absence of a market and the existence of alternative sources of food or livelihood.

3.4.3 Utilisation and economic significance of roots and tubers

Table 3.6 presents the uses of the plants specified according to the plant parts used.

As shown in Table 3.6, it is not only the enlarged underground roots, tubers and corms that are used for food but other plant parts as well, like the young leaves or shoots of the cassava and sweet potato, the young stalk, leaves, petioles and runners of taro and tannia, and the young pod of the wild yambean. Roots, tubers, corms, cormels (smaller corms usually produced by larger corms of taro or tannia, also used as planting material), rhizomes, vines, stalks and leaves are also used for animal feed, sold to provide additional cash income for the family, or bartered in exchange for household necessities.

Among the *Kapampangan* and *Ilocano* in Central Luzon, cassava, sweet potato and yam bean are planted as a cash crop to sell to traders or industrial buyers. Among the *Kalanguya, Aeta, Bago* and *Kankana-ey*, sweet potato, taro, greater yam and potato are important cash crops. Any excess or a portion of the harvest is kept for household use for food or feed. Only few households domesticate or gather wild roots and tubers for food, feeds and for sale. The *Aeta* informants claimed that only a few of these wild roots and tubers remain but they gather these to eat as staple instead of rice if they want to save cash, or for cash sale. The wild

Table 3.6. Utilisation of roots and tubers and plant parts used by indigenous people in Northern Philippines.

Roots and tubers	Part of plants used according to use			
	Human food	Animal feed	Cash sale	Barter
Cassava	Roots, young leaves or shoot	Roots	Roots	
Sweetpotato	Roots, young leaves or shoot	Roots, vines, leaves	Roots, shoots and vines	
Yambean	Roots			
Yacon	Roots		Roots	
Wild yambean	Roots, young pod			
Greater yam	Tubers	Tubers	Tuber	
Lesser yam	Tubers		Tubers	Tubers
Potato	Tubers		Tubers	
Arrowroot	Tubers	Tubers		
Canna	Tubers			
Aerial yam	Tubers			
Nami	Tubers			
Elephant yam	Young stalk or petiole			
Wild yam (3 species)	Tubers	Tubers		
Wild lesser yam	Tubers			
Taro	Corms, stalk, leaves, petiole, runners	Corms, stalk, leaves	Corms, stalk, leaves, petiole, runners	
Tannia	Cormels, young stalk and leaves	Corms, stalk, leaves	Corms	
Wild taro (3 species)	Young stalk, leaves, petiole	Stalk, leaves	Young stalk, leaves, petiole	Young stalk, leaves, petiole
Giant taro		Stalk, leaves		

roots and tubers are gathered from September to December when they go to their swidden farm or to the forest to hunt. The *Iyattuka, Isneg, Kalinga* and *Tingguian* harvest wild taro during summer in April or anytime of the year as alternate feed for pigs if sweet potato is unavailable. Among the *Tingguian*, the wild taro called *pikaw* is harvested for vegetable, sometimes sold or bartered with *bagoong* (fish sauce). Sometimes people from the capital town of Abra (Banguet) buy *pikaw* (PhP5-10 per bundle) as *palwad* or token for visitors during weddings. The informants described *pikaw* as growing near the creeks or in rocks and as sensitive to other plants. Wild species of taro are usually gathered as viand and feed for swine. Otherwise, households do only occasionally harvest wild roots and tubers, for example when children go dig-up the wild yambean (*sagsag-ot)* during summer vacation or when a household likes to cook yam for a gathering. The *kalot* requires thorough processing, thus is rarely gathered from the wild.

Starchy roots and tuber crops play a pivotal role in the human diet as they add variety to the diet in addition to containing a number of bioactive constituents, such as phenolic compounds, saponins, bioactive proteins, glycoalkaloids, and phytic acids responsible for numerous desirable nutritional and health benefits such as antioxidative, hypoglycemic, hypocholesterolemic, antimicrobial, and immunomodulatory activities (Chandrasekara and Kumar, 2016). The literature mentions the potential benefits of indigenous practices of mixed cropping for dietary diversity (Koppmair *et al.*, 2017; Kuhnlein, 2007), although this potential is not always realised (Sibhatu *et al.*, 2015). When cross-tabulating parameters of production, malnutrition and poverty in Benguet, Northern Philippines, three municipalities proved to have high sweet potato production and low malnutrition, but the same study also found many of these households living below the poverty line (Tenorio and Gonzales, 2013). Scott *et al.* (2000) claim that over two billion of especially very poor people use roots and tubers for food, feed or income.

Most indigenous people consider roots and tubers healthy foods and a special food for the sick, pregnant women and infants. Eating roots and tubers allegedly delays hunger, is more filling, prolongs the lifespan, is good for cleansing and that those who eat roots and tubers regularly are not sickly. There are claims that roots and tubers have anti-cancer properties, i.e. preventing colon cancer, that their consumption invigorates red blood cells and is good for the malnourished, constipated, anaemic and for those having hypertension, diabetes, haemorrhoids, diarrhea, asthma and skin allergies. More specifically, yam bean and lesser yam are considered a healing diet for those having ulcers and yam bean is supposed to relieve arthritis. The aerial tubers of greater yam are eaten raw for liver cleansing. Boiled sweet potato broth or *sabeng* is good for those having ulcers; juice prepared from purple sweet potato leaves is good for those with dengue fever and the tops for the anaemic. Sweet potato is also a dewormer if eaten raw. The *Tingguian* claim that the juice of the *alla-at* variety of sweet potato is good for the heart. Those who regularly consume taro vegetables are allegedly thin and live longer. The *Kankana-ey* claim that young stalks of *pikaw* or wild taro stalks cure those suffering from appendicitis and cleanse kidney stones. Decoction of wild *pikaw* corms is good for people with hepatitis. *Kamangeg* (wild yam) is good for those with appendicitis. The *Kalinga* use roasted *lidoy* stalk as skin patch against itchiness. Raw sweet potato roots are given to stroke patients to chew and exercise their jaws. Touching the cheeks of babies with young taro leaves will stop them from continuous crying. *Hebeng* or *sabeng* from sweet potato is an indigenous vinegar and if regularly consumed is believed to prolong life. To the *Applai*, *bakkay* serves as an alternative medicine. The *Bago* and the *Kankana-ey* use grated *karot* poultice as a cure to centipede bites and boils, and the *karot* juice as anthelmintic for carabao. The sap of cassava is used to stop blood flow in flesh wounds and crushed cassava leaves to treat wounds of cows.

The sharing of root crops planting materials and harvest during family and community celebrations, during wakes of relatives and neighbours, and during group work, i.e. *am-among* or *og-ogbo*, promotes social harmony. Root crops are also used as barter goods for household necessities, and as gifts or tokens, and as offerings to appease spirits and dead ancestors.

Factors that directly and indirectly affect valuation, i.e. landscapes, cropping systems, farm size, plant population, crop yields, farm price, external input use and unit cost, labour use and wage rates, use of farm tools, work animal or machinery and source of planting materials, were taken into account in the cost-benefit analysis (cf. Gayao *et al.*, 2014). Further, comparing the socio-economic importance of root and tuber crops farming between the majority *Kapampangan* and *Ilocano* with that of minority indigenous peoples such as the *Ibaloi* and *Bago*, also takes into account the disparities in resources. The former plant up to four hectares of one or two kinds of roots and tubers (cassava, sweet potato, taro or yam bean) in plain areas as mono crops, mainly for commercial purposes. The latter plant roots and tubers in small farms in mountainous or hilly areas, mixed with legumes, rice, ginger, banana and other crops. The *Kapampangan* and *Ilocano* earn most from the cash sale of taro (US$6,116/ha), followed by sweet potato (US$5,234/ha), cassava (US$3,667/ha) and yambean (US$670/ha) after a cropping period of 3-4 months at net benefit-cost ratio ranging from 75 to 733%. The *Ilocano* respondents said: 'We plant rice for our food, and sweet potato to earn cash.' Household food benefit is less than 1% for all crops; animal feed benefit is 9.4% for cassava and about 4% for sweet potato. Benefit in terms of planting materials ranged from 1.4% for cassava to 29% for yambean (Table 3.7).

As Table 3.8 shows, the total benefit generated from nine kinds of roots and tubers harvested by an *Ibaloi* farm household for a period of 3-10 months from more or less 1000 m^2, was computed at US$778, which is higher than the US$512 earned by a *Bago* farm household from six kinds of roots and tubers harvested. Since family labour is used in root and tuber crop farming, self-employment was the highest contribution of root and tuber crops farming among the *Ibaloi* at 41.2%, followed by cash sale at 35.9%, household food at 14.3%, planting materials at 5.5%, animal feed at 2.3% and gifts at 1.4%. Among the *Bago*, the self-employment benefit (24.2%) is second to cash sale primarily from greater yam and sweet potato (25.5%), followed by household food contribution at 20.4%, animal feed at 12.9%, gifts at 11% and planting materials at 4%.

Table 3.7. Benefits and costs of root and tuber crop production among the Kapampangan and Ilocano farmers in Central Luzon, Northern Philippines.

Details	PhP/ ha					
	Cassava	Sweetpotato[1]			Taro	Yam bean
		A	B	C		
# of hills	10,000	42,500			40,000	40,000
Benefits						
Sale	168,800	251,400	100,500	31,200	298,500	49,800
Food	900	1,400	600		1,500	200
Feed	17,800	10,500	4,200			
Planting materials	2,800	6,800	6,800		12,800	20,400
Total	190,300	270,100	112,100	31,200	312,800	70,400
Costs						
Labour	15,600	20,050	20,050	1,200	17,150	18,900
Fertilizer/insecticide	5,960	6,775	6,775	1,744	6,664	-
Tools depreciation	944	944	944		944	944
Planting materials	2,800	6,800	6,800		12,800	20,400
Total	25,304	34,569	34,569	2,944	37,558	40,244
Net benefit/ha (PhP)	164,996	235,531	77,531	28,256	275,242	30,156
(US$)	3,667	5,234	1,723	628	6,116	670
Net benefit-cost ratio (%)	652	681	224	960	733	75
Net benefit/hill (PhP)	16.50	5.50	1.80	0.70	6.90	0.75
% contribution to total benefits						
Sale	88.7	93.1	89.6		95.4	707
Food	0.5	0.5	0.5		0.5	0.3
Feed	9.4	3.9	3.6			
Planting materials	1.4	2.5	6.1		4.1	29.0

[1] A = during high price months at PhP14/kg; B = during low price months at PhP5.50/kg; C = clean planting materials/ vine cuttings product at PhP130/500pcs.

These results show that the importance of roots and tubers among the resource-poor indigenous people is more encompassing than providing household food, feed and cash income but also inculcates the virtue of work, industriousness and self-reliance.

Table 3.8. Benefits and costs summary from root and tuber crops among the Ibaloi and Bago people in Northern Philippines.

Details	Total benefit (PhP)		% contribution to total benefits	
	Ibaloi	*Bago*	*Ibaloi*	*Bago*
Benefits				
Food	10,083	6,635	14.3	20.4
Sale	25,257	8,285	35.9	25.5
Feed	1,610	4,200	2.3	12.9
Planting materials	3,483	1,305	5.5	4.0
Gifts	955	3,590	1.4	11.0
Employment	29,029	7,856	41.2	24.2
Total	70,417	32,446		
Costs				
Labour	29,029	7,856		
External inputs		324		
Tools depreciation	2,885	384		
Planting materials	3,483	856		
Total	35,397	9,420		
Net benefit (PhP)	35,020	23,026		
(US$)	778	512		
Net benefit/cost ratio (%)	99	244		

3.4.4 Retention and loss of traditional root crop knowledge

Conceptual knowledge pertains to knowledge on the diversity and variety of roots and tubers, roots and tubers landscapes, cropping practices and uses, and local terms of the aforementioned. The practical skills component pertains to cultivation and utilization practices including the local terms for said cultivation and utilization practice. Table 3.9 presents the test results on conceptual knowledge and skills according to age group and gender for selected IPs.

Between the middle-aged and younger group, all IP groups showed a loss of TRK as indicated by the intergenerational rates of less than one. Between the middle-aged group (35-56 years) and the oldest-aged group (57-77 years), there was retention of conceptual TRK as indicated by the intergenerational rate of retention (1.00) especially among the male *Iyattukas* (1.10). Their higher score was attributed to the fact that it is mostly the men who oversee land preparation and harvesting of rice and cash crops. There was also retention of

Table 3.9. Intergenerational rate of traditional root crop knowledge retention according to age groups and gender of selected indigenous peoples in Northern Philippines.

Age group	IP group	Rate of retention			
		Concept	Skills	Male	Female
15-35 years	*Ibaloi*	0.68	0.71	0.62	0.75
	Bago	0.69	0.83	0.80	0.64
	Iyattuka	0.72	0.54	0.70	0.80
	Tingguian	0.48	0.69	0.38	0.64
36-56 years	*Ibaloi*	1.19	1.00	1.15	1.17
	Bago	0.96	0.80	0.93	0.89
	Iyattuka	1.00	0.80	1.10	0.90
	Tingguian	0.94	0.87	0.89	0.96
57-77 years		1	1	1	1

practical skills and an increase in the conceptual knowledge among male and female *Ibalois* as shown by the respective intergenerational rates of 1.15 and 1.17. *Ibaloi* men and women when not employed elsewhere, are most knowledgeable on collecting, planting, harvesting, cooking and preserving the different kinds of roots and tubers. Small quantities of roots and tubers are exchanged or sold by housewives and elders for kitchen necessities and cash needs Among the middle-aged *Bago* and *Tingguian,* the results show a loss of TRK (rate below 1). In the case of the *Bago* IPs where men have higher intergenerational retention rate (80-93%) as compared to the female retention rate of 64-89%; and this is because of greater yam cultivation which is now a cash crop included in upland rice farming.

Vandebroek and Balick (2012) claim that younger people can have the same amount of knowledge as older people and that knowledge is not necessarily affected by age. Our results show a mixed picture. The higher retention rates of the middle-aged *Iyattuka* and *Ibaloi*, the female *Tingguians* and young female *Iyattukas* can be due to the fact that they were more exposed to root crop cultivation and consumption in their younger years than the others. It is also the unemployed who are left at home to take care of young children, oversee caretaking of farm and forestlands, and are in-charge of day-to-day household food provision who have ample opportunity to absorb knowledge on roots and tubers. Since, traditional knowledge and skills are passed on orally or by experience (Kothari, 2007), there is less transmission of knowledge on indigenous roots and tubers to young people who are studying or employed in non-agriculture jobs. A similar study comparing traditional plant knowledge between herders and students in Northern Kenya showed that young herders have more knowledge of traditional plants than students do (Bruyere *et al.*, 2016).

TRK test scores among the IP groups showed highly significant differences *(sig.013)*, in both the conceptual and practical skills component (Table 3.10). There was weak relationship of IP group to TRK test scores. The *Iyattukas* showed the highest total test score for both concept (56.21%) and skills component (62.91%), followed by the *Ibalois* at 62.37% in the skills test.

Table 3.11 shows the annualized rate of change according to IP group, age group and gender, and by concept and skills component.

The younger age groups showed a negative trend. The *Bago* and *Tingguian* had a sharper annual loss rate of >2% than the *Ibaloi* and *Iyattuka* at an annual loss rate of between 1% and 2%. Among the middle-aged group a significant TRK loss of 1% annually is shown only in the practical skills component, specifically by the male *Tingguian* and the female *Bago* and *Iyattuka*. The middle-aged group of *Ibalois*, however, exhibited not a loss but an annual 1% increase in TRK knowledge.

The declining trend according to age group was confirmed by the TRK test scores with the oldest age group having the highest mean scores (55.24%) followed by the middle-aged (53.15%), and the youngest age group (36.83%) for both the conceptual and skills component (Table 3.12). In addition, females score slightly higher than males but the difference is not statistically significant (Table 3.13). There appears to be only a weak association between age and sex and test scores.

Table 3.10. Traditional root crop knowledge (TRK) scores among selected indigenous people in Northern Philippines.

IP group	TRK mean scores (%)		
	Concept	Skills	Both
Ibaloi, n=38	42.50	62.37	46.79
Bago, n=42	41.52	38.86	41.12
Iyattuka, n=42	53.67	62.91	56.21
Tingguian, n=36	48.33	56.94	49.83
Pearson's chi-square value	0.023	0.03	0.022
Sig	0.01	0	0.013
Pearson's R value	0.180	0.041	0.153
Sig	0.240	0.606	0.055

Table 3.11. Annual rate of change in traditional root crops knowledge of selected indigenous people in Northern Philippines.

Age group	IP group	Rate of change (CA)			
		Concept	Skills	Male	Female
15-35 years	*Ibaloi*	-0.01	-0.01	-0.01	-0.01
	Bago	-0.02	-0.02	-0.02	-0.02
	Iyattuka	-0.01	-0.01	-0.01	-0.02
	Tingguian	-0.03	-0.02	-0.03	-0.02
36-56 years	*Ibaloi*	0.01	0.00	0.01	0.01
	Bago	0.00	-0.01	0.00	-0.01
	Iyattuka	0.00	-0.01	0.01	-0.01
	Tingguian	0.00	-0.01	-0.01	0.00
57-77 years		0.00	0.00	0.00	0.00

Table 3.12. Difference and relationship of traditional root crop knowledge (TRK) scores according to age group.

Age-group (years)	TRK mean scores (%)		
	Concept	Skills	Both
15-35, n=52	34.04	46.21	36.83
36-56, n=53	52.51	54.72	53.15
57-77, n=53	52.83	63.98	55.24
Pearson's chi-square value	0.016	0.013	0.013
Sig	0.012	0	0.195
Pearson's R value	0.406	0.271	0.405
Sig	0	0.001	0

Indigenous knowledge on roots and tubers is learned from formal and informal sources. The results on the source of indigenous knowledge are presented in Table 3.14. Schools, church and *barangay*-based extension services (in this case mostly health and nutrition workers) were the most mentioned single source of TRK knowledge, which means that knowledge TRK could be learned regardless of ethnic group, age, gender, education and occupation (Table 3.14). The difference in retention or loss of traditional knowledge among IPs, age

Table 3.13. Difference and relationship of traditional root crop knowledge (TRK) scores according to gender.

Gender	TRK mean scores (%)		
	Concept	Skills	Both
Male	45.41	51.65	46.91
Female	45.56	56	48.66
Pearson's chi-square value	63.95^{ns}	43.81^{ns}	51.88^{ns}
Sig	0.443	0.205	0.765
Pearson's R value	0.058	0.119	0.08
Sig	0.481	0.138	0.315

Table 3.14. Traditional root crop knowledge mean scores according to source of knowledge.

Source of knowledge	TRK mean scores (%)		
	Concept	Skills	Both
1 – Kin, relatives, friends	44.24	53.29	46.37
2 – School, church, extension services	48.00	100.00	62.00
3 – Media (print, radio, TV)	49.50	66.75	54.25
4 – 1,2 and 3	53.03	56.79	53.82
5 – 1 and 2	36.94	52.44	40.17
6 – 1 and 3	56.20	58.20	56.43
Pearson's chi-square value	0.032	0.022	0.031
Sig	0.667	0.063	0.287
Pearson's R value	0.125	0.046	0.107
Sig	0.118	0.568	0.180

groups and gender domains validates the general conclusion from a case study by Zent (2010b) that an understanding of traditional knowledge retention or loss requires multiple measures disaggregated by semantic domain and socio-demographic group.

3.5 Conclusions and recommendations

Twenty kinds of roots and tubers consisting of 10 cultivated species and more than 10 wild species as identified by indigenous peoples in the Northern Philippines were documented. The socio-economic importance of these roots and tubers among resource-poor indigenous people is evident in self-employment from utilising and selling root and tuber crops and among the groups who have wide plain farmlands, root and tuber crops are a major source of cash income. The non-monetary benefit of roots and tubers is in providing variety and quality to human diets that could have lasting returns in terms of health and well-being. Although this study and other studies have already pointed at the value of roots and tubers as functional or medicinal foods, their potential as functional foods, nutraceutical ingredients and in industrial applications needs to be explored further.

There is retention or little loss of traditional knowledge on roots and tubers among the indigenous people born before the eighties. The loss of knowledge is a reality only among the younger generation born after the 1980s, as evidenced by the continuing decline of planting and consumption of roots and tubers. This again reiterates the need to share indigenous practices on roots and tubers and raise awareness on the value of indigenous knowledge as well as enhance external knowledge in the area of nutrition and health value of roots and tubers.

Building on the traditional knowledge of root and tuber crops should be considered to improve food security and diversity, well-being, and the development and adoption of environment-friendly or organic farming practices and livelihood opportunities. Traditional knowledge is important baseline information for those planning to go into production, marketing and processing of these crops, and for the research and extension personnel whose mandate is the development of technologies in the field. The Northern Philippines Root Crops Research and Training Center at Benguet State University, for example, could start by publishing a source book of roots and tubers resources in the Northern Philippines, and a summary of indigenous production and utilisation techniques on roots and tubers that could guide farmers and future entrepreneurs. The centre could be a learning venue on indigenous production and utilisation practices in Northern Philippines and elsewhere in the world.

The continuing promotion and transfer of traditional knowledge, and the development of improved production, processing, and market utilization of the root and tuber resources is needed to ensure the conservation, sustained production and the increased consumption of roots and tubers, which may reduce rice imports, improve quality of food intake and provide additional sources of household income. Awareness about the important role of roots and tubers in household food security and income should start by teaching children to regularly eat roots and tubers. Most roots and tubers are not yet museum items but some species are candidates.

Acknowledgement

The researchers are indebted to the real authors, the key informants and respondents who shared their knowledge and practices in honour of their parents and grandparents who taught them to live off the mountains when the roads were not yet developed, and as a legacy they could hand down to the younger generation. The researchers are also grateful to the collaborating researchers from the local government units of Sugpon, San Jose City and Bauko, from DA-Region 3, NCIP-Nueva Vizcaya, Apayao State College, Batanes State College, Kalinga-Apayao State College, and Abra State Institute of Science and Technology, especially thanking their agency heads and directors for their support to the project and the succeeding activities. The researchers are most thankful to the Neys-Van Hoogstraten Foundation for the grant that enabled the conduct of the fieldwork and workshops. Lastly, the researchers thank their supervisors and colleagues at the Benguet State University and Northern Philippines Root Crops Research and Training Center for their support during the conduct and presentation of the research at in-house reviews, symposia and conferences.

References

Agada, M. and Igbokwe, E., 2016. Influence of food culture and practices on household food security in North Central Nigeria. *Journal of Food Security* 4(2): 36-41.

Balaki, E. and Solimen, J., 1991. The role of sweet potato in the diet, culture and ecology of an ethno-community. In: *Sweet potato cultures of Asia and South Pacific. Proceedings of the 2nd Annual UPWARD International Conference*. Los Baños, Philippines, pp.113-119.

Bareja, B.G., 2010. *Root crops, tuber crops and corm crops distinguished, list of examples*. Available at: http://www.cropsreview.com/root-crops.html.

Briggs, J., 2005. The use of indigenous knowledge in development: problems and challenges. *Progress in Development Studies* 5(2): 99-114.

Bruyere, B.L., Timarco, J. and Lemungesi, S., 2016. A comparison of traditional plant knowledge between students and herders in Northern Kenya. *Journal of Ethnobiology and Ethnomedicine* 12: 48.

Camacho, L.D., Gervana, D.T., Carandang, A.P. and Camacho, S., 2016. Indigenous knowledge and practices for the sustainable management of Ifugao forests in Cordillera Philippines. *International Journal of Biodiversity Science, Ecosystem Services and Management* 12: 1-2; 5-13.

Chandrasekara, A. and Kumar, T.J., 2016. Root and tuber crops as functional foods: a review on phytochemical constituents and their potential health benefits. *International Journal of Food Science* https://doi.org/10.1155/2016/3631647.

Coulibaly, I., 2011. *The global food crises: the right to decide what we eat*. The Guardian 2 June 2011. Available at: http://tinyurl.com/y88u264d.

Davies, M., 1998. *Biological diversity and indigenous knowledge*. Research paper #17. 1997-98. Parliament of Australia. Available at: http://tinyurl.com/ybp4d8vb.

Dayo, H.F., Labios, J.D. and Wagan, A.M., 1998. Rootcrop agriculture in Batanes: diversity and transformation of an island food system. In: Prain, G. and Bagalanon, C.P. (eds) *Conservation and change: farmer management of agricultural biodiversity in the context of development*. UPWARD, Los Baños, Laguna, Philippines, pp.14-44.

FAO, 2017. Report on the inter-centre review of root and tuber crops research in the CGIAR. Available at: https://cgspace.cgiar.org/handle/10947/1205.

Fu, X. and Soete, L., 2011. The role of foreign technology and indigenous innovation in the emerging economies: Technological change and catching-up. *World Development* 39(7): 1204-1212.

Fuglie, K.D., Oates, C.G. and Jiang Xie, 2006. *Root crops starch and agro-industrialization in Asia*. Social Sciences Working Paper No. 2006 – 3. International Potato Center (CIP), Lima, Peru. Available at: http://cipotato.org/wp-content/uploads/2014/09/AN67288.pdf.

Gadgil, M., Berkes, F. and Folke, C., 1993. Indigenous knowledge for biodiversity conservation. *Ambio* 22 (2/3): 151-156.

Gartaula, H., Patel, K., Johnson, D., Devkota, R., Khadka, K. and Chaudhary, P. 2017. From food security to food wellbeing: examining food security through the lens of food wellbeing in Nepal's rapidly changing agrarian landscape. *Agriculture and Human Values* 34(3): 573-589.

Gayao, B.T., Meldoz, D.T. and Backian, G.S., 2013-2014. *Traditional roots and tubers knowledge series*. #1-14. Northern Philippines Rootcrop Research and Training Center, La Trinidad, Benguet, Philippines.

Gollin, M.A., 2008. *Driving innovation: intellectual property strategies for a dynamic world*. Cambridge University Press, Cambridge, UK.

Hens, L., 2006. Indigenous knowledge and biodiversity conservation and management in Ghana. *Journal of Human Ecology* 20(1): 21-30.

Horton, D., 1988. *Underground crops. long-term trends in production of roots and tubers*. Winrock International, Little Rock, AR, USA.

International Plant Genetic Resources Institute (IPGRI), 2002. *Neglected and underutilized plant species: strategic action plan of the International Plant Genetic Resources Institute*. IPGRI, Rome, Italy.

IRRI, 2008. Why does the Philippines import rice? International Rice Research Institute, Los Baños, Philippines. Available at: http://irri.org/news/hot-topics/why-does-the-philippines-import-rice.

Koppmair, S., Kaseie, M. and Qaim, M., 2017. Farm production, market access and dietary diversity in Malawi. *Public Health Nutrition* 20(2): 325-335.

Kothari, A., 2007. *Traditional knowledge and sustainable development*. International Institute for Sustainable Development, Winnipeg, MB, Canada. Available at: https://www.iisd.org/pdf/2007/igsd_traditional_knowledge.pdf.

Kuhnlein, H.V., 2007. Indigenous peoples' food diversity and food security. *The FASEB Journal* 21(5): A672-A673.

NAARAP, 2009. The importance of indigenous/traditional knowledge in agriculture. NGO for association agricultural research in Asia-Pacific. Available at: http://tinyurl.com/y8skjewu.

Niehof, A., 2010. *Food, diversity, vulnerability and social change: research findings from insular Southeast Asia*. Wageningen Academic Publishers, Wageningen, the Netherlands.

Sano, E., Fang-asan, M., Gayao, B. and Alupias, E., 1991. Sweetpotato as a buffer in crisis situations: the case of Amlimay, Buguias during the 1990 earthquake crises in Benguet. In: *Sweet Potato Cultures of Asia and South Pacific. Proceedings of the 2nd Annual UPWARD International Conference.* UPWARD, Los Baños, Philippines, pp. 120-125.

Scott, G.J., Best, R., Rosegrant, M. and Bokianga, M., 2000. *Roots and tubers in the global food system: a vision statement to the year 2020.* A co-publication of the International Potato Center (CIP), Centro Internacional de Agricultura Tropicana (CIAT), International Food Policy Research Institute (IFPRI), International Institute of Tropical Agriculture (ITA) and International Plant Genetic Resources Institute (IPGRI). Lima, Peru: CIP. Available at: http://tinyurl.com/y95z2teq.

Shawki, N., 2012. The 2008 food crises as a critical event for the food sovereignty and food justice movements. *International Journal of Sociology and Agriculture and Food* 19(3): 423-444.

Sibhatu, K.T., Krishna, V.V. and Quaim, M., 2015. Production diversity and dietary diversity in smallholder farm households. *PNAS* 112(34): 10657-10662.

Solimen, J., Gayao, B. and Prain, G., 1998. Community-based knowledge systems in sweet potato genetic resources management among four ethno-linguistic groups in the northern Philippines. In: Prain, G. and Bagalanon, C.P. (eds) *Conservation and change: farmer management of agricultural biodiversity in the context of development.* UPWARD, Los Baños, Philippines, pp.120-186.

Sthapit, B., Rao, R. and Jarvis, D., 2001. *Processes of implementation of in-situ conservation of agro-biodiversity on-farm in the perspective of sweetpotatoes.* Paper for the Asian Network for Sweetpotato Genetic Resources (ANSWER) Workshop on exploring the potentials of in-situ (on-farm) conservation of sweet potato genetic resources in Asia. 2-4 October 2001. Bali, Indonesia.

Sunderland, T.C.H., 2011. Food security: why is biodiversity important? *International Forestry Review* 13(3): 265-274.

Tenorio, A.L. and Gonzales, R., 2013. *Geographic information system (GIS) in identification and monitoring root and tuber crops production areas in selected barangays in the Province of Benguet, Cordillera Administrative Region.* In: Proceedings of the 2nd Philippine Geomantics Symposium 2013.

Thiagarajan, S., 2008. *Importance of indigenous technology.* The Hindu 12 October 2008, updated 10 October 2016. Available at: http://tinyurl.com/y7z56r62.

Vandebroek, I. and Balick, M.J., 2012. Globalization and loss of plant knowledge: challenging the paradigm. *PLoSONE* 7(5): e37643. https://doi.org/10.1371/journal.pone.0037643.

Verdonk, I. 1991. The role of sweetpotato in the Philippine diet. In: *Sweet potato cultures of Asia and South Pacific.* Proceedings of the 2nd Annual UPWARD International Conference. UPWARD, Los Baños, Philippines, pp.429-439.

Zent, S. 2010a. A VITEK case study. In: *Vitality index of traditional environmental knowledge: an introductory guide.* Terralingua, Salt Spring Island, BC, Canada. Available at: http://terralingua.org/wp-content/uploads/2015/07/VITEK_Report.pdf.

Zent, S. 2010b. *Resilience and vitality of traditional econological knowledge.* Proceedings of the 1st National Conference on Biodiversity, Gender and Indigenous Knowledge. October 6-8, 2010. Agricultural Training Institute, Benguet State University, La Trinidad, Benguet, Philippines.

4. Gathering practices and actual use of wild edible mushrooms among ethnic groups in the Cordilleras, Philippines

D. Corazon M. Licyayo

Ifugao State University, The Philippines; dmlicyayo@yahoo.com

Abstract

This study aimed at documenting the gathering practices and utilisation of wild edible fungi in the Cordillera Region, Philippines. Thirty species from 23 genera of wild edible fungi were identified. Most of these species are found in Benguet and Mt. Province areas where the coniferous type of forest is conducive to the growth of wild edible mushrooms. The abundant fungal diversity in Benguet and Mt. Province created a stronger tradition of mushroom gathering than in Apayao, Kalinga, and Ifugao where non-coniferous types of forests prevail. Primarily male members of the family are responsible for gathering edible mushrooms in the wild; women for preparing the mushrooms for family consumption and for selling mushrooms. As yet, the contribution to the household income of mushrooms sales is very little, but the edible fungi in the region could be a promising source of income to local households. Gathering and utilisation of wild edible mushrooms is influenced by the ecological environment and ethnicity.

Keywords: fungi, food security, fungal diversity, mushroom collection

4.1 Background of the study

In 2012, poverty incidence in the Philippines was estimated by the Philippine Statistics Authority at 25.2%, and in the Cordillera Administrative Region (CAR) at 22.8% (PSA, 2015). Still, except for Benguet, the Cordillera provinces belong to the 20 poorest provinces in the country. In the Cordillera region of the northern Philippines, the highest and largest mass of mountains in the country can be found. It has six provinces: Benguet, Mountain Province, Ifugao, Kalinga Apayao, Abra, and the city of Baguio. The indigenous peoples by major ethnolinguistic identity are Ibaloy, Kankanaey, Bontoc, Ifugao, Kalinga, Isneg (Apayao), Tinggian, and Kalanguya.

The region is rich in biodiversity and natural resources, such as rivers and hot springs, minerals (mainly gold and copper), lowland and temperate agricultural areas, and some remaining forests. The native inhabitants of the Cordillera are indigenous peoples who have occupied and sustained their territories since times immemorial. They have maintained their culture, distinct languages, and socio-political systems. Not having been effectively colonized by Spain spells the difference between minorities like the peoples of the Cordillera and the

DOI 10.3920/978-90-8686-864-3_4, © Wageningen Academic Publishers 2018

majority of Filipinos. This difference persisted even throughout the American colonization and succeeding Filipino regimes (Tapang, 2001).

The indigenous peoples live in highly fragile ecosystems and comprise small-scale farmers who cultivate land received through agrarian reform, landless workers and fishers. Although they are all poor, there are substantial differences in the level of poverty between the regions and provinces. The poverty gap between urban and rural areas is widening. People living in the CAR uplands and in Mindanao are among the poorest in the country. Causes of rural poverty are a decline in the productivity and profitability of farming, smaller farm sizes and unsustainable practices that have led to deforestation and depleted fishing waters. Rural areas lag behind in economic growth and have higher underemployment. Poor people have few non-farm income-generating activities and little access to productive assets, business opportunities, microfinance services and affordable credit. Additionally, indigenous peoples have high illiteracy rates and are affected by the encroachment of modern technology onto traditional practices. Fishers face continuing reduction of their catches and they have few opportunities or skills outside fishing. Women's roles are mainly confined to family responsibilities and marketing own produce (IFAD, 2002).

In Asia, mushrooms are generally regarded useful foods (Chang and Buswell, 1996; 1999). In countries like Japan and China mushrooms are part of the everyday diet, because they can be grown easily in the backyard throughout the year and because of their nutritional value. *Volvariella volvacea*, for example, contains both standard and non-standard amino acids that also exhibit antihypertensive and antidiabetic properties (Eguchi *et al.*, 2008). In addition, non-food uses are important. Reyes *et al.* (2009) have demonstrated the ability of *Schizophyllum commune* to produce schizophyllan, a jelly that is soluble in water and can be utilised in skin care products.

The Philippines is rich in mushroom diversity. Popularly grown cultured mushroom species in the Philippines are the *Pleurotus* sp. (*P. ostreatus/P. sajor caju*) and *Volvariella* species, due to their adaptability to the environmental conditions of the country. However, there are still many unidentified mushrooms in the wild. If properly documented and studied, these species can be used as direct sources of protein and bioactive compounds to enhance population food and nutrition security. While Filipinos are known to be consumers of mushrooms, the commodity remains an expensive food for the ordinary Filipino family. Its high price and the fact that it is not always available, make the mushroom a special food. However, local people who have access to wild edible mushrooms can enjoy the gift of nature.

The CAR has a rich fungal biodiversity. The Cordillerans usually gather edible fungi in the forests to feed their families, especially during rainy season. Some of these forests have been undergoing changes for the past years, brought about by factors such as the wood carving industry, land conversion, encroachment and neglect of forest areas. This calls for fungal biodiversity management, yet there is scant information about fungi quantity and distribution. The practice of mushroom hunting in the area can be traced generations back and was not influenced by foreigners or missionaries entering the region. In the CAR, the most common edible mushrooms that are grown in growing houses include the *Pleurotus* strain and the *Volvariella volvacea* varieties. Government agencies such as the Department of Science and Technology (DOST) and the Department of Agriculture (DA) assist small-scale mushroom growers, while large-scale mushroom producers are private companies usually found near urban areas like Baguio and La Trinidad. The use of wild edible mushrooms by indigenous peoples in the Philippines like the Aeta in Pampanga, Tarlac and Zambales (De Leon *et al.*, 2012) and the Gaddang in Nueva Vizcaya (Lazo *et al.*, 2015) has been documented. About the gathering and utilisation of wild edible fungi in the Cordillera Region, however, little knowledge is available. This study examined the role of wild edible mushroom gathering practices in food and nutrition security and in poverty reduction in the Cordillera Region, taking into account the intersection of these practices with factors such as gender, ethnicity, and geography.

4.2 Conceptual framework of the study

Figure 4.1 shows the interplay of environment, people and resources for attaining food and nutrition security. Adequate nutrition is the prime requirement for ensuring the normal physical and mental development of children, as well as the health and productivity of adults.

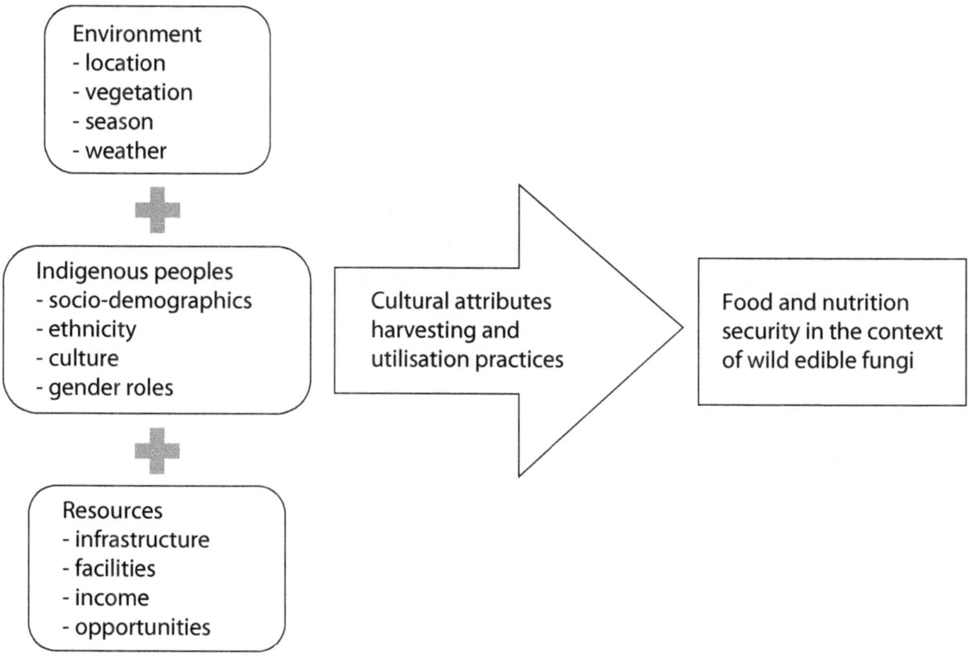

Figure 4.1. Conceptual framework of the study.

Ecosystems can contribute greatly to food security, especially in areas where indigenous peoples live. The indigenous peoples' dependence on agriculture for survival is rooted in their environment. Ecological biodiversity plays a crucial role in human nutrition through its influence on food production, as it ensures the sustainable productivity of soils and provides the genetic resources for crops, livestock, and marine species harvested for food.

However, environments and ecosystems are deteriorating fast due to natural and man-made factors. Climate change is affecting the agricultural production systems due to changes in temperature and rainfall (IFAD, 2013). This is exacerbated by human activities such as conversion of ecological landscapes for economic reasons. Indigenous peoples are closely associated with the forests, rivers and agricultural lands in their environments. They are, in fact, the stewards of the natural resources surrounding them, and they have their own methods to sustainably preserve, conserve and protect their environment, such as the *pinugo* or *muyong* system of the Ifugao (cf. Jang and Salcedo, 2013; Camacho *et al.*, 2015). Indigenous territories are often situated in landscapes that affected by factors such as logging, mining, and the immigration of farmers and ranchers. Deforestation and forest fragmentation increasingly encroach on indigenous lands as road networks expand into wilderness areas. These changes will affect the indigenous peoples' interaction with the natural environment and their culture.

Ecological sustainability of the environment as source of food depends on how people manage their communities and on the roles individual men and women perform in these activities. Indigenous knowledge on food sources, environment conservation and agricultural productivity has been proven to be effective for sustainable agriculture. Hence, it is important to consider culture, ethnicity, indigenous knowledge, and gender roles in practices aiming at household food security.

As outlined by Niehof (2016), key practices in food security include food procurement (crop production, buying, borrowing and exchanging food, and gathering of wild food), and food processing and storage. Nutrition security practices involve selective food procurement for food safety; proper processing, storage and preparation; and safeguarding food hygiene. These practices are contained in the arrow in Figure 4.1, in which harvesting and utilisation practices are highlighted.

4.3 Research area and methodology

The study was conducted in five provinces of the Cordillera Administrative Region. The province of Abra was not included in the survey due to its geographical location. We used mixed methods, including survey, focus group discussion, key informant interview, and observation.

4.3.1 Research instruments

An interview guide was used to gather the demographic profile of the respondents and wild edible fungi utilisation practices of the respondents. Focus group discussions and individual interviews were done. Photo-documentation of wild edible mushrooms as identified by the respondents was performed as well.

4.3.2 Respondents

Thirty respondents were selected in each province, amounting to a total of 150 respondents for the entire region. Respondents were found by purposive sampling. Initially, the author collected information from the community to identify the usual mushroom collectors in each area. Focus group discussions were also used to collect this information. In several locations, key informants knowledgeable about wild edible mushrooms were identified to help in the identification of mushrooms in the forest. In the FGD activity, the respondents' answers were gathered collectively. Gender balance was observed in choosing the respondents.

4.3.3 Collection and identification of specimen

The field research for seasonal mushrooms was conducted from May 2012 to October 2012 and repeated in the same months of 2013. These months are considered the mushroom season in the region. For the non-seasonal mushrooms, surveys were conducted the whole year round. Wild edible mushrooms were gathered from the study sites and were identified by the key informants. For the whole two-year duration of the study, the researchers were able to gather data through close contact with the key informants. In the mushroom gathering season, the researchers took the opportunity to see and join mushroom gathering.

The identification of the mushrooms took place during the actual mushroom gathering. In the Benguet and Mountain Province areas where mushroom hunting is part of the local culture, key informants did the identification. In the provinces of Ifugao, Kalinga, and Apayao, common wild edible mushrooms were identified by the survey respondents. These three provinces have common wild edible mushroom species, are due to their similar forest characteristics and climatic conditions. The mushrooms were collected with the head of the family (key informant) who usually gathers wild edible mushrooms. The collected mushrooms were then photographed and important information was recorded.

4.4 Results and discussion

4.4.1 Demographic and socio-economic profile of respondents

Tables 4.1 presents the demographic profiles of the respondents per province. Households commonly depend on their men for collecting mushrooms in the forest, where men also go to collect firewood. Men predominate in gathering wild mushrooms, which explains that they are more knowledgeable than women regarding the identification of wild edible mushrooms. However, the respondents were selected among the usual mushroom collectors and Table 4.1 shows a rather balanced picture in terms of gender. Table 4.1 also shows that the majority of the respondents belongs to the age bracket of 41-60, and is married. Only in Mt. Province, one third of the respondents belong to the youngest age group.

The socio-economic characteristics of the respondents are presented in Table 4.2.

The data on income was based on the monthly income of the respondents. Most respondents have a monthly income of less than Php 10,000.00 per month. The income derived from selling wild edible mushrooms is considered off-farm income. The average household size ranges from six to ten. The larger households comprise an extended family that includes the married child who is the recipient of a bigger part of inheritance from the parents.

Table 4.1. Demographic profile of the respondents (n=150).

Province	Sex (%)		Age (%)			Marital status (%)		
	Male	Female	21-40	41-60	>61	Single	Married	Widowed
Apayao	33	67	7	67	27	0	87	13
Benguet	43	57	13	60	27	0	97	3
Ifugao	60	40	0	50	50	0	97	3
Kalinga	47	53	10	33	57	3	97	0
Mt. Province	63	37	33	43	23	0	93	7

Table 4.2. Socio-economic profile of the respondents (n=150).

Province	Monthly income PhP (%)		Household size (%)			Occupation (%)		
	≤10,000	10,000-20,000	<6	6-10	11-15	Farmer	Employee	Skilled worker
Apayao	97	3	33	64	3	97	3	0
Benguet	87	13	23	70	7	93	7	0
Ifugao	100	0	33	57	10	97	0	3
Kalinga	100	0	16	77	7	100	0	0
Mt. Province	93	7	47	53	0	90	3	7

The majority of the respondents reported farming as their major occupation with their daily subsistence coming from farming-related activities. When their farm labour demand is low, the respondents work in infrastructure construction or as daily wage earners in farm-related activities.

4.4.2 Ethno-linguistic affiliation

The CAR is ethnically and linguistically a diverse area. Table 4.3 shows the ethno-linguistic affiliations of the respondents. Ethnicity is determined by a person's origin (parents) and the language spoken. The *Isnegs* are from the province of Abra. The *Ilocanos* emanated from the Ilocos region, therefore are considered migrants to the CAR. The *Ayangans* are mostly from the eastern part of Ifugao, the *Tuwalis* from Lagawe, Kiangan, and Hingyon in Ifugao province. The *Ibalois* are from Benguet, and the *Kalingas* from the province of Kalinga.

4.4.3 Gathering practices

The earliest recorded publication on wild edible mushrooms in the Philippines by Edwin B. Copeland about the *Polypodiciaea* of the Philippine Islands dates from 1873 (Copeland, 1905). Prill-Brett (2004) reports details about the indigenous resource management system among the Bontoc in the Cordillera, where communal forests are open areas for gathering wild vegetables, fruits, and mushrooms. Based on our interviews, the Cordillerans have been practising the culture of wild edible mushroom gathering from times immemorial.

Table 4.3. Ethno-Linguistic Affiliation.

Ethno-linguistic affiliation	Areas					n	%
	Apayao	Benguet	Ifugao	Kalinga	Mt. Province		
Isneg (Apayao)	27	0	0	0	0	27	18.00
Ilocano	3	0	2	5	0	10	6.66
Ayangan (Ifugao)	0	0	13	0	0	13	8.66
Tuwali (Ifugao)	0	0	15	0	0	15	10.00
Kankanaey (Mt. Province)	0	0	0	4	30	34	22.66
Ibaloi (Benguet)	0	30	0	0	0	30	20.00
Kalinga (Kalinga)	0	0	0	21	0	21	14.00
Total	30	30	30	30	30	150	100.00

In the Cordillera region, it is men usually who gather the wild edible mushrooms. They combine this with gathering firewood and attending to their swidden farms in the forest. Nonetheless, gathering edible mushrooms is not an exclusive male activity. Women and children also collect mushrooms, but their foraging activities are done nearer to their houses. The women have to prepare the wild edible mushrooms for household consumption and do the selling of marketable mushrooms. Among the Kankanaey from Mt. Province and the Ibaloi from Benguet province, families go to the mountains every mushroom season to look for mushrooms for home consumption or trade. The mountain vegetation of Benguet and Mt. Province is mostly coniferous (*Pinus* sp.), which offers diverse edible mushroom species.

Time of the year and location for gathering wild edible mushrooms

Wild edible mushroom availability differed throughout the two-year study period. Depending on the weather conditions, the mushrooms appear at different times. The volume of production is not consistent per specie. The peak of mushroom gathering is the rainy season between May and October; hence, wild edible mushrooms are mostly a seasonal food. Most respondents reported that the wild mushrooms grow during the same time of the year, and usually at the same spot or nearby where they were previously gathered. According to the respondents, locating particular species is challenging. They have to familiarize themselves with the 'usual spot' where they gather mushrooms. The collectors are cautious about the toxicity of unfamiliar mushroom species and do not collect those.

Interview data shows that different kinds of wild edible mushrooms are mostly gathered in their natural habitat during the rainy season, particularly in the months of June to August. These species are found in the forest areas on rotten fallen trees, in the swidden farms (called *uma* in the Cordilleras), and in the forest floor areas. However, there are also species that are found any time of the year, depending on the substrates (local natural growing material) where they naturally grow. Examples are the *Auricularia* sp. and *Schizophyllum* sp. that are abundant on dead tree trunks commonly found in decaying woods near residential areas.

Edibility of wild mushrooms

There is no simple test for determining wild mushroom edibility. The scientific literature is the best objective source of advice, but local practices and preferences can also reveal useful information. Empirical evidence is the ultimate basis of whether or not a species is edible (Boa, 2004). Knowledge of fungi was passed on from one generation to the next, as part of oral traditions. Burke (1983) reported that the Indians of the Pacific Northwest and British Columbia tended to be suspicious of mushrooms and considered them best used for medicine rather than food. The Cordillerans in the Philippines seem to have been using fungi for food and not so much as medicine. This is also the case with the Gaddang in Nueva Vizcaya and the Aetas in Central Luzon. (De Leon *et al.*, 2012).

The respondents claimed that their knowledge in determining the edibility of wild mushrooms derived from their grandparents, parents, elders, and neighbours with experience of eating wild edible mushrooms. They did not mention obtaining knowledge from government extension workers. Transfer of knowledge occurs in practice during daily routine activities in the field, by showing the sites where edible fungal species abound and pointing out the physical appearances of the commonly gathered edible mushrooms. Unrecognized mushroom species are not gathered, reflecting a cautious attitude towards unfamiliar species. More specifically, the following principles were followed to assess edibility:

- Avoid mushrooms with rings because these are thought poisonous or not edible.
- Only collect mushrooms with similar characteristics as those of previously gathered edible mushrooms.
- Similar kind of mushroom that have sprung in the same area during the same time of the year where edible mushrooms were gathered previously may be edible.
- Gathered mushrooms are not eaten until the elders have verified edibility.

Strong and weak gathering traditions

There are cultural differences in the approach to gathering wild mushrooms. In the provinces of Mt. Province and Benguet, mushroom hunting is traditionally practised every year when the wild edible mushrooms are in season. Then the members of the community go into the forest to search for the mushrooms, sometimes as a family activity. We classified this group as having a strong mushroom hunting tradition. In the other parts of the Cordillera region particularly in Ifugao, Kalinga and Apayao, there is no such tradition. The respondents said that they do not intentionally look for wild edible mushrooms in the forest, but only for two mushroom species *Termitomyces* sp. that are found on the riverbanks and in grasslands. This weak mushroom gathering tradition was observed in places with lower geographical elevation (Ifugao, Apayao and Kalinga). Edible mushroom species such as *Volvariella volvacea, Pleurotus* sp., *S. commune, Galiella celebica* and *Auricularia* sp. abound in these areas, but are underutilised because people are unaware their edibility and fear of toxicity.

So, groups with a weak mushroom gathering tradition do not intentionally go for mushroom 'hunting', whereas the groups with a strong mushroom gathering tradition are those that intentionally go for mushroom hunting during the mushroom season. The latter live in an environment with high mushroom diversity, such as coniferous forests, and know the places where they can gather mushrooms in the wild. The groups with a weak gathering tradition are exposed to fewer wild edible species. Table 4.4 shows the wild edible mushroom species that are consumed by groups with weak a hunting tradition. Table 4.5 lists those of groups with a strong hunting tradition.

Table 4.4. Wild edible mushrooms (WEM) consumed by respondents (n=150) with a weak gathering tradition.

Species	Local name	Description	Location	Ethnic group	n respondents consuming WEM
Termitomyces sp.	O-ong bunton	Long stalk, brownish cap, grows on soil mounds made by termites, solitary.	Apayao, Ifugao, Kalinga	Isneg	10
				Tuwali/Ayangan	30
				Kalinga	10
				Total	50
Termitormyces sp.	O-ong	Smaller stipes, brownish-greyish cap, grows on soil mostly on cogonal areas, by colony.	Apayao, Ifugao, Kalinga	Isneg	10
				Tuwali/Ayangan	30
				Kalinga	10
				Total	50
Collybia reinikeana	O-ong dir-an	White capped, can grow as big as 10 cm (cap), grows on soil with decayed wood.	Ifugao	Tuwali/Ayangan	15
Unidentified sp.	Kalaw	Yellow funnel shaped cup, white stype. Can grow as big as 12 cm (cap).	Ifugao	Tuwali/Ayangan	18
Volvariella sp.	O-ong saba,	Brown, grows in under the banana shade, soil-based.	Ifugao, Kalinga Apayao	Tuwali/Ayangan	15
				Kalinga	30
				Isneg	15
				Total	60
Pleurotus sp.	Kulat	Brownish with stem, quite hard to chew even when cooked. Grows in dead wood.	Ifugao, Apayao Kalinga	Tuwali/Ayangan	10
				Isneg	13
				Kalinga	22
				Total	45
Pleurotus sp.	Kulat Folong/ Fullong	Brownish white, grows in dead wood.	Ifugao, Kalinga Apayao	Isneg	10
				Tuwali/Ayangan	5
				Kalinga	10
				Total	25
Schizophyllum commune	Kudjel/Kudi / Kudet, Kaysop Uyhop	White-brownish white, 1-1.5 cm, like a small fan, grows on dead wood.	Ifugao, Kalinga, Apayao	Tuwali/Ayangan	5
				Kalinga	15
				Isneg	10
				Total	30

Table 4.4. Continued.

Species	Local name	Description	Location	Ethnic group	n respondents consuming WEM
Auricularia sp.	Laplapayag, Ballutak, Gargarot	Brownish, sometimes brown reddish, very soft.	Kalinga, Apayao, Ifugao. Mt. Prov.	Kalinga	7
				Isneg	5
				Tuwali/Ayangan	7
				Kankanaey	11
				Total	30
Coprinus comatus	O-ong mais/O-ong pagay	Small mushrooms found in decaying rice straw or corn leaves mound.	Ifugao, Kalinga Apayao	Tuwali/Ayangan	5
				Kalinga	13
				Isneg	17
				Total	35
Galiella celebica	Lusi-lusi, fofotoy	Dark brown, soft, mostly found under the mango trees on dead wood. Can be eaten raw.	Kalinga Apayao	Isneg/Ilocano	13
				Kalinga	12
				Total	25
Polyporous squamosus	Bannog	Watery flesh light brown, grows vertical on dead wood.	Ifugao	Tuwali/Ayangan	10
Termitomyces sp.	Lam-laming O-ong billit	Small, brownish white, gathered under pine tree vegetation.	Mt. Province	Kankanaey	15
			Ifugao	Tuwali/Ayangan	30
			Kalinga	Kalinga	25
				Total	70

Table 4.5. Wild edible mushrooms (WEM) consumed by respondents (n=150) with a strong gathering tradition.[1,2]

Species	Local name	Description
Unidentified sp.	Damayan	Slimy cap, light brown to dark brown, found under pine trees.
Sclerodermataceae	Bu-u (Puffball)	Yellowish to brown, round, soil grown, grows as big as 5 cm in diameter, found under exposed areas, soft watery tissues when opened.
Russula virescens	Kapputan; Ofot	Dirty white with greenish colour at the cap surface, soil grown and found under the pine trees. Can be eaten raw.
Hygrophorous russula	Lumsek	Pinkish-dirty white cap colour, grows on pine tree vegetation.
Boletus sp.	Dangkiyan	Yellowish cap and yellowish stipe. Found in pine forest.
Strobilomycetaceae	Kuyyopan	Brown to black cracky cap.
Tricholoma robostum	Panggayya	Brown cap and stype, excellent flavour.
Rammaria botrytis	Pansit-pansitan	Brown to light brown colour, grows on ground but associated with tree roots.
Albatrellus	Shapannan	Velvety, pale yellow, grows on the base of the trees. Very leathery, even when cooked.
Boletus speciousus	Sinabog	Pinkish cap, with pinkish stipe. (Can be mistaken for the poisonous Boletus species that has brownish cap with pinkish stipe).
Amanita javanica	Bagel	Yellowish mushroom, egg-like shape when immature. (Can be easily confused with deadly members of the Amanita family.)
Sparassis crispa	Mala-koli	Unknown to the community as a medicinal mushroom (alleged anti-cancer components). Used for home consumption only.
Cantharellus	Bagelni Bishing	Yellow-funnel shaped cap.
Lactarius volemus	Gatas-gatasan	Thus named because it produces white milky (*gatas*) sap.
Unidentified sp.	Botaytay	Shiny brownish cap, grows up to 4 cm.
Lyophillum fumosum	Bagel	Mushrooms that grows in bunch with 0.5-2 cm brownish cap.

[1]Except for the last item, 30 Ibaloi and 30 Kankanaey respondents reported consuming the species. For *Lyophillum fumosum* consumption was reported only by 15 Ibaloi respondents.

[2]Locations of the reported species are Benguet (Ibaloi) and Mt. Province (Kankanaey), except for *Lyophillum fumosum*, which was only reported in Benguet.

4.4.4 Post-harvest handling and cooking practices

Results of the data gathering from the different places show similar practices in post-harvest handling and cooking of wild edible mushrooms. There are no particular post-harvest practices for the wild edible mushroom except for the thorough washing and removal of soiled and spoiled parts. Drying the mushrooms for future use is generally not done.

The mushrooms are prepared by plain boiling or by sautéing with other vegetables or legumes. Respondents claimed to appreciate eating mushrooms because of their delightful 'mushroomy' taste and aromatic smell as well as because the mushrooms are nutritious. When mushrooms are sold, women take on the role of mushroom vendors.

4.4.5 Reasons for gathering wild edible mushrooms

Sixty-four percent of the respondents believed that mushrooms are beneficial and important as source of food as well as income. The mushroom gatherers intimated that when they have the opportunity to gather them, they prefer to bring the harvest home for family consumption. If the harvest is abundant, they sell the surplus in the neighbourhood.

Almost all key informants were aware of the health and medicinal benefits that can be derived from the mushrooms. However, they just pick edible mushrooms in the wild for food consumption and are not very mindful of their nutritional or medicinal value. A key informant who had been a mushroom gatherer for almost three decades, shared that it is only recently that local gatherers are becoming aware of the nutritional and medicinal benefits of mushrooms. An example is the popular mushroom *Ganoderma lucidum* from which a medicinal beverage can be made. This information is spread by mushroom buyers and through media advertisements, and can enhance the appreciation of the utility of the mushrooms among the indigenous peoples of the CAR. Better information on the benefits of wild edible mushrooms can also correct wrong ideas among the gatherers and consumers, such as the common (unfounded) fear that mushrooms have very high uric acid content.

4.4.6 Socio-economic significance of wild edible mushrooms

Wild edible mushroom collection has been considered as a recreational activity, intended only for family consumption. However, with the increase of population and the need to augment household income, wild edible mushroom collection has also become a way to earn additional income. Several wild edible mushrooms are now commonly sold to the public. During the harvest season, the wild edible mushroom pickers sell at the roadside or in the wet market. According to the respondents, most buyers are those who have experienced eating the kind of mushroom being sold. People who do not have this experience but are interested to buy, would often first secure the seller's assurance that the mushroom is edible.

In Ifugao province, during the peak season of the *Termitomyces* species, which usually lasts for two weeks only, mushrooms are gathered and sold along the roadside. The mushrooms are carefully tied using bamboo strips and displayed in conspicuous places. In Benguet province, different species of fresh mushrooms are all mixed and put together in one container, to be sold along the roadside. To strengthen their marketing potential, the packaging should be improved. Benguet respondents said that selling wild edible mushrooms is a recent phenomenon. In the municipality of Bokod, selling of wild edible mushrooms along the newly concreted national highway started only about two to three years ago. Interestingly, buyers are identified as experienced consumers of wild edible mushrooms. Interviews with key informants from Sagada and Besao, Mt. Province, disclosed that mostly wild edible mushroom collectors bring home the mushrooms for family consumption, but that there is now a demand for wild edible mushrooms in local restaurants. This prompts these collectors to start selling the mushrooms to restaurants, where it is a delicacy especially for tourists.

Respondents were interviewed on the contribution of the sale of wild edible mushroom to their household income. Those from the provinces of Ifugao, Apayao and Kalinga derive only a small income from selling mushrooms, which does not exceed 500 pesos (about €7-8) per season. For respondents from the provinces of Benguet and Mt. Province the income per season is 1,200 pesos (about €20) on average. Based on the respondents' average monthly income of 10,000 pesos (€167), the additional income from mushrooms is very little. We found no significant relationship between income derived from selling mushrooms and overall income of the respondents. The disparity in income derived from the sales of wild edible mushroom between those with a weak and a strong gathering tradition can be explained by the fact that compared to the groups with a weak tradition, groups with a strong mushroom hunting tradition have access to more diverse species of mushrooms to sell. The disparity does not imply that the latter are poorer. The economic status of the respondents does not seem to affect their attitude to the consumption and utilisation of wild edible mushrooms, which is more influenced by knowledge, culture and geographical location.

In Benguet province, household size turned out to be positively associated with income derived from mushroom hunting. The bigger the household size the more members can engage in gathering, which eventually contributes to the household income. Furthermore, it has to be noted that, in the absence of good preservation and processing technologies, the income derived from mushroom gathering is only seasonal, lasting for only two weeks to one month in a year. Finally, the income generating potential of wild edible mushrooms depends on availability of selling points and markets.

4.5 Conclusion

The environmental conditions and ecological diversity of the Cordillera region contribute to the availability of wild edible mushrooms. The dipterocarp and coniferous types of forest in the region offer abundant macro-fungi that are potential sources of food and income. In this study, we identified 30 wild edible mushroom species from 23 genera as being consumed by the indigenous peoples.

The knowledge on fungal biodiversity among the respondents in the Cordillera region varies. The Kankanaeys of Mt. Province and Ibalois of Benguet have a stronger wild edible mushroom hunting tradition than the people from the provinces of Ifugao, Apayao and Kalinga. Ecological characteristics strongly influence the gathering tradition. The research revealed that there are more mushroom species gathered in coniferous forests found in Benguet and Mt. Province than in the deciduous type of forest that is common in Apayao, Ifugao and Kalinga.

The collection of wild edible mushrooms in the Cordillera provinces is primarily for household food and nutrition security. Additional income comes second. Some species have economic viability; others are a valuable addition to the food supply of the community. As yet, the commercial demand for wild edible mushrooms is low and the income derived from it is rather insignificant. However, there is potentially a promising commercial market.

Underutilisation of wild edible mushroom is attributed to lack of proper information and knowledge on their use and benefits. According to Yorou and De Kesel (2002), the importance of wild edible fungi in developing countries may have gone unnoticed for the reason that collection is mainly done for household consumption. Nevertheless, the potential of wild edible fungi for the diet and income of rural people should not be underestimated. However, knowledge on the utilisation and benefits of wild edible mushrooms among the Cordilleran people is still poor. Mushroom education would be a way of addressing this issue, particularly through expanding knowledge on collection and consumption in localities with a weak mushroom gathering tradition. Additionally, the management of the forest resources needs to be considered, to ensure sustainable sources of wild edible mushrooms for food and nutrition security and as a source of income.

Acknowledgment

The author gratefully acknowledges the financial support of the Neys-Van Hoogstraten Foundation and the Ifugao State University (IFSU) for making this research possible.

References

Boa, E., 2004. *Wild edible fungi: a global overview of their use and importance to people*. Food and Agriculture Organization of the United Nations, Rome, Italy. Available at: http://tinyurl.com/yal357oy.

Burke, W., 1983. Puffball usages among the North American Indians. *The Journal of Ethnobiology* 3(1): 55-62.

Camacho, L., Gevana, D., Carandang, A. and Camacho, S., 2015. Indigenous knowledge and practices for the sustainable management of Ifugao forests in Cordillera, Philippines. *International Journal of Biodiversity Science Ecosystem Services & Management* 12(1-2): 5-13.

Chang, S.T. and Buswell, J.A., 1996. Mushroom nutriceuticals. *World Journal of Microbiology and Biotechnology* 12: 473-476.

Chang, S.T. and Buswell, J.A., 1999. *Ganoderma lucidum* – a mushrooming medicinal mushroom. *International Journal of Medicinal Mushrooms* 1: 139-146.

Copeland, E.B., 1905. *The Polypodiciaea of the Philippine Islands. new species of edible Philippine fungi.* Issue 28. Bureau of Government Laboratories, Manila, Philippines.

De Leon, A.M, Reyes, R.G. and De la Cruz, T.E.E., 2012. An ethnomycological survey of macrofungi utilized by Aeta communities in Central Luzon, Philippines. *Mycosphere* 3(2): 251-259.

Eguchi, F., Yoshimoto, H., Reyes, R.G., Kikukawa, T. and Matsuda, T., 2008. *The change of component and functionality in the growth stage of* Volvariella volvacea. Proceedings of the 5th Meeting of the East Asia for Collaboration on Edible Fungi. September 17-20, 2008, Kyushu University, Fukuoka, Japan.

IFAD, 2002. *Rural poverty in the Philippines*. Available at: http://tinyurl.com/y7m7za7w.

IFAD, 2013. Smallholders, food security and the environment. Available at: http://tinyurl.com/yd6bqvyf.

Jang, J. and Salcedo, S., 2013. *The socio-political structure that regulates the Ifugao forest maintenance.* Fourth International Conference on Biology, Environment and Chemistry. November 23-24, 2013. Phuket, Thailand. Available at: http://tinyurl.com/ybgawcj3.

Lazo, C.R.M, Kalaw, S.P. and De Leon, A.M., 2015. Ethnomycological survey of macrofungi utilized by Gaddang communities in Nueva Vizcaya, Philippines. *Current Research in Environment and Applied Mycology* 5(3): 256-262.

Niehof, A., 2016. Food and nutrition security as gendered social practice. *Apstract* 10(2-3): 59-67.

PSA, 2015. *Philippine poverty statistics report*. Philippines Statistical Authority Manila, Philippines.

Prill-Brett, J., 2004. Gender relations and gender issues on resource management in the Central Cordillera, Northern Philippines. *Review of Women's Studies* 14(1). Available at: http://journals.upd.edu.ph/index.php/rws/article/view/2919.

Reyes, R., Kalaw, S., Bellere, A., Yoshimoto, H., Higaki, M., Kikukawa, T. and Eguchi, F., 2009. Status and prospects of medicinal mushroom cultivation in Central Luzon region of the Philippines. *Mushroom Science and Biotechnology* 17(2): 87-89.

Tapang, B.P. Jr., 2001. *Economic transaction flows in a typical Cordillera village.* Research Report 1: Perspectives on Resource Management in the Cordillera. Cordillera Studies Center, University of the Philippines, Baguio City, Philippines.

Yorou, S.N. and De Kesel A., 2002. *Ethnomycologic knowledge of Nagot in Central Benin (West Africa).* Proceedings of the 16th AETFAT Congress, Brussels 2000.

5. Household food security, food care, child nutrition, and wellbeing of ethnic Tengger families in East Java, Indonesia

D. Hastuti

Division of Child Development, Department of Family and Consumer Sciences, College of Human Ecology, Bogor Agricultural University, Indonesia; tutimartianto@yahoo.com.

Abstract

This study was conducted to profile family life among the Tengger people in the Mount Bromo area in East Java, Indonesia, in terms of their household's food security, food caring practices, children's nutrition and family wellbeing. Most Tengger families in this study felt food secure, but based on energy intake 81.7% of the mothers proved to be food insecure, 36.7% severely and 45.0% moderately food insecure. At the same time, 5.5% of the mothers was obese and 25.8% overweight. Exclusive breastfeeding is only 40%, and food caring for children can be considered moderate. The Tengger children's energy and nutrients reached adequacy levels of more than 90%, but based on anthropometric measurement still 39.2% of the Tengger children under six years of age is stunted. Subjectively, the Tengger families are quite satisfied with their economic situation, but 19.2% of the families is not satisfied with their family's health status, and 20.8% of the children under six years old experienced illness that lasted for seven days or more during last three months. Although the research yielded a mixed picture, overall the results are more positive than expected given that the Tengger people are an indigenous community living in a relatively remote area.

Keywords: food security, child nutrition, wellbeing, indigenous community

5.1 Introduction

5.1.1 Background

The Tengger people are an indigenous ethnic group in Indonesia. Their living area is on the slopes of Mount Bromo in the district of Probolinggo in East Java. They have limited access to modern life in terms of health, nutrition and education services. Hence, one could be expect that the nutrition status of their children is relatively low and their morbidity status relatively high. In other indigenous communities in Indonesia, such as the Mentawai in West Sumatera and the Baduy in West Java, there are diverse linkages between culture, environment and the food system in the area (Niehof, 2010). Mentawai culture, for example, has two types of rituals: community or clan rituals and household rituals. During these rituals, chicken and pigs are killed and consumed. In Mentawai, men are hunters, women fish. For the Baduy people the relationship between man and nature should be harmonious;

DOI 10.3920/978-90-8686-864-3_5,

damaging nature is considered a great sin. The Baduy see tilling the dry land as worship and laziness as sinful. In term of gender relations, the Baduy people treat man and wife equally, but they have different responsibilities. Both man and women are expected to work hard but women can do lighter work in agriculture than men. In terms of their wellbeing, the Baduy have low literacy rates since they see formal education as a potential threat to their culture (Khomsan *et al.*, 2009).

Many studies have documented aspects of the relationship between socioeconomic status and household food security in Indonesia. For example, Martianto *et al.* found a significant relationship between socioeconomic status as measured by household expenditures and household food security in Banjarnegara, Central Java. Food insecure households implement coping strategies such as changing staple food to a cheaper, more common and available food, borrowing money and food, selling livestock, or finding an alternative or temporary job outside the village (Martianto *et al.,* 2006, 2007, 2008). In a study conducted in West Sumatra by Zein (2007), the coping strategies of women to address a food insecure situation included borrowing from neighbours and family or borrowing from the local shop at an undisclosed rate of interest. In her study about a fisher village in Madura, East Java, Niehof (2010) found that women used a variety of strategies to meet the food needs of their households, such as selling gold or jewellery, selling other goods, pawning gold or other jewellery, finding additional sources of income (work), changing the menu, reduce food intake, and borrowing from the moneylender or boat owner. The Tengger people rely on seasonal agriculture. Presumably, they have specific strategies to cope with household food insecurity as well, in particular during the hard times of the eruptions of the Mount Bromo volcano in 2010-2011.

Children's welfare and wellbeing are nationwide concerns in Indonesia, but among indigenous ethnic groups such as the Tengger people, these are understudied. Since children are the most vulnerable members of a family in terms of nutrition and health, this study investigated the impacts of household food security status and food caring practices on overall family wellbeing and children's nutritional status in Tengger families.

5.1.2 Objectives and research questions

This study was conducted to profile family life among the Tengger people in East Java, in terms of their household's food security, food caring practices, children's nutrition and family wellbeing. Specifically it aims to:

- Identify family characteristic, family resources and access to health and nutrition services.
- Estimate prevalence of household food insecurity and document coping strategies to deal with it.
- Identify food caring practices for children and assess children's nutritional status.
- Determine the factors related to family wellbeing of Tengger families, subjectively and objectively.

Based on these aims, the following research questions were addressed:

- What is the food availability situation and how is food produced, processed and stored?
- How do natural resources, the seasonal calendar, and community resources contribute to achieving household food security?
- What coping strategies were/are used in relation to food insecurity, in particular during the Mount Bromo eruptions?
- How do food security, food caring practices, access to health services and nutrition influence family wellbeing objectively and how do families subjectively perceive their wellbeing?
- What are the relations between dietary habits, food caring practices, family wellbeing and the nutritional status of children?

5.1.3 Relevance of the study

Malnutrition prevalence in Indonesia is still considered high. The Ministry of Health Report of 2008 (MoH, 2008) indicated that malnutrition prevalence among children under five years old in Indonesia was 18.4% (5.4% severe wasting and 13.0% wasting). In the province of East Java, the under-five malnutrition prevalence was 17.4%, which is slightly lower than the national figure. However, the province of East Java is still facing acute malnutrition with underweight and severe underweight status higher than 10%, indicating a serious malnutrition problem. Besides, there are still seven out of 38 districts in the province of East Java that have high malnutrition prevalence, which include the four Madurese districts and the district of Probolinggo where the Tengger people live (BPS, 2011).

As an indigenous ethnic group living on the slopes of Mount Bromo, the Tengger people would have more limited access to health and nutrition services, and education than most people in the province, and their income, household expenditure, and socio-economic status would be lower. Therefore, it can be expected that the Tengger people have a low socio-economic status, score low in terms of household food security, and that their children have a higher risk to be malnourished.

In this situation, it is relevant to study how a group like the Tengger people provides for its food with the help of their agricultural system, social organization and community resources. The cultural system of the Tengger people will affect food utilisation and distribution among family members, which is why it is also relevant to document their food habits and feeding practices. Improvement of the latter could be a component in government interventions to increase food consumption and improve the nutritional status of the Tengger children.

5.2 Conceptual framework

The concept of food security has a long history. FAO (2013) describes food security as a situation in which all people having access to sufficient, safe and nutritious food that meets their dietary needs and food preferences for an active and healthy life. Safe food is understood in two ways. First, in the sense of containing no harmful chemicals, second of being free of microbiological contamination (caused by inadequate hygiene) that may obstruct the bodily uptake of nutrients. In actual cases, food preferences may be at odds with dietary needs (e.g. food taboos for children or pregnant women). Food security is anchored at the household level, since households are resource-managing, family-based groups that are geared towards meeting the primary and daily needs of their members (cf. Niehof, 2011). In this study, we shall emphasize food availability and access, food habits and preferences (culture), health and hygiene, food caring practices and nutritional status of children.

Many studies have documented the relations between household food security, socio-economic status and household income, showing that food insecurity has its roots in poverty.

In studies conducted in Nusa Tenggara Province, it is shown that household food security is related not only to household income but also depends on the mother's knowledge of and access to health and nutrition services. The latter was related to the mother's educational attainment and to her exposure to *posyandu* activities. The *posyandu* is a village-level health post that regularly (usually monthly) provides health, nutrition and family planning information and services. It is supervised by the health centre in the sub-district capital. The mothers who regularly attended the *posyandu* proved to be more knowledgeable and had healthier and better-nourished children than mothers who did not do so (Martianto *et al.*, 2006, 2007, 2008). We expect a similar pattern in our case. Additionally, among the Tengger families drought and the eruption of Mount Bromo may have led to difficulties with meeting food needs, which could result in malnutrition. Further, it is assumed that apart from household income and food security, good health and sanitation and adequate caring practices increase family wellbeing and children's nutritional status (Figure 5.1).

Below it is briefly explained how the key concepts (Figure 1) were measured.

- Household food security is a situation when every household member has access to the food needed for a healthy life, as measured by food consumption intake and nutrient adequacy level, and by the subjective perception of satisfactory food availability and consumption.
- Nutrition and health practices are the maternal practices to take children to the *posyandu*, to get a health card, to prepare and handle the children's food consumption, to look after the children's sanitary and hygienic needs and to take proper steps when a child has diarrhoea.
- Caring practices refer to the habit of the mother to provide food (food caring) and seek access to health services (health caring). Food caring practices refer to administering the food, including the mother's approach in the selection, provision and delivery of food, and the mother's attitude to child feeding.
- Family wellbeing in the objective sense is the condition that reflects socio-economic status, measured by the level of household income and asset ownership, adequate energy

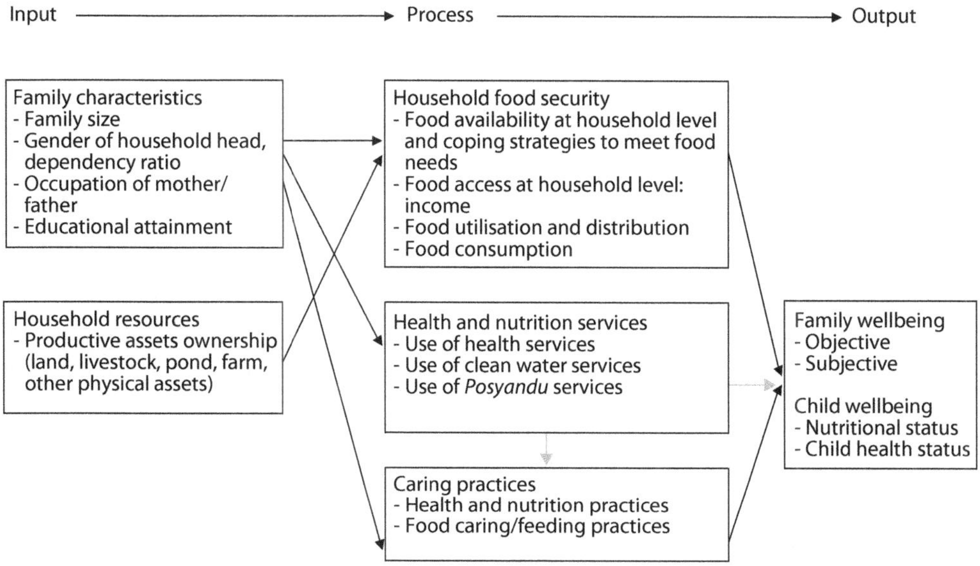

Figure 5.1. Conceptual framework of study.

consumption and nutritional status of family members. Subjective family wellbeing is measured in terms of satisfaction with income and expenditure (economic dimension), satisfaction with household necessities (material dimension), and satisfaction with social relationships with other family members and the community (psychosocial dimension).

- Food consumption of mothers and children under six is about the intake of nutrients through food to meet the nutritional needs for an active and healthy life, the level of nutrient intake being compared with recommended dietary allowances (RDA) of Indonesia (LIPI, 2008).
- Children's objective wellbeing is measured by the child's energy consumption and nutritional status, and using child's health status. Nutritional status of children is the state of the child's body resulting from the use of nutrients as measured by the results of measurements of weight and height. There are three indicators of nutritional status used in this study: the indexes of weight for age, height for age, and weight for height. Child health status refers to the absence of any bodily illness during the three months prior to the survey.

5.3 Research methodology

5.3.1 Research design

The study was conducted among ethnic Tengger families in the Mount Bromo area, in Probolinggo district, East Java province, Indonesia. The Tengger people live in four districts:

Probolinggo (in the villages of Ngadas, Wanatara, Jetak and Ngadisari), Pasuruan (villages of Wanakersa, Ledokombo, Pandansari, Tosari, Baledono, Sedaeng, Wonokitri, Ngadiwono, Kandangan, Mororejo, Keduwung, Ngadirejo, and Ledok Pring), Malang (Ngadas village) and Lumajang (Ranupani village). Since prevalence of malnutrition in the district of Probolinggo is higher than in the other districts, the study focused on Probolinggo.

5.3.2 Selection of study site and time frame of the study

Through a preliminary survey in the district of Probolinggo, we selected the sub-district of Sukapura and the villages of Ngadas, Wanatara, Jetak and Ngadisari, as places of predominant Tengger residence. The study was conducted during the period September 2011- January 2012.

5.3.3 Methods of sampling and data collection

A selection of 120 children was randomly selected from 156 children in the four villages. In this study, primary data and secondary data were collected. Secondary data are about, the agricultural and socio-demographic features of the study area, primary data about family characteristics and household resources, food security, health and nutrition services, caring practices and family wellbeing (Figure 5.1). Data on family characteristics and household resources, health and nutrition services, caring practices, and family wellbeing were collected using a questionnaire. Health and nutrition practices, caring practices and family wellbeing were assessed using a Guttman scale with 'yes' or 'no' answers. Data on food security (i.e. food availability, food access, and distribution) were collected using a structured interview schedule. For data on food consumption, the 2 × 24 hours recall method was used. Nutritional status was assessed by anthropometric measurements, using a weighing scale and a microtoice.

5.3.4 Data analysis

Data about family characteristics and household resources such as family size and educational attainment are presented by frequency distributions, means and standard deviations. A scoring system was used to assess nutrition, health and caring practices. Based on the scoring system, frequency distribution, means and standard deviation are applied to describe the variables under study. The following categorization was applied: low <60%; moderate 60-80%; and high >80%.

Food consumption data measured daily food intake in terms of energy (calories), protein (g), Calcium (mg), Fe (mg), Vitamin A (RE) and Vitamin C (mg). The data are presented by means and standard deviation, and compared to the recommended dietary allowances (RDA) of Indonesia. RDA performance was classified according to four levels: <70%, 70-90%, 90-120%, >120%.

Household food security was objectively measured by food intake and nutrient adequacy level, and subjectively by using a Guttman scale to measure perceptions of food availability and consumption. Data on the nutritional status of children is classified based on: weight-for-age z-score, height-for-age z-score, and weight- for-height z-score indexes. Children with a z-score lower than -3SD were classified as having severe malnutrition, a z-score of -3SD up to -2SD as having malnutrition, a z-score of -2SD up to +2SD as normal, and children with a z-score higher than +2SD were classified as being overweight. Cross tabulation of the results was done and correlation analysis applied, to determine association between the relevant variables.

5.4 Results and discussion

5.4.1 Geographic location and socio-economic conditions in the study area

Probolinggo's location around the equator causes the area's climate to consist of two annual seasons: the dry season (April-October) and the rainy season (December-March). The rainfall in the rainy season can amount to ±360 mm. Between the two seasons there is a short transitional season. The Mount Bromo volcano is the nature conservation area of the Bromo Tengger Semeru National Park. This National Park, located between 1000 and 3,676 meters above sea level, is quite spectacular and attracts many visitors. The last eruptions took place in 2010. Crops were damaged, and plants, backyards and houses were covered with dust.

The sub-district of Sukapura has a surface area of 102.08 km^2 and is located in the southern part of the Tengger Mountains, 30 kilometres from Probolinggo capital. In 2010, the population of Sukapura was 20,275, 9,920 males and 10,355 females, spread over 12 villages. The population density was 199 people/km^2. The livelihoods of most residents in the district are based on farming, whether as farmer or farm worker. Some people worked as traders, civil servants, industrial labourers, or in the army. Tourism also provides employment. Most adult residents in the Sukapura District had graduated from junior high school (66.3%), 28.9% graduated from senior high school, and 4.8% had a university degree (BPSP, 2010).

Sukapura is located at an altitude of 650-1,800 meters. According to key informants and secondary sources (BPP, 2011), due to minerals derived from the volcanic eruptions, the land of Sukapura is fertile and contains sand, stone, and mud turned into yellowish clay. The soil is suitable for vegetables cultivation. Planting is done in the rainy season only. Crops planted are potatoes, maize, cabbage, scallion, carrot, and tomatoes. Rice can be planted in a few places, and only in the rainy season. There is no irrigated rice cultivation in the area.

5.4.2 The Tengger cultural heritage

The Tengger people have four types of rituals: (1) life cycle rituals such as at birth, reaching puberty and marriage; (2) ecological rituals, around the harvest and the mountain; 3) calendar rituals such as the Friday ritual; and (4) a religious ritual dedicated to the (Hindu) goddess Saraswati. The Saraswati ceremony is celebrated twice a year. It symbolises a homage to the Hindu books the *Wedas*. The celebration takes place for the adults in the temple and for kids at school. It must be performed by all communities and, according to key informants, is considered to strengthen social cohesion.

The biggest day of the Tengger people is Karo, an annual religious ritual performed in the month Karo. Tengger families celebrate it by preparing and offering special foods (*sesajen* and *sedekah*). They visit each other and pray for the community to be safe and free from evil. Families will invite persons who pass by to join them and enjoy the food which has been prepared. A regular ritual is the weekly Friday *legi* (*legi* meaning sweet), at which food is prepared to honour the ancestors. Usually the food is placed in several corners of the house, such as in the kitchen, near doors, or in front of the bathroom. It is intended that the ancestors are honoured in every place, to protect the family and keep the peace in the home.

Food taboos for pregnant and lactating women are common. Examples are eating chili straight from the plate, eating legged fish, consuming birds, or eating a half-peeled banana. Breaching these taboos would negatively affect the appearance of the child or the process of childbirth. Lactating women should not eat foods or drink beverages with a sharp smell, because the effect on the mother's milk will make the child unwilling to drink. Most mothers believe that spicy food causes the child's lips to become hot. Lactating mothers should also refrain from drinking or ice water because it can cause children to be cold. Young women should not eat banana because it is believed that it will cause difficulties in finding a marriage partner. Children should not eat beef and drink soda. For men there is a taboo on eating ham.

5.4.3 Household resources and expenditures

Agricultural production, specifically food crop cultivation, is the main occupation of the Tengger people and serves as a source of food and income. Most men (93.3%) work as farmer. A few of them work in tourism, as a horse-driver or car-driver to transport tourists going to Mount Bromo. Most women (95.0%) also work as farmer, helping their husbands in the field.

According to Central Bureau of Statistics (BPS, 2011), in 2011 the poverty line for East Java Province was Rp 206,275. The average household income of the Tengger families in our sample was Rp 644,140. Therefore, most of them (79.2%) can be considered non-poor. Only 21% of the families proved to be poor according to this standard.

Household expenditures were divided into food and non-food expenditures. Average monthly household expenditure was Rp 611,692. Food expenditures amounted to Rp 289,909 and non-food expenditure to Rp 321,783; hence, the proportion of food expenditures (47.4%)

was slightly less than that of non-food expenditures (52.6%). Based on Engle *et al.* (1997) the wealthier the people are, the lesser the proportion of expenditures on food. Wealthier people spend less than 30% of their income on food. The highest food expenditures were those on sources of animal protein and snacks or street food. For non-food expenditures, the biggest expenditures were those on education and social fees, including taxes.

5.4.4 Household food security

Food crop production and household food availability

The largest agricultural production is from horticultural crops. The vegetable providing the highest yield is cabbage, with an average annual production of 6,018.8 kg. Potatoes come second (4,696.2 kg), chives third (3,094.0 kg). The annual production of maize only amounts to 221.8 kg. Families also planted some horticultural crops, mostly onion, in their backyard for daily consumption or for a small-scale sale.

Food security history and coping strategies to overcome food shortages

Coping strategies are the effort made by households to overcome difficulties in accessing staple foods. Among the Tengger the staple is rice (99.2%), sometimes in combination with maize. Most families (94.2%) did not have difficulties in getting their staple food in the past year and during the Mount Bromo eruptions. Those who did, deployed several strategies to meet their staple food needs (Table 5.1). The most common first course of action is to

Table 5.1. Type of coping strategies according to priority.[1]

Coping strategies	P1	P2	P3	n
Change of staple food	1	0	1	2
Decrease daily food frequency	0	1	0	1
Selling non-productive assets	1	1	0	2
Selling productive assets	2	0	0	2
Men go outside area/abroad for work	1	0	0	1
Borrow money from relatives	9	4	0	13
Borrow money from non-relatives	11	3	0	14
Other effort[2]	5	1	0	6
Total	30	10	1	41

[1] P1 = first order priority; P2 = second order priority; P3 = third order priority.
[2] Other effort = search for another job, take savings, sell the crops.

borrow money from non-relatives. Borrowing money from relatives comes second. This reflects the existence of mutual help among villagers and between kin. Third comes searching for another job, using savings or selling the crops. The other options are quite rare.

Levels of household food security

The level of household food security was measured subjectively and objectively. Subjective measurement was done by asking about perceived food security in terms of changes in staple food and frequency and size of meals, as well as weight loss during the past two months. Only in few families (<1%) it was felt that family members could be hungry or food insecure. Adequacy of nutrient intake levels as a percentage of RDA was the objective measurement.

The results of the analysis of dietary intake in relation to RDA levels shows that the proportion of families who experienced moderate to severe food insecurity was 64.2% (Table 5.2). This proportion is lower than the proportion of families experiencing this in the Banjarnegara study (Martianto *et al.*, 2009) where it reached up to 97.3%.

When applying a subjective approach, most villagers do not feel they have a food insecurity problem because they get food easily and never experienced difficult times in getting food. However, the objective assessment shows that mothers who feel food secure often have a nutrients intake below the recommended dietary allowances (RDA) levels. Their food intake also compares unfavourably with that of their children. Mothers are more often severely or moderately food insecure and less often lightly food insecure or scoring above adequacy levels. It could be observed that mothers prioritise their child(ren) and husband when serving lunch while working in the field together with the husband and having small child(ren) with them.

Table 5.2. Distribution of family members by food security level based on dietary intake.

Food security level	Mother (n=120) (%)	Children under six (n=120) (%)	Average (%)
Severely food insecure (<70%)	36.7	20.0	28.3
Moderately food insecure (70-90%)	45.0	26.7	35.9
Lightly food insecure to food secure (90-120%)	17.5	33.3	25.4
Above adequacy (>120%)	0.8	20.0	10.4

Social security network in the community

Unlike the situation in rural areas in the district of Nusa Tenggara Timor (Martianto *et al.*, 2008, 2009) where villages have a special place to store food for emergency situations such as famine, the Tengger communities does not have a form of food storage. However, among the Tengger people the following forms of collecting rice as a social safety net were found:

- Collection of seven spoons of rice per household every week. This rice is collected monthly to be sold to underprivileged villagers or people struck by disaster at a very low price or given as a loan to these villagers to meet their food needs.
- Collection of 2.5 kg of rice every week from every household. The collected rice is sold or donated directly to families in need.
- Collection of ¼ cup of rice each month, After three months the rice is redistributed to communities in need.

5.4.5 Health and nutrition

Health and nutrition services access

The coverage of health and nutrition services in the community determines family access to health and nutrition services (Table 5.3). A family's access to and use of health and nutrition services indirectly influences children's nutritional status (Myers, 1990; Engle *et al.*, 1997).

Table 5.3. Households' use of services by kind and place of nutrition and health service.

Kind of health service	*Posyandu*		Midwive		Public health centre		*Dukun bayi*		Hospital	
	n	%	n	%	n	%	n	%	n	%
Children weighing	120	100.0	27	22.5	13	10.8	0	0	0	0
Vitamin A	100	83.3	6	5.0	3	2.5	0	0	0	0
Iron (Fe)	40	33.3	9	7.5	0	0	0	0	0	0
Contraception	5	4.2	107	89.2	10	8.3	0	0	2	1.7
KMS (health card)	113	94.2	1	0.8	0	0	0	0	0	0
Immunization	117	97.5	7	5.8	1	0.8	0	0	0	0
Medical treatment	8	6.7	95	79.2	44	36.7	6	5.0	10	8.3
Other services[1]	9	7.5	4	3.3	2	1.7	54	45.0	4	3.3

[1] Services of the *dukun bayi* (TBA) include body massage and herbal treatment (for pregnant women, mothers and babies).

In the study area, health and nutrition services delivered by midwives and the *posyandu* are available in the villages, whereas not in every village there is a traditional birth attendant (*dukun bayi*) and a public health centre. Apart from assisting at childbirth, the *dukun bayi* provides herbal treatment and massage for mothers and babies. She also uses spells and mantras.

Health environment and sanitation

All families (100%) in the study area get potable water from spring water from the mountain, a water supply system managed by local leaders. Seventy percent of the families in the study area has a house with brick walls, 21% of the houses is entirely or half made from wooden board, and 9% of the houses is made of bamboo. In half of the houses most of the floor is made of ceramic tiles, 45% has floors of cement or wooden boards, and 5% of the houses has an earthen floor. Nearly all families (99.2%) in the study area have their own toilet. Most families (81.7%) handle waste by discharging it into the abyss without any treatment. The others collect it in an empty field and immediately burn it (15.8%) or throw it into the river (2.5%).

Family health care practices

Of all families 78.3% have moderate hygiene and health practices, 7.5% have good hygiene and health practices, meaning that the bodily health of children under six years of age is maintained by bathing, washing hands, brushing teeth, clipping nails, hair cuts, and using footwear. More than 75% of the children under six years of age habitually bath with soap, use shampoo, brush their teeth, cut their nails, and use footwear. However, only 35% of the children regularly wash their hands before eating, 33% seldom do so. Less than 40% of families wash their hands with soap.

5.4.6 Food care

Household food consumption pattern

According to key informants, up to the 1970s, the staple food among the Tenggerese was white maize, known as 'maize rice'. Maize is stored as food reserve, then it is processed and crushed into flour. Currently, the staple food is purchased rice (99.2%). The animal food consumed is mainly fish (salted or fresh) (75.0%) and eggs (45.8%). Vegetables consumed are mostly cabbage (52.5%) and green/white mustard (46.7%). Orange, mango, watermelon, and apple are the most frequently consumed fruits. Excepting the vegetables, all of these are purchased.

Breastfeeding

Although only half of mothers is given their baby immediately after birth, 79% of the mothers said to give the colostrum to their infants. As much as 60% of the mothers reported giving complementary foods to their baby before it had reached the age of six months. Mothers do give breast milk substitutes, i.e. bottle-feeding, for example when they work in the fields. Foods fed to infants are meshed bananas, meshed green beans, and rice cake.

Feeding practices

More than two-thirds of the mothers affirmed to plan and prepare meals for the children. More than half of the mothers said to monitor the food eaten by their children, assist the children and eat together with the children. The rest of the mothers did so irregularly or never (Table 5.4), which could have a negative effect on the children's food consumption.

We found that 70% of the mothers always respond adequately when the child refuses to eat, but that 71.7% of the mothers never respond adequately when the child does not finish its food and that only 27.5% of the mothers always introduce new food in a positive way. The mothers' overall positive attitude when feeding the child shows a mixed picture in terms of frequency: never: 22.5%; sometimes: 45.8%; always: 31.7%. The convenience of the situation when the child is eating shows a similar nuanced, though slightly more positive, picture; the situation when the child is eating is always convenient in 41.7% of the cases.

Table 5.4. Frequency distribution of the mothers' feeding care practice.

Feeding care	Never		Rarely		Always	
	n	%	n	%	n	%
1. Mother plans the child's meals	8	6.7	32	26.7	80	66.7
2. Mother prepares the child's meals	0	0	21	17.7	99	82.5
3. Mother provides menu of vegetables to the daily child's meals	2	1.7	50	41.7	68	56.7
4. Mother monitors the child's meals	10	8.3	48	40.0	62	51.7
5. Mother chooses the child's snacks	39	32.5	57	47.5	24	20.0
6. Mother assists the child when eating	6	5.0	43	35.8	71	59.2
7. Tradition of family meals	7	5.8	47	39.2	66	55.0
8. Mother gives a varied meals to the child	11	9.2	66	55.0	43	35.8
Min-Max (score)	44.0-100					
Mean ± SD (% score)	71.98±1.35					

Studies by Myers (1990), Engle *et al.* (1997) and Zeitlin (2000) showed that the attitude of the mother while feeding the child affects the child's appetite and food consumption. Feeding the child with affection will have a positive effect on the efficiency of digestion, absorption of nutrients and utilisation by the body. Figure 5.2 shows that the level of feeding care provided by the mothers is predominantly low to moderate.

5.4.7 Family wellbeing

Family income and asset ownership

The average monthly income is Rp 644,140 per capita (± Rp 562,810). Compared to the poverty line of East Java province of Rp 206,275, 20.8% of the households can be classified as poor, 79.2% as non-poor. Almost all households (97.5%) owned agricultural land and less than half of them also owned livestock such as chicken or ducks (37.5%) and cow, buffalo or horse (49.2%). They mostly owned gold with an average ownership of 4 grams, and home appliances such as television (92.5%), cellular phone (83.3%) and a rice cooker (85%). Additionally, 78.3% of the households owned a motor cycle and 20.8% a car.

Satisfaction about economic status

Using subjective measurement of wellbeing, 69.2% of the families is satisfied with their household income, and 90.8% is satisfied with the quality and quantity of their food consumption (Table 5.5). Two-thirds of the families feel able to donate money to charity causes. However, 43.3% and 59.2% of the families are dissatisfied with the increasing prices of fuel and food, respectively.

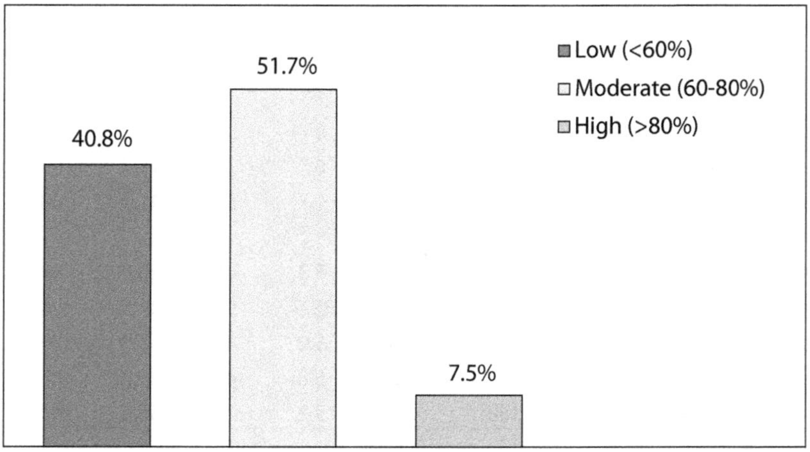

Figure 5.2. Distribution of levels of feeding care by mothers to children under six years of age.

Table 5.5. Responses to statements on subjective economic wellbeing of the household.

Economic wellbeing of the household[1]	No		Yes	
	n	%	n	%
1. Earn sufficient income	37	30.8	83	69.2
2. Food consumption is sufficient	11	9.2	109	90.8
3. Experiencing difficulties with health expenses*	108	90.0	12	10.0
4. Experiencing difficulties with children's education*	102	85.0	18	15.0
5. Able to meet the education needs of family members	23	19.2	97	80.8
6. Experiencing difficulty with living costs of children*	98	81.7	22	18.3
7. Able to meet the daily needs	16	13.3	104	86.7
8. Finding the fuel prices heavy at this time*	68	56.7	52	43.3
9. Feel burdened by the price of current goods*	49	40.8	71	59.2
10. Feel satisfied with the social support system	39	32.5	81	67.5
11. Difficult family*	103	85.8	17	14.2
12. Able to assist the poor / neglected children / the elderly	39	32.5	81	67.5
means ± sd (index)	74.858±17.238			

[1] * = negative statements.

Satisfaction on material wellbeing

The majority of the families (74.2%) feel that their house is appropriate and facilities are adequate (Table 5.6). This may be related to the fact that 79.2% of the households can be classified as non-poor (see above) and can adequately meet family necessities. More than 85% of the respondents expressed satisfaction with health and nutrition services (*posyandu*, midwife, health centre, medical doctor, and hospital) in the area.

Table 5.6. Response to statements on material wellbeing.

Responses to statements on subjective material wellbeing	No		Yes	
	n	%	n	%
1. House is now owned	31	25.8	89	74.2
2. House is comfortable for the family	31	25.8	89	74.2
3. Clothing is sufficient	23	19.2	97	80.8
4. Easy to obtain pharmaceutical drugs when needed	10	8.3	110	91.7
means ± SD	80.208±29.330			

Satisfaction on psycho-social dimension

In this study, the psychosocial dimension is the third dimension of subjective family wellbeing. Assessment was done using indicators on religious freedom, security, social activities, and social harmony. The majority of the families expressed satisfaction in response to the statements on psycho-social wellbeing. About having health problems that interfere with daily activities the level of dissatisfaction is highest, but still only 19% (Table 5.7).

Overall, the families feel satisfied (28.3%) to very satisfied (64.2%) with their economic status and material and psychosocial condition, only 7.5% of the families did not. Contrary to expectation, this study reveals that most families are not poor, have moderate access to health and nutrition information and services, have a moderate quality of housing with often a lavatory of their own, and are quite satisfied in terms of the psychosocial dimension of wellbeing.

Factors related to family wellbeing

In a study among families in rural Tapanuli Tengah (Sumatra), Butar Butar (2008) found that socio-economic aspects, including education, family size, type of occupation and family assets ownership, significantly contributed to poverty status in the village. In a study in Ireland on the association between living at risk of poverty and family wellbeing (Swords *et al.*, 2010) family wellbeing was conceptualised as comprising the dimensions of parent wellbeing (health), child wellbeing (health, happiness, school performance, no behaviour problems) and positive family relationships (parents' responsiveness and low level

Table 5.7. Responses to the statements on the psycho-social dimension.

Statements on the psycho-social dimension	No		Yes	
	n	%	n	%
1. Having health problems that interfere with daily activities	97	80.8	23	19.2
2. Feel free to worship according to own religion	3	2.5	117	97.5
3. Enjoy the holiday season for happiness	0	0	120	100
4. Feel safe from crime such as hold-up, robbery, extortion	5	4.2	115	100
5. Satisfied with current job	13	10.8	107	89.2
6. Involved in environmental activities	12	10.0	108	90.0
7. Having a harmonious marital relationship	5	4.2	115	95.8
8. Happy with the number of children they have	8	6.7	112	93.3
9. Satisfied with the communication with the partner	4	3.3	116	96.7
Means ± sd	92.875±9.370			

of conflict). Families who lived at risk of poverty had lower scores for all three dimensions of wellbeing than families not doing so.

We did a statistical analysis on the relations between different variables indicating aspects of objective and subjective family wellbeing, which yielded a number of statistically significant associations. As Table 5.8 shows, there is a significant and positive correlation between length of mother's education and the material dimension of family wellbeing (r=0.230), implying that in families of better educated mothers the needs regarding housing, clothing, and health needs are better met. As could be expected, family income is associated positively with economic wellbeing. Finally, better access to health and nutrition services leads to more psycho-social wellbeing (such as harmonious family relationships), which reflects the importance of such access for feelings of security and safety.

Table 5.8. Relations between selected variables of subjective and objective family wellbeing.

Variables	Dimensions of wellbeing[1]		
	Economic	Material	Psycho-social
Mother's education	0.097	0.230*	0.129
Monthly income	0.193**	0.125	-0.112
Access to nutrition and health services	0.078	0.114	0.206**
Nutrition knowledge	-0.066	0.015	0.145
Feeding practices	0.074	-0.041	-0.029
Breastfeeding practices	0.069	0.022	-0.131
Healthy practices	0.102	0.110	0.142

[1] *** = significant at 0.01 level; ** = significant at 0.05 level; * = significant at 0.1 level (2-tailed).

5.4.8 Children's nutrition

Energy and nutrients intake

The average energy intake of children under six years of age was 282.7±396 kcal. Their protein intake was 34.07±15.67 g, calcium intake 454.64±561.88 mg, Fe intake 6.63±2.63 mg, and Vitamin C 11.17±11.89 mg. Vitamin A intake was high (1,335±669 RE).

The children's energy and nutrients adequacy levels reach 91.0% for energy, 96.2% for protein, 90.9% for calcium, and 73.6% for Fe. The lowest adequacy level is that for vitamin C: only 24.8%. The Tengger children's energy adequacy level is higher than that of the children in Baduy Luar (88.15% RDA) and Baduy Muslim (76.9% RDA) but lower than that of the children in Baduy Dalam (97.3% RDA) in the study by Khomsan *et al.* (2009) on the Baduy in West Java. Compared to the Baduy children, especially vitamin A adequacy is much higher for the Tengger children (269.7% RDA), the figures for the children in Baduy Luar, Baduy Dalam and Baduy Muslim being 90.27% RDA, 152.8% RDA and 102.6% RDA, respectively. Table 5.9 presents the distribution of the four categories of the energy and nutrient adequacy levels.

Children's nutritional status

The three indicators of nutritional status used are the index weight-for-age, the index height-for-age, and the index weight-for-height. Data were analysed using the WHO Anthroplus 2005 version 3 programs and WHO anthropometric measurement standards (WHO, 2005).

Index weight-for-age

Table 5.10 shows that based on weight-for-age, the prevalence of underweight is 18.3% and of severe underweight 4.2%. WHO (MoH, 2010) classifies prevalence of underweight higher than 10% as a public health problem. Figure 5.3 shows that the median weight-for-age index of the children in this study lies to the left of the WHO standard (WHO, 2005), meaning that according to this indicator the nutritional status of the children is below WHO standards.

Table 5.9. Distribution of under six years children by nutrition adequacy level.

Nutrients	Nutrition adequacy level (% of recommended dietary allowance)			
	<70	70-90	90-120	>120
Energy (Cal)	20.0	26.7	33.3	20.0
Protein (g)	24.2	21.7	25.8	28.3
Calcium (mg)	60.8	5.0	10.0	24.2
Iron (mg)	48.2	27.5	17.5	6.7
Vitamin A (RE)	2.5	1.7	5.0	90.8
Vitamin C (mg)	91.7	5.0	2.5	0.8

Table 5.10. Children under six years of age by nutritional status (weight-for-age).

No	Nutritional status	Total	
		n	%
1	Overweight	4	3.3
2	Normal	89	74.2
3	Underweight	22	18.3
4	Severe underweight	5	4.2
	Total	120	100.0
	Min-Max	-4.57-3.75	
	Means ± std	-0.90±1.44	

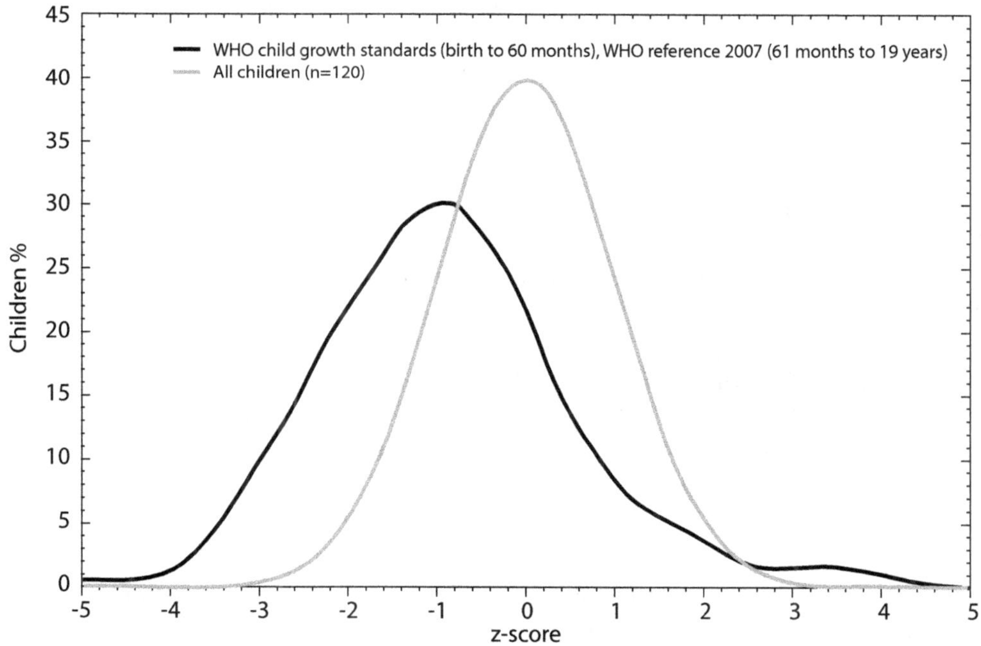

Figure 5.3. Distribution of children under six years of age by z-score weight/age.

Figure 5.4 shows the average value of the index weight-for-age by age groups in months. It can be seen that children aged 12-23 months already show the occurrence of underweight.

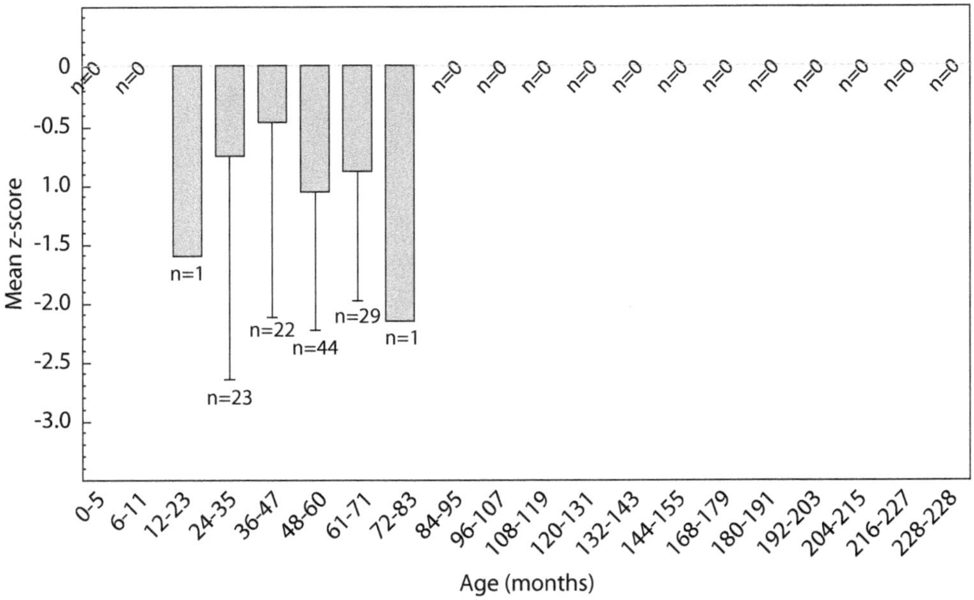

Figure 5.4. The index weight-for-age by age group.

Index height-for-age

Based on the index height-for-age, we found that the prevalence of stunting was 39.2% (Figure 5.5).

According to WHO criteria (MoH, 2010), prevalence of stunting higher than 20% is a public health problem. Figure 5.6 shows that the median of height-for-age index of the children under six in this study lies to the left of the WHO standard (WHO, 2005). The average value of the z-score was -1.23 standard deviations, which is considerably below the WHO standards.

Figure 5.6 shows the average value of the height-for-age index by age group in months. Children aged 12-23 months had an index close to -2.5, indicating a stunting problem in this age group. Table 5.11 shows that height-for-age is associated with mother's education, household income, and mother's nutrition knowledge, and that the latter correlated significantly with mother's education (r=0.299), access to services (r=0.375) and health practices (r=0.231).

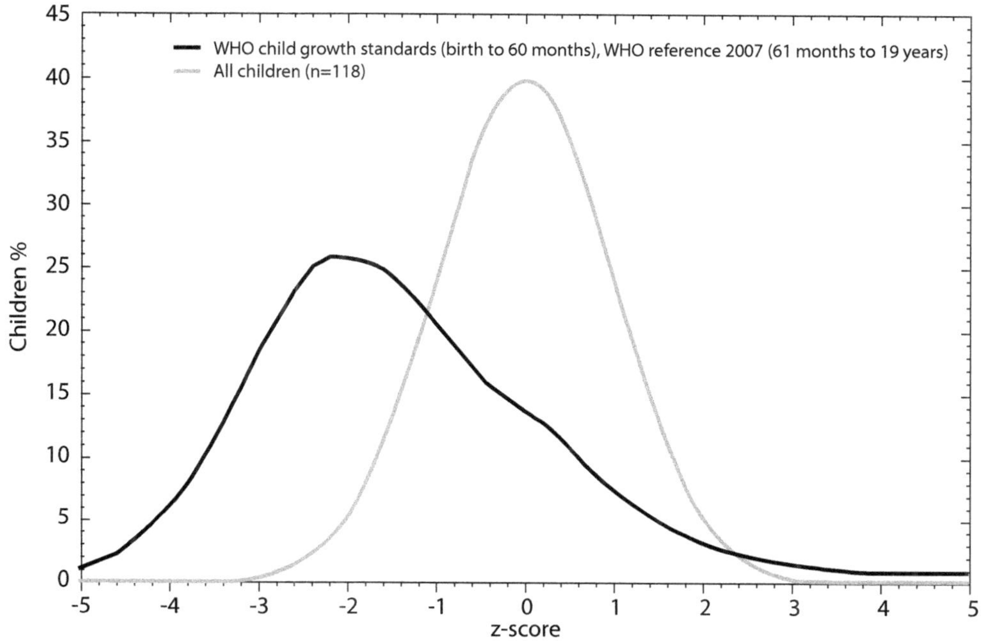

Figure 5.5. Distribution of children under six years of age by z-score height/age.

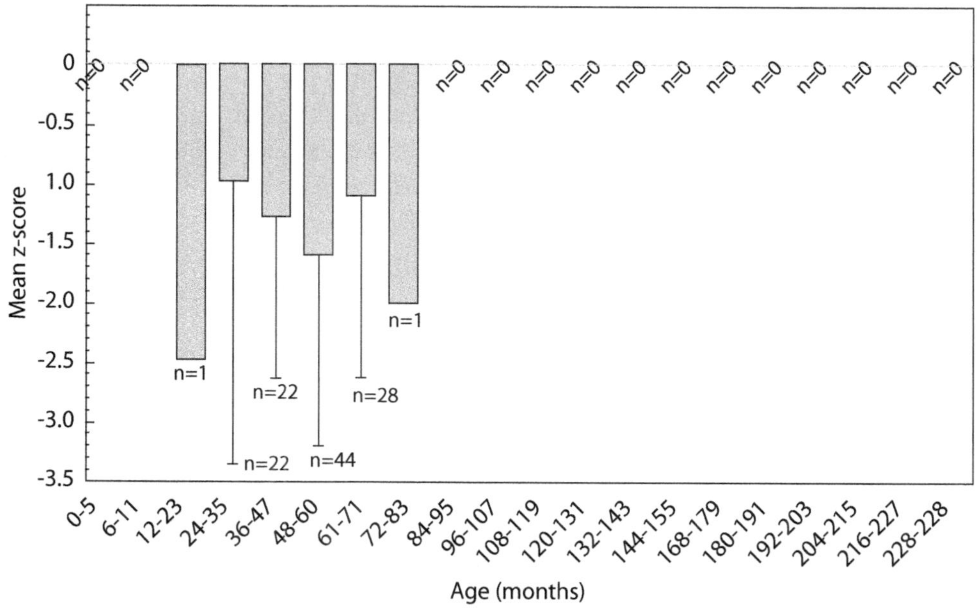

Figure 5.6. The index height-for-age by age group.

Table 5.11. Relations between selected variables and nutritional status of children (height per age).

Variables	Height per age[1]
Mother's education	0.224**
Monthly income	0.173*
Access to nutrition and health services	0.085
Nutrition knowledge	0.318***
Feeding practices	0.075
Breastfeeding practices	-0.057

[1] *** = significant at the 0.01 level (2-tailed); ** = significant at the 0.05 level (2-tailed); * = significant at the 0.1 level (2-tailed).

The findings imply that families with better-educated mothers are likely to have better access to health and nutrition services and better nutrition knowledge. A study conducted in Congo (Emina *et al.*, 2009) similarly found that maternal education was associated with lower prevalence of simultaneous multiple-malnutrition. The authors conclude that national policies that integrate education, access to food and use of health services are pivotal to improve child health and nutrition.

Index weight for height

Low weight-for-height (wasting) is a sensitive indicator of acute malnutrition. Prevalence of wasting of 10.1-15.0% is considered a serious health problem and if more than 15.0% a critical health problem (MoH, 2010). Hence, the wasting prevalence of 17.4% (including 5.8% severe wasting) among the Tengger children indicates a serious public health problem. Figure 5.7 shows that the median weight-for-height index of children under six lies to the left of the WHO standard (WHO, 2005), the average z-score for this indicator being -0.13 standard deviations.

When the values of the weight-for-height index are calculated by age group in months, we found that the z-scores deviation starts at the age of 12 months.

Children's health status

Morbidity is an important factor affecting children's nutritional status and wellbeing. Figure 5.8 shows that in the three months prior to the survey, 35.8% of the children did not get sick. During the same period, 20.8% was ill for more than seven days. The average duration of illness experienced by the children was 6.2±1.2 days.

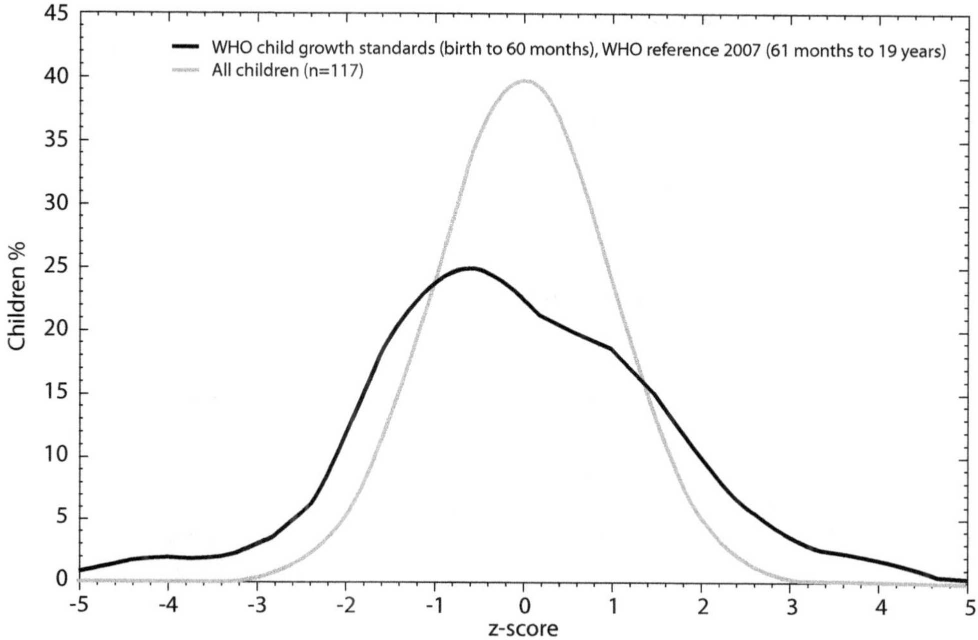

Figure 5.7. Distribution of children under six years of age by weight-for-height z-score.

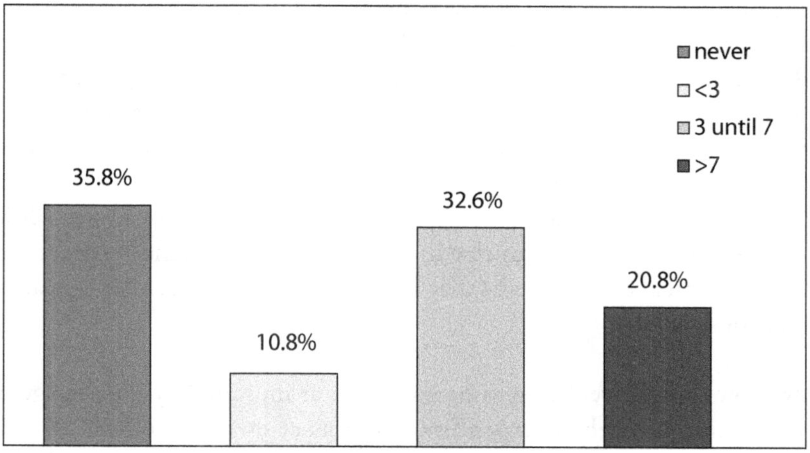

Figure 5.8. Distribution of children under six by illness duration in days (past three months).

The children suffered mainly from infectious diseases, such as (reportedly) cold/influenza (41.0%), fever (33.3%), and cough and cold (33.3%). Child morbidity is affected by many factors, one of them being access to nutrition and health service facilities. In this study, most mothers (60.8%) had been pregnant once. Almost all mothers (95.8%) never had a miscarriage, but 7.5% of the mothers reported infants who had died in the womb (4.2% prenatal and 3.3% postnatal), in 33.4% of the cases allegedly because of fatigue during pregnancy caused by the mother's working in the fields.

5.5 Conclusion and recommendations

The average income of the Tengger households in our study was well above the provincial poverty line of East Java (BPS 2011), although the monthly household incomes ranged from a minimum of Rp 300,000 to a maximum of Rp 2,850,000. Most Tengger families in this study (79.2%) are not poor according to the East Java standard. This is a better result than expected. The Tengger households mainly engage in farming (horticulture) and they own on average 1.2 hectares of agricultural land. Perhaps the fact that Mount Bromo area is a national park that attracts many tourists also contributes to the relatively good economic status of the Tengger households. Tourism generates employment and a market for agricultural produce (such as cabbage). People may also provide paid lodgings for tourists and some of them cook meals for their guests, especially during the peak season of religious celebrations such as Saraswati or Karo. This situation is quite different from that of a more isolated indigenous ethnic group like the Baduy (cf. Khomsan *et al.*, 2009).

Most Tengger families have good access to health and nutrition services and they only have one or two children, as advocated by the national family planning program. They seem to follow government programs quite well, whether these programs are about family planning, health, nutrition or education. However, child-feeding practices are considered to be at a low (40.8%) to moderate (51.7%) level. The mothers do not respond appropriately when a child refuses to eat or is unable to finish the food. Forty percent of Tengger mothers practice exclusive breastfeeding of infants under six months old, which is more or less at the level of the national rate. According to the 2012 Indonesian Demographic and Health Survey (CBS, 2013), the level of exclusive breastfeeding in Indonesia increased from 32% in 2007 to 42% in 2012. Promotion of exclusive breastfeeding is needed, since also the Tengger mothers are exposed to television and other media that advertise giving formula milk instead of breast milk to very young infants.

This study shows family wellbeing to be moderate, as indicated by the average monthly household income, household expenditure and asset ownership. Still, 20.2% of the households are poor according to the East Java poverty line. Subjectively, the respondents feel moderately to very satisfied with their economic situation. The only discontent is about the increasing prices of food and fuel. In terms of socio-psychological wellbeing, it was revealed that in general the families feel secure and are satisfied with their social relationships and their marital relationship. At the same time, 19.2% of the families is not satisfied with their family's health status and 20.8% of the children under six years old experienced illness that lasted for seven days or more during the three months prior to the survey.

The majority of the families in the study felt food secure. According to the village head and the local informal leader, the people never experienced household food insecurity, also not at the time of the Mount Bromo eruptions. At community level there is a local social security system in the form of rice collections and redistribution, from which poor and food insecure people can benefit, especially at times of disaster. There is also the institution of the *arisan*, an informal savings and credit group, which is found all over Indonesia (cf. Hospes, 1996).

In spite of the people's positive own assessment of their food security situation, this study found a double malnutrition problem in the Tengger community. More than 70% of the mothers have a deficient protein, Calcium, Fe, and vitamin C intake. Energy deficiency proved to be 36.7%. At the same time, there is obesity (5.8%) and overweight (25.8%) among the mothers. While the majority of the mothers rated themselves as food secure, based on energy intake the prevalence of food insecurity among mothers could be estimated at 81.7% (36.7% severely and 45.0% moderately food insecure). This might be related to cultural gender norms according to which women are expected to give priority to feeding the other family members, even if at the expense of themselves.

The Tengger children's energy and nutrients adequacy levels reach more than 90% RDA, which compares favourably to the children of the Baduy community in West Java (Khomsan *et al.*, 2009). Still, based on anthropometric measurement using the height-per-age indicator we found that 39.2% of the Tengger children under six years of age is stunted. Stunting was significantly associated with access to health and nutrition services and with the mother's educational attainment. The latter association could explain the overall better nutritional status of the Tengger children compared to the Baduy children, since more than 60% of the Tengger mothers finished junior high school whereas more than 85% of the women in the Baduy study were illiterate (Khomsan *et al.*, 2009).

Even though in many respects the Tengger families were doing much better than expected, still the following recommendations can be formulated based on this study:

- The traditional local social security system should be sustained and it is suggested to maintain food stocks at the household or community level.
- Nutrition knowledge should be improved by involving the respected traditional spiritual experts (*dukun adat*), traditional birth attendants (*dukun bayi*) and health staff (doctor and midwife), especially to change food beliefs that have adverse effects.
- The efficacy of *posyandu* nutrition activities could be enhanced by installing a desk for nutrition counselling by the midwife and health centre staff. Mothers need to be provided with materials on subjects like local nutritious food sources, exclusive breastfeeding for six months – although this is a problem in Indonesia (cf. Wibowo *et al.*, 2008) – and proper supplementary feeding afterwards, which could all help to reduce malnutrition.
- There is a need to (1) help ease waste pollution and litter in the area, especially from plastics; (2) improve facilities for public waste disposal; (3) educate people on waste

handling starting at home by separate disposal of organic and inorganic waste; and (4) establish a machine for production of organic fertiliser from organic waste.

- There is a need to educate the Tengger mothers on reducing children's street foods consumption, especially those that are unsafe for consumption.
- Family welfare can be enhanced through existing programs on women's empowerment and poverty alleviation, and by setting up village- or sub-district level cooperatives to provide soft loans or grants for small enterprises. Since the Bromo area is a tourist destination, more households could be involved in income generating activities.

References

BPP, 2011. Agricultural extension program of Sukapura district. Sukapura, Probolinggo: Badan Penyuluhan Pertanian [Agency for Agricultural Extension].

BPS, 2011. The poverty rate in East Java. Surabaya: Badan Pusat Statistik Provinsi Jawa Timur. Surabaya: Central Official of Statistics of East Java Province.

BPSP, 2010. Sukapura in figures. Probolinggo: Badan Pusat Statistik Kabupaten Probolinggo [Central Official of Statistics of Probolinggo].

Butar Butar, D., 2008. Analisis sosial ekonomi rumah tangga kaitannya dengan kemiskinan di pedesaan (studi kasus di Kabupaten Tapanuli Tengah) [Soci-economic status and its relationship to rural poverty (a case study in the District of Tapanuli Tengah] *Jurnal Perencaan & Pengembangan Wilayah* 4(1): 6-16.

CBS, 2013. Indonesia demographic and health survey 2012. Jakarta: Central Bureau of Statistics, BKKBN, Ministry of Health, and AUSAID.

Emina, J.B., Kandala, N.B., Inungu, J. and Yazoume Y.E., 2009. Maternal education and child nutritional status in the Democratic Republic of Congo. *Journal of Public Health and Epidemiology* 3(12): 576-592.

Engle, P.L., Menon, P. and Haddad, L., 1997. *Care and nutrition. concept and measurement.* International Food Policy Research Institute, Washington DC, USA.

FAO, 2013. *The state of food security in the world.* Rome: Food and Agriculture Organization of the United Nations.

Hospes, O., 1996. *People that count: changing savings and credit practices in Ambon, Indonesia.* Amsterdam: Thesis Publishers.

Khomsan, A., Anwar, F., Sukandar, D., Riyadi, H., Mudjajanto, E.S. and Wigna, W., 2009. *Socio-cultural aspects of nutrition and food systems of the Baduy.* Bogor: Faculty of Human Ecology, Bogor Agricultural University.

LIPI, 2008. *Widya Karya Nasional Pangan dan Gizi.* Jakarta: Lembaga Ilmu Pengetahuan Indonesia.

Martianto, D., Riyadi, H., Damayanthi, E., Hastuti, D. and Murtilaksono, K., 2007. *Food and nutrition security: situation analysis and alternative programs to strengthen household food security and improve child nutritional status in Kefamenanu district, East Nusa Tenggara province.* Bogor: Bogor Agricultural University and PLAN International

Martianto, D., Riyadi, H. and Hastuti, D., 2008. *Food and nutrition security situation analysis and alternative programs to strengthen household food security and improve child nutritional status in Timor Tengah Selatan district, East Nusa Tenggara province.* Bogor: Bogor Agricultural University and PLAN International.

Martianto, D., Riyadi, H., Hastuti, D. Briawan, D. and Alfiasari, 2006. *Food and Nutritional Assessment in Lembata, Nusa Tenggara province.* Bogor: Bogor Agricultural University and PLAN International.

Martianto, D., Riyadi, H., Hastuti, D. and Alfiasari, 2009. *Household food security, family resource allocation and impact on child development in Banjarnegara, Central Java province, Indonesia.* Bogor: Faculty of Human Ecology, Bogor Agricultural University, in collaboration with the Neys-Van Hoogstraten Foundation (The Hague).

MoH, 2008. *Report on basic health research Indonesia year 2007.* Jakarta: Centre of Research and Development, Ministry of Health.

MoH, 2010. *Basic health research (Riskesdas) in 2010.* Jakarta: Ministry of Health.

Myers, R.G., 1990. *The twelve who survive: strengthening programs of early childhood development in the third world.* Michigan: High Scope Press.

Niehof, A., 2010. *Food, diversity, vulnerability, and social change: research findings from insular Southeast Asia.* Wageningen: Wageningen Academic Publishers.

Niehof, A., 2011. Conceptualizing the household as an object of study. *International Journal of Consumer Studies* 35(5): 488-497.

Swords L., Merriman, B. and O'Donnell, M., 2010. Family wellbeing and a limited income: a study of families living at risk of poverty in Ireland. Dublin: Children's Research Centre, Trinity College.

WHO, 2005. WHO anthro 2005 for personal computers manual. Software for assessing growth and development of the world's children. Geneva: World Health Organization.

Wibowo, Y., Frebruhartanty, J., Fahmida, U. and Roshita, A., 2008. A formative research of exclusive breastfeeding practice among working and non-working mothers in an urban setting. [Unpublished report.] Jakarta: SEAMEO-TROPMED and Universitas Indonesia in collaboration with the Neys-Van Hoogstraten Foundation (The Hague).

Zein, A., 2007. The role of fisher-women in food security: a case study of traditional fisherman households, West Sumatera, Indonesia. [Unpublished report.] The Hague: Neys-Van Hoogstraten Foundation.

Zeitlin, M., 2000. Toddler nutrition in developing countries: the role of parenting. Jakarta: LIPI, Positive Deviance for Nutrition Program in WKNPG VII.

6. The role of social capital and institutions in food security and wellbeing of children under five for resettled households in Central Java, Indonesia

M. Quetulio-Navarra[1], E. Frunt[2] and A. Niehof[3]*

*[1]Presidential Commission for the Urban Poor, Quezon City, Metro Manila, the Philippines; [2]Bouw en Van de Meerendonk, Vleuten, the Netherlands; [3]Professor Emeritus, Wageningen University, Wageningen School of Social Sciences (WASS), the Netherlands; *melisnavarra@gmail.com*

Abstract

The study investigated the role of institutional inputs and social capital in achieving food security and wellbeing of children under five years old among households in an involuntary resettlement site in Central Java, Indonesia. Mixed methods were applied in data collection. The results reveal that in spite of flawed institutional inputs like basic services and public places, household food security was better than anticipated based on the literature. One explanatory factor is that because households were resettled within a relatively short distance from their original residence in the mountains, they could continue vegetable gardening. More importantly, however, the fast rebuilding of social capital in the resettlement site had a positive impact on food accessibility. Households in general relied on their new strong ties for securing food. The fact that culturally underpinned practices of food exchange and assistance were continued had a similar positive effect. Since household food security is a necessary condition for the wellbeing of children, they could benefit as well. Additionally, the institutional context after resettlement proved to be more favourable for the wellbeing of under-fives. The physical layout of the new site allowed for better child supervision, and health care facilities and schools were now closer to their homes.

Keywords: involuntary resettlement, social capital, food security, under fives well-being, Indonesia

6.1 Introduction

The number of people involuntarily displaced by natural and man-made disasters has been steadily increasing. In 2010, the number was 19.3 million, in 2014 more than 43 million (Ascher and Mirovitskaya, 2016). Additionally, around 10 million people are involved in infrastructure-induced involuntary displacement and resettlement each year (Cernea, 2000). The World Bank (2016) asserts that 'involuntary displacement occurs when the decision of moving is made and imposed by an external agent and when there is no possibility to stay.'

A. Niehof *et al.* (eds) **Diversity and change in food wellbeing – Cases from Southeast Asia and Nepal**
DOI 10.3920/978-90-8686-864-3_6, © Wageningen Academic Publishers 2018

Dislocation usually pushes the affected families into impoverishment (Cernea, 1985; Cernea and McDowell, 2000). Parasuraman and Cernea (1999) stress the role of institutions in displacement and resettlement through policies and programs for resettlement outcomes. The international community applies the World Bank standard that involuntary resettlement is to 'improve' (get ahead) or at least 'restore' (get by) the economic and social base of the displaced (World Bank, 2016). In his Impoverishment, Risks and Reconstruction (IRR) Model, Cernea (2000) posits that impoverishment results from neglecting nine interlinked potential risks that are inherent to displacement: loss of land, employment, shelter, and access to common resources; economic marginalisation; increased morbidity and mortality, food insecurity, educational loss, and social disarticulation. Of these, this study focused on the less investigated risks, i.e. social disarticulation, food insecurity, and morbidity. We investigated these risks at two levels: social disarticulation and food insecurity at household level, morbidity at the level of children under five years old in these households.

Cernea and Mc Dowell (2000: 30) assert that 'forced displacement tears apart the existing social fabric' through social disarticulation, which disperses and fragments communities, dismantles patterns of social organization, and scatters kinship groups. Life-sustaining informal networks of reciprocal help and local voluntary associations are disrupted. This loss of valuable social capital compounds the loss of natural, physical, and human capital. Additionally, 'the social capital lost through social disarticulation is typically unperceived and uncompensated by the programs causing it and this real loss has long-term consequences.' They also warn that forced displacement increases the risk of chronic food insecurity. Green (2000) points at the physical side of food security, i.e. the sufficient quantities of food that have to be procured, moved, and delivered to avoid critical conditions among the resettlers. Another threat is increased morbidity among the weakest segments of the demographic spectrum, such as children under five. Unsafe water supply and improvised sanitation systems associated with displacement may not only induce social stress but also parasitic and vector-born diseases such as malaria, schistosomiasis, and chronic diarrhoea (Cernea, 2000).

Resettlement studies have pointed out the significant role of social capital and institutions for the wellbeing of the resettled families after involuntary resettlement. Social networks and trust built after displacement were indispensable in building a 'new social world' for refugees in Australia (Westoby, 2008) and were a crucial resource among the 'new Canadians' (Lamba and Krahn, 2003). A large body of literature indicates that social capital facilitates positive outcomes for health and for children's wellbeing (e.g. Ferguson, 2006; Helliwell and Putnam, 2004; Mohnen *et al.*, 2015). On the other hand, Parasuraman and Cernea (1999) observed that resettlement outcomes are significantly influenced by institutions through their inputs in the form of resettlement policies and programs.

The studies referred to above highlight the value of both social capital and institutions for the wellbeing of involuntary resettlers. However, we applied a longitudinal perspective by looking at how social capital and institutions affect the resettlement community in its early years and later on, in terms of household food security and wellbeing of the young children. In this way, the present study can contribute to theory by assessing the role of social capital and institutions at different points in time in a resettlement trajectory. The study also aims at yielding insights that are relevant for government-managed resettlement sites in a developing country like Indonesia. Notably, Indonesia features prominently in Centre on Housing Rights and Eviction (COHRE) reports in terms of incidence of forced demolition and resettlement and violations of the right to adequate housing (cf. COHRE, 2008).

Specifically, this study addresses the following research questions:

1. How did involuntary resettlement affect the growth and quality of institutional inputs in the resettlement community as well as the households' stock of social capital and state of food security?
2. What are the effects of institutional inputs and social capital on the change in food security among the resettled households?
3. What are the effects of social capital and institutional inputs on the under-fives' wellbeing after the displacement?

6.2 Food security and under-fives through the lens of social capital and institutions

Using the double lens of institutions and social capital, we assume that institutional inputs and social capital have an impact on the resettled households' food security and the wellbeing of under-fives from their first year of residency and several years later. The Indonesian Bureau of Statistics (BPS, 2012) defines households as 'a person or a group of people living in a physical/census building or part thereof who make common provision for food and other essentials of living.' Niehof (2011: 490) sees the household as 'a co-residential unit [...] which takes care of resource management and primary needs of its members.' In this research, the number of people belonging to the household, including the children, is counted 'de facto', i.e. referring to all persons staying in that household at the time of the survey.

6.2.1 Social capital

Social capital has been described as a resource accessed through social networks (Ferlander, 2007). Woolcock and Narayan (2000: 243) note that 'social relations provide opportunities for mobilizing other growth-enhancing resources, that social capital does not exist in a political vacuum, and that the nature and extent of the interactions between communities and institutions hold the key to understanding the prospects for development in a given society.' A longitudinal study done by Curley (2009) involving 28 women from three relocation groups revealed that their homophilous ties (in terms of gender and socioeconomic situation) were the bedrock of their social networks, from which they derived emotional and instrumental support. Similarly, Ferlander (2007) pointed out that social capital has been empirically linked to reduced mortality as well as increased self-rated health.

In a study related to this one, Quetulio-Navarra *et al.* (2014) measured social capital in the new community in different forms and periods. They distinguished bonding, bridging and linking social capital and the different determinants of these, and looked at the state of the resettled families' social capital before and after the forced relocation. The present study uses a similar approach.

6.2.2 Institutional inputs

It is assumed that marginalisation in terms of food security and wellbeing of the resettled occurs when the administrative structures and health care institutions at the 'new' local level are inadequate. Bang and Few (2012) stress that relocated people should not become more vulnerable to socio-economic risks upon involuntary resettlement and that the institutional context has to be supportive to enable the relocated people to sustain their livelihood. Contractor (2008) states that the non-provision of health care facilities leads to severe health consequences for involuntary resettled people. In this study, the institutional inputs were measured by the availability of basic services and public places, administrative institutions and health care institutions at the local level.

The relevant health institutions operate on district level and below. In the formal health sector, health services are delivered by midwives, general practitioners, the *posyandu* (monthly health post for integrated mother-and child health care services), and hospitals. Apart from these formal institutions, there are informal health care providers who also play a significant role in the health and wellbeing of the people, such as traditional birth attendants and traditional healers.

The *posyandu* provides supplementary foods for under-fives, vitamin A capsules for infants 6-11 months old and under-fives, nutrition and health extension, immunisation of children, and family planning and reproductive health services. It links the local community to the rural health centre at sub-district level, to which cases of severe malnutrition can be referred. The *posyandu* works with volunteers from the community, with technical services being provided by doctors and midwives from the sub-district health centre. It has been shown that regular attendance of the *posyandu* has a positive influence on the health of under-fives (Niehof, 2010).

To improve population health and wellbeing, the national government has created a number of support programs: *Raskin, PKH, Jamkesmas,* and *Jamkesda. Raskin* is a rice support program for poor households. *Program Keluarga Harapan* is a cash transfer program in the form of allowances given to chronically poor households to assist them bringing up their children (Arif *et al.*, 2010). *Jamkesmas* and *Jamkesda* are health insurance programs for the poor. The first is funded by the national government, the second by the provincial government.

6.2.3 Wellbeing of under-fives

'The true measure of a nation's standing is how well it attends to its children – their health and safety, their material security, their education and socialization, and their sense of being loved, valued and included in the families and societies into which they were born' (Bradshaw, 2011: 1). Pollard and Lee (2003) distinguish five domains of wellbeing: physical, psychological, cognitive, social, and economic. The indicators in this study are limited to physical and economic (through food security) wellbeing. Bronfenbrenner and Morris (2006) claim that human development takes place through the dynamic interaction of the four principal components of the bioecological model of human development known as: Process, Person, Context and Time. The context consists of the surrounding layers of social groupings of family, relatives, friends, school, neighbourhood, and work. In this study, we focused specifically on the contextual layers of the household (family) and neighbourhood (physical layout and institutions).

6.2.4 Food security

Food is a basic need for all individuals and a universal dimension of household and personal wellbeing (Bickel *et al.*, 2000). Inadequate food intake induces mental and physical changes that may lead to harmful effects on learning, development, productivity, physical and psychological health, and family life (Bhattacharya *et al.*, 2004). Food security exists when all people, at all times, have physical and economic access to sufficient safe and nutritious food to meet their dietary needs and food preferences for an active and healthy life (FAO, 2013). For a rural community, access to food also depends on its geographical location and the likelihood of natural disasters occurring there. Cultural factors play a role as well. Niehof (2010) and Den Hartog *et al.* (2006) mention the *slametan* in Indonesia as a cultural food mechanism. The *slametan* is a communal meal staged at individual life course transitions (birth, marriage, death) and at important community events, to promote harmony (*rukun*) and social cohesion in the community and family (Guinness, 1986). According to Engle's' Law of Consumption, poor households typically spend three fifths of their income or more on food, leaving little for other necessities like health care, education, and recreational needs of the household members (FAO, 2016). In this study, we measured household food security by the reported food expenditures as proportion of total household expenditure for the year before resettlement, the first year in the site after resettlement, and the full year prior to the survey. Additionally, we asked the respondents whether they had experienced food shortages in the household.

6.3 Study design and methodology

The fieldwork was conducted during February-December 2012 in Bantarpanjang Translok (BT), a government-managed resettlement site situated in the village of Bantarpanjang, sub-district Cimanggu, district Cilacap in Central Java. BT covers an area of 3.1 ha. There are

97 housing plots of about 45 square metres each. The households do not own the house and the land they occupy. The site was built for households that lost their homes to landslides. It has been accommodating households since 2001. Data collection and analysis were carried out at two levels: household level and children under five.

6.3.1 Data collection at household level

A survey was conducted utilising a household composition sheet and a tailored-calendar tool. The sample comprised all the 70 households who were resettled in the area between 2001 and 2005. Data collected at household level are the basis for addressing research questions one and two.

The survey questions on social capital were a modification of the tool developed by Krishna and Shrader (2000). The Life History Calendar (LHC) method was used to strengthen reliability of the retrospective data (cf. Quetulio-Navarra *et al.*, 2015). For a proper application of the method, the enumerators were trained to ask the questions in a flexible way, adapted to the life course of the respondents. The life course events (landmarks) were used to create an accurate recall. During the interview, the enumerators could use the calendar in which the dates of children born and relocation were highlighted. The survey used a grid format to graphically display the sequencing and timing of life events. The following topics were addressed: (1) respondent's profile; (2) children; (3) household features; (4) physical features of the community; and (5) respondent's social engagement. The periods covered by the questions comprised a year before resettlement, first year in the site, and the last full year prior to the survey. Qualitative methods included key informant interviews, observation, in-depth interviews and focus group discussions.

Social capital was measured by questions on seven types of ties made by a household head in each year, comprising friends and acquaintances made in public places and during community activities in the location, individuals living in the community they would engage with frequently, people they knew from the government, non-government organization/s (NGO) and church, and the individuals who would help them in the areas of food security and health of the under-fives. Data on these ties were collected by asking for numbers of new friends and acquaintances made per year (number generator), names of persons the respondents would frequently engage with outside the family, and of persons they know from the government, NGOs and the church (name generator), and the name(s) of the person(s) who help(s) to meet specific needs such as food and childcare (combined name and resource generator). Names were converted into numbers and summed. Subsequently, the ties were grouped according to strong and weak ties.

Institutional inputs were measured through three indicators: number of basic services (i.e. power, water, school, day care centre) present in years covered by the study; number of public places (i.e. sidewalks, mosque, stores) present in each year covered by the study; and the respondent's annual frequency of attendance in government-convened meetings each year covered by the study. Answers of the respondents would vary since they transferred to the site in different years.

Food security was measured by securing reports from the respondents on the share (in percentage) of their household's total expenditure devoted to food for the year before resettlement, the first year in the site after resettlement, and the full year prior to the survey.

Details of these measurements are given in Table 6.1. The profile of the household heads, social capital, food security, and institutional inputs (basic services, public places, government-convened meetings) were based on reports of the respondents.

6.3.2 Data collection about under-fives

To address the third research question on the wellbeing of children under five years old, qualitative and quantitative methods of data collection were used. The research started with exploratory semi-structured interviews, to improve the survey tool and to understand the research context. Observation and non-structured interviews with residents in BT were conducted as well. The structured survey with the household heads was the principal method for obtaining quantitative data. It contained socio-demographic questions on household size, sex, age, marital status, relationship with the household head, educational level and employment status.

The wellbeing of the under-fives was measured by asking questions about immunisation, receipt and utility of the booklet with health information from the *posyandu*, vitamin supplements, ability to afford medical treatment, number of relations supporting the household in terms of food for under-fives, and the nature of these relationships. Respondents were also questioned regarding the institutions used in case of child injury, how often they visited the *posyandu*, and whether they had a *Jamkesmas* insurance. Exploratory semi-structured interviews were conducted with health and administrative personnel. Observation was done in conjunction with short non-structured interviews with residents. Lastly, in-depth interviews with BT residents were carried out to fully understand the circumstances in the settlement.

The sample comprised all involuntarily resettled households with under-fives between 1999 and 2011. In case a household had more than one child under five within the reference period, the two under-fives covering most years in the reference period were selected. This yielded a sample of 58 households. The key informants for the exploratory and in-depth interviews were purposively selected. The sample size for the qualitative part of the study was determined in the field, using the saturation principle. In the end, 18 exploratory interviews, two non-participant observations and four in-depth interviews were conducted.

Table 6.1. Table of variables studied.

Variable	Explanation
Dependent variable	
Food security	Percentage of household income spent on food
Independent variables	
Friends	Sum of no. of friends made in public places and in community activities
Acquaintances ties	Sum of no. of acquaintances made in public places and in community activities
Support ties	No. of individuals whom they rely on for support in different areas (money, health, etc.)
Food ties	No. of individuals whom they rely on for food support
Strong ties	Sum of no. of friends, close individuals, and family members, neighbours, and friends whom they rely on for support after resettlement
Weak ties	Sum of no. of acquaintances, ties from church, individuals who are not family or friends whom they rely on for support after resettlement
Linking ties	Sum of no. of ties with government (local, national, central level) and NGOs
After resettlement profile	
Individual attributes	
Place of origin	Cities of origin were ranked based on distance from site
Came from Bantarmangu	Dummy variable, where Bantarmangu=1
Year resettled	Entry year in the community
Gender	Dummy variable, where Male=1
Age	
Living in RT1/RT2	Dummy variable RT1=1; where RT2=1
Educational level	
Household attributes	
Household income	
Household size	Number of family members in the household
Wife employment status	Dummy variable, where Employed=1
Husband employment status	
Percentage of household income spent on food	
Institutional inputs	
No. of basic services in the community	
No. of public places in the community	
No. of attendances in government meetings	

In developing the Life History Calendar (LHC) tool, the need for collecting retrospective data regarding the under-fives in the community was taken into account.

6.3.3 Cohort approach and data analysis

The quantitative data were entered into Excel and analysed using SPSS 22 and STATA 9. Indicators for the yearly (a year before resettlement, first year in site, year when survey was conducted) food expenses profile, institution-related profile, and social capital profile were reported in numbers, frequencies, means, and percentages. These data were analysed for patterns regarding their trajectory over time.

T-tests were done on the differences between years. A correlation test was done on the association between percentage of food expenditure and institution-related and social capital variables. Lastly, regression analysis was conducted to determine the effects of the two sets of variables on the dependent variable of the yearly percentage of household food expenditures during the three periods. The independent variables in the regression model included some control variables and variables representing individual and household attributes, institutional factors, and the stock of social capital (food ties, strong ties, and weak ties) during the first year and in the last full year. The institutional factors were basic services, public places and attendance in government meetings (Table 6.1).

Qualitative data were recorded, transcribed and translated from Indonesian into English, and content analysis was applied. The analysis was directed at evaluations of social relations as well as the process of social capital development. The results were triangulated with the quantitative data.

As explained above, the overall study design integrated a time dimension (comparing outcomes at different points of historical time), an age dimension (comparing outcomes per cohort, per age), and a cross-sectional dimension (describe outcomes of the population at one point in time) in order to show outcomes from different angles and trace developments over time. In short, we applied a cohort approach; a combination of a longitudinal and a cross-sectional perspective (cf. Yuan, 2010). Altogether, the 58 households that participated in the survey had 95 children under five years of age. These children were divided over three cohorts: children born before 2001 (before resettlement) were assigned to Cohort 1 (40 children), those born between 2001 and 2006 to Cohort 2 (35 children), and those children born between 2007 and 2011 to Cohort 3 (20 children).

6.4 Results

6.4.1 General picture

Table 6.2 reveals that the majority of the respondents is male (94.3%). Most of them (68.6%) are in the age bracket of 41-60. Almost all respondents are married (97.1%). The majority (74.3%) only has elementary education or less, only 5.7% has high-school level education. Over half of the respondents are either doing elementary jobs (31.4%) or farming (25.7%). All are Muslims; most of them (68.6%) affiliated with Nahdatul Ulama (NU), the rest with Muhammadiyah. All are ethnic Sundanese. The average household size is 3.96 and the average annual household income is IDR 10,975,006.58 (915.773 USD). The site is divided into three blocks, RT1, RT2 and RT3, which comprise 67 households. For personal reasons, three households moved to RT5, a block that belongs to another ward (RW) but is adjacent to RT3. The landslide happened in 2000, but only in 2001 the resettlement site became minimally habitable for its victims. Sixty-four households resettled in BT during 2001-2003, six transferred in 2004 and 2005.

The Bantarpanjang resettlement project is managed by the Cilacap Provincial Government and the Cilacap Department of Transmigration. The interview with the officer in Cilacap who was directly involved in the supervision of the project, revealed that there was actually no earmarked budget for the affected families. However, given the urgency of the case, the Ministry reallocated some of their funds for regular transmigration activities to the development of the site. To acquire the land for the resettlement project, the Provincial Government entered into a land exchange contract with the forestry agency Perhutani in which the Cilacap Provincial Government compensated Perhutani for its land with a property twice the size of the target land.

Table 6.2. Profile of the respondents (n=76).

Variables	%	Frequency
Gender		
Male	92.11	70
Female	7.89	6
Age		
20-30	3.95	3
31-40	23.68	18
41-50	34.21	26
51-60	30.26	23
61-more	7.89	6

Table 6.2. Continued.

Variables	%	Frequency
Civil status		
Single	2.63	2
Married	93.42	71
Widowed	3.95	3
Education		
Never been to school	3.95	3
Elem. school or less	71.05	54
Junior high school	19.74	15
High school	5.26	4
Islam practice		
Nahdatul Ulama	67.00	51
Muhammadiyah	33.00	25
Occupation		
Farmer	23.68	18
Entrepreneur	13.16	10
Elementary occupation	31.58	24
Govt/private employee	7.89	6
Housewife	5.26	4
Others	15.79	12
Household size		
1-3	34.21	26
4-6	63.16	48
7-9	2.63	2
Household income		
<4,560,000	25.00	19
≥4,560,000 and <9,000,000	27.63	21
≥9,000,000 and <14,850,000	22.37	17
≥14,850,000	25.00	19
Present address		
RT 1	26.32	20
RT 2	36.84	28
RT 3	32.89	25
RT 5	3.95	3
Year resettled		
2001-2004	89.47	68
2005-2008	6.58	5
2009-2011	3.95	3

The provincial government still owns the land and the houses and there is no option for the resettlers to acquire it. To date, the Indonesian government still has no national policy for resettlement. The Transmigration Ministry follows the presidential decree Keppres 55/1993 on the transmigration of families, but that only covers the aspect of land acquisition, not the provision of basic services and public facilities (Zaman, 2002).

6.4.2 Institutions, social capital, and food security profile

The significant increase in the number of basic services and public places in BT compared to the location of origin (Table 6.3) is not surprising. Most of the new residents are farmers and came from mountain communities that had neither piped water connections nor permanent electric power or public places like mosques, day-care centres or sidewalks. Many of them did not have a private toilet and bathroom. Instead, they were using the mountain streams.

The resettlement took place before most of the basic services and public places had been constructed. The official in charge of the project explained in an interview that the provision of basic services and public places was delayed due to lack of budget, despite the emergency situation (see above). In 2000, the first two basic facilities in BT were a water container for collecting mountain water and eight public toilets. Subsequently, public wells and a day-care centre were added. Group electricity (one source for 10-11 households) was installed in 2003 and piped water connections were installed in 2005. Sidewalks and streets were built in 2005, mostly privately funded by the households with a little support from the government. As a respondent said, you could tell the richest RT by just looking at the streets

Table 6.3. Institutional inputs, social capital, and food consumption profile by year.

Institutions	Before	1st year	Diff. Bef & 1st	t-stat	*P*-value[1]	Last year	Diff. 1st & last	t-stat	*P*-value[1]
No. of basic services	3.36	3.63	0.27	-4.466	0***	3.71	0.08	-0.815	0.418
No. of public places	3.84	4.09	0.25	-1.79	0.078*	4.74	0.65	-1.962	0.054*
No. of government meetings	2.46	2.37	-0.09	0.349	0.728	2.15	-0.22	1.081	0.283
Social capital									
Ave. support ties	13.09	13.26	0.17	-1.797	0.077*	13.73	0.47	-2.46	0.016*
Ave. food ties	1.53	1.53	0			1.59	0.06	-1.653	0.103
Ave, strong ties	25.41	20.88	-4.53	4.583	0***	15.78	-5.1	5.095	0**
Ave. weak ties	61.92	49.88	-12.04	3.895	0***	33.93	-15.95	6.721	0**
Food expenditures	59.89	59.96	0.27	-0.3	0.765	59.59	-0.37	0.332	0.741

[1] Significant at *$P<0.10$; ** $P<0.05$; ***$P<0.01$.

and sidewalks. The only mosque is located in RT3 and was built by a Muhammadiyah foundation in 2005. It caters only for the RT3 residents since the other RTs are NU. Facilities like the health centre, NU mosque, market and schools are outside the community but are easily accessible.

The attendance of government meetings did not significantly change over the whole study period and was consistently low; 50-66% of the household heads failed to attend any meeting called by their RT and RW (ward) leaders or the village head. Only the RT leaders and older men would attend the few meetings that they themselves convened. Respondents said that they did not go to RT meetings because they were already tired from work and would rather stay at home and rest. They claimed that during 12 years in BT, their RW leader only organised two meetings. RT and RW leaders and the elders would attend the meetings organized by the village head. The most pressing issue in these meetings was how to acquire legal ownership of the lots, to avoid another displacement in the future.

Table 6.3 shows a statistical change in the households' social capital in all years studied. There was a consistent increase in support ties (13 to 14 ties on average) from before resettlement until 2011. On average, the respondents had around two persons who could assist them in securing food. One mother shared that she was happier in BT because she now had a bigger support group than in her previous residence where she could only rely on a few neighbour-relatives and friends. Another one added that moving to BT gave her access to credit that she did not have previously. She could now buy items from sellers in the community and pay later. A single father disclosed that he managed to raise his child alone due to the continued assistance of his neighbours who would look after the child while he was at work. Rather than relatives, especially family members, friends, neighbours and co-workers in the farm constitute the residents' support network in the new site. Both widowers and widows in the community even articulated that they heavily depend on their neighbours for food and for information on farm labour jobs. One of the women said that it was hard to contact their relatives because they do not live nearby anymore. Hence, people would rely on their neighbours to take care of them when sick, for emotional support, and for borrowing money without interest. One mother said that the friends and relatives lost during the displacement were replaced with many more friends and neighbours after the resettlement.

Strong and weak ties of the households steadily increased from the pre-relocation phase until the year 2011 in the new location. The pool of potential weak ties (acquaintances) would get bigger with the coming of new resettlers and as a result of participation in community activities, such as a *slametan*, increased. Women explained that even if they would go to the well to fetch water more often, they would have little time to connect with other people due to the long queue. At parties, however, they feel relaxed and have time to talk with other people while cooking, eating, or watching entertainment. Also the men considered attending community activities, particularly the informal ones, as a better way of making new friends and acquaintances than going to public places where they would just say hello and go. The people's way of practicing Islam through following either Nahdatul Ulama or Muhammadiyah, similarly stimulated acquaintanceship and friendship among the resettlers. Overall, the respondents were happier in BT than in their previous place of residence because of the opportunities to make new friends.

The households did not forge government ties apart from those with their RT and RW leaders (also resettlers) after their move to BT. Government officials in charge only visited the community when they needed to roll out a program, which rarely happened. The site's project manager revealed that he visits the site twice a year, but during the first year of resettlement twice a week. The government connections of RT and RW leaders are limited to the village head, staff in *Perhutani*, and the people who were working in government facilities like the health centre. The resettlers consider their RT leader as the only government tie valuable for their household. The RT and RW leaders considered the village head most important because of his power and money. They cannot join any program or project without his signature or invite NGOs to BT without his approval. They also need his approval and funding for activities such as *Raskin* (rice assistance programme), deep well construction, or burials. Almost all respondents had ties with the *imam* (priest) or the *ustadz* (chaplain) of the mosque. Of these, the *imam* is deemed more important because he is a spiritual guide as well as instrumental in keeping the peace in the community. Sometimes, people request his intervention in resolving community issues.

The share of household income spent on food (around 60%) did not significantly change over the years. This result affirms Engle's Law of Consumption that it is common for poor households to allocate as much as 60% of their income for food, leaving only 40% for other necessities (cf. FAO, 2016). Women farm labourers told us that although most of their land was swallowed by the landslide, they were able to find new spots for gardening on the same mountain they were once living and could ask the help of their relatives who were still living there. This was possible because of the relatively short distance (one to three kilometres) between BT and the previous residence. Others were able to make an arrangement with forestry officers for access to a farming lot in a nearby forest for a small rental fee. In both cases, the crops they harvested were either sold for extra income or consumed by the household.

6.4.3 Relationship of institutions and social capital with household food security

The correlation results for food security during first year of residency depict diverse relationships. Being employed, fewer public places in the community, more basic services, more support and strong ties, and less food ties are associated with food security among the households. Whereas the positive association of an employed household head with food security is logical, the negative association of 'number of public places' with food security needs to be explained. An explanation could be the fact that the construction of public places in BT, like sidewalks and streets, was paid for by the personal contributions of the households, at the expense of their food budget.

Providing the household with more permanent basic services like water and electricity is important for household food security. A housewife in BT explained how she spent most of her time fetching water at the deep well at the expense of time for farm labour and gardening. A husband complained about the temporary power system in the community wherein a group of residents had to share one electric metre and pay an equal amount regardless of amount of usage. He said that he was paying too much for electricity, leaving less for other more important items such as food.

Forging more new support, food, and strong ties was found to contribute to the attainment of food security among the newly resettled households. For food security, people would rely on the individuals they frequently engaged with. Considering that most of the resettlers are farm labourers and some even maintained their own farm, food exchange and assistance among neighbours was common practice. Friends and relatives in the mountain area who were unaffected by the landslide were helping tending food gardens.

Women belonging to the same RT could be seen gathered for small talk on the sidewalk in front of their houses. The women were keen on participating in activities that could support their family, like savings and credit groups (*arisan*). There were different *arisan* in the different RTs, which would vary in the contribution asked from the members (money or vegetable crops). The contributions were pooled and used for parties or to help a household in need of assistance. Friendship among women was also kindled by the community practice of *slametan*, which entailed cooking and eating together on different occasions. One housewife stressed that attending a *slametan* and contributing to it (in the form of money or assistance in cooking) are social obligations and that declining an invitation is unacceptable. *Arisan* and *slametan* are part of Sundanese and Indonesian culture, hence also practiced before resettlement. Religion also played an important role in binding new residents in BT and building their new stock of social capital. As explained above, this was either through affiliation with Nahdatul Ulama or Muhammadiyah. Clearly, these cultural institutions contribute to household food security in a resettlement context through increasing the stock of social capital.

The regression results presented in Table 6.4 show that the two institutions variables (public places and basic services) and two variables pertaining to social capital (strong ties and weak ties) had an impact on the households' food security during the first year. In the last year, however, their influence vanished, leaving only household income as a positive predictor of food security 12 years after settling in BT.

Table 6.4. effects of institutions and social capital on household food security (n=67).

Variables	First year		Last year	
	Coeff.1	St. error	Coeff.1	St. error
Age	4.21	3.17	4.52	3.37
Education level	-1.48	5.66	-4.49	6.16
Occupation	0.67	2.09	-0.18	2.04
Year resettled	14.16	15.84	17.60	15.21
Present address (RT1)	-6.68	6.44	-1.95	7.00
Married	0.68	8.20	1.56	8.59
Employed	5.59	10.59	14.10	11.07
Household income	-4.99	4.02	7.26*	4.24
Household size	-8.81	5.33	-5.86	5.62
No. of public places	-10.43*	5.60	0.78	4.79
No. of basic services	35.23*	18.19	2.58	14.62
No. of gov. meetings	-1.89	1.92	-0.70	2.36
No. of food ties	1.93	2.52	1.80	2.60
No. of strong ties	0.87**	0.37	-0.47	0.46
No. of weak ties	-0.41**	0.18	0.28	0.23
R^2	37.6%		26.9%	

[1] Significant at *$P<0.10$; ** $P<0.05$.

6.4.4 Institutional inputs, social capital, and under-fives' wellbeing

For a small fee, monthly health checks on the children were conducted at the *posyandu*. Monthly visits to the *posyandu* were highly recommended. The village midwife would write the check-up results in a small booklet kept by the mother. Almost all respondents in each cohort had this booklet.

The results reveal a mode of 12 visits per year across all cohorts and age groups. Noteworthy is that after resettlement, the frequency of *posyandu* visits decreased among the cohorts, presumably, because in BT there were more child health care options available. There were no significant differences per age group between the different cohorts. However, irrespective of cohort, an overall significant trend of a slightly decreasing number of visits as the age of the child increases is visible, which was to be expected. This might be another reason for the observation that the frequency of *posyandu* visits decreased among the cohorts. According to the booklet data, nearly every child in Cohort 1 was immunised, and all children in Cohort 2 and 3 were immunised. Additionally, most of the children received vitamin supplementation, both before and after resettlement. We found no changes over time or differences according to age.

The t-test analysis shows that the ability to pay for medical treatment for under-fives does not significantly change with age. The descriptive statistics show some differences between before and after resettlement, but these are not significant. A one-way ANOVA test showed that among all age groups, the ability to pay for medical treatment does not significantly vary between cohorts. Likewise, the ability to pay for medical treatment between cohorts among children aged 0 and 1 did not significantly differ. An analysis of the records in the booklets did not show a notable difference regarding *posyandu* attendance, vitamin supplementation, and immunisation coverage between before and after resettlement.

In case of disease or injury of a child, there are several options to get help. In the survey and in the exploratory interviews the following were mentioned: traditional healers, the traditional birth attendant, the auxiliary nurse, the village midwife, the sub-district health centre, the *posyandu*, the general practitioner, and the hospital. The answers to the question where the parents would bring a child in case of injury were mostly in the categories 'missing' and 'other' (between 65% and 80%) in all cohorts. In Cohort 1, before resettlement the traditional healer appeared to be the most chosen option, and after resettlement the popularity of the traditional healer even increased. Before relocation, In Cohort 2, before resetllement the sub-district health centre was the most chosen health institution in case of injuries among under-fives. After resettlement, the traditional healer and the village midwife became the most chosen options. In all cohorts, the village midwife became more popular over time. In Cohort 3, she appeared to be by far the most favoured option in case of injuries among under-fives, with the traditional healer coming as second. The hospital is not popular, because of the distance, costs, and bureaucracy involved. The general practitioner was rarely mentioned at any time in any cohort. Across all ages, the village midwife and the traditional healer appeared to be the most important persons in the institutional health care system for reasons of cost effectiveness and accessibility.

The health insurance scheme for poor families in Indonesia (*Jamkesmas*) could contribute to under-fives' wellbeing. However, most of the households in BT were not included in this insurance scheme, and in all cohorts the number of beneficiaries even decreased after resettlement. Qualitative research revealed that the institutional context in terms of health services was more favourable after resettlement. One respondent said there was no midwife in the previous village. Household had to borrow money to go to the sub-district health centre in Cimanggu, which was relatively far. Before resettlement, they would get traditional herbs from the religious leader. In BT, they could now buy medicine in a shop or go to the traditional healer.

About 75% of the respondents from households with under-fives indicated that they rarely or never experienced a food shortage situation. Regarding food shortage, the one-tailed t-test revealed only negligible differences between before and after resettlement. Also over time, trends are not striking. Cohort 1 shows a higher average number of relations supporting the under-fives' households in terms of food information before resettlement. The opposite is true for Cohort 2. The t-test reveals that across all cohorts, the number of contacts supporting the household in terms of food information for under-fives does not significantly differ between one year before and one year after resettlement. In Cohorts 1 and 2, most respondents said they had none or one relation before as well as after relocation in terms of providing food information. Most respondents of Cohort 3 mentioned two contacts. However, the number of contacts supporting the household on food information for under-fives does not vary significantly between cohorts among the different age groups, including among infants. On the other hand, the cohort analysis shows remarkable differences across some age groups for the same variable. T-tests show a significant increase in number of contacts between children of 0 to 3 years old (P=0.040), between children of 0 and 4 years old (P=0.040), and, lastly, between children of 0 and 5 years old (P=0.040). The statistics indicate that when the child gets older the number of contacts increases significantly. The support mentioned would mostly come from family. It is remarkable that, across all cohorts, the category 'friends and neighbours' form a much bigger support group after resettlement than before. In Cohort 1, the support from family dramatically decreased after resettlement. This is not the case for Cohort 2.

The regression analysis shows that the number of relations supporting the household in terms of food information for under-fives significantly predicts the ability of the household to afford medical treatment for their under-fives before (P=0.010) as well as after resettlement (P=0.004). The number of relations in the periods before and after resettlement can explain the significant proportion of the variance in ability to afford medical treatment. This applies in the year 2001 (P=0.015), but not for the years 2006 and 2011.

Qualitative research revealed that prior to resettlement, neighbours and relatives would always lend money in case a sick child was in need of medical treatment. Sometimes the religious leader could also be asked for traditional herbs. After resettlement, help in the form of money and advice were frequently provided in the community. Clearly, support came increasingly from friends and neighbours and less from relatives. This indicates a shift in type of bonding social capital. The supervision of under-fives improved after resettlement due to the population density in BT. Linking social capital helped the BT residents in obtaining *Jamkesmas* and connecting them with health care institutions. The RT leader was responsible for finding women volunteers to assist in the monthly *posyandu*. These women volunteers would establish linking ties in the *posyandu* and the sub-district health centre.

6.5 Conclusions and discussion

Before formulating a general conclusion about the case of Bantarpanjang resettlement site, we shall first summarise the answers we found to the research questions.

> Question 1. How did involuntary resettlement affect the growth and quality of institutional inputs in the resettlement community as well as the households' stock of social capital and state of food security?

Although there was a delay in construction of basic services, which was inconvenient, the transfer to the resettlement site after the landslide provided opportunities for the households to benefit from better access to basic services, health facilities, and public places. The resettlement site provided water and sanitation facilities, whereas in the previous location the people used the mountain streams for their personal hygiene and sanitation. Additionally, since the resettlement project was a government program, the community was considered a priority site for social programs such as the rice assistance program.

Despite going through disruption in their stock of social capital due to physical displacement, the new residents were able to rebuild their social capital, which in the new place was composed of friends and neighbours instead of relatives. Their social networks improved in terms of substance and distance. After the transfer, their ties comprised their neighbours and block mates, who were all important social resources for meeting various needs. The cultural practices of *slametan* and *arisan* and the religious affiliation with either *Nahdatul Ulama* or *Muhammadiyah* encouraged the creation of social capital among previous strangers as well as the replacement of support ties of relatives by neighbours and friends. Linking ties with government officers and officials remained weak.

A surprising outcome is that household food security did not seem to have deteriorated as a consequence of resettlement. Involuntary resettlement did not change the food share of the households' total expenditure in the first year after resettlement and several years after. This was to a large extent due to the households' ability to adapt in the new situation by forging new support ties. The fact that the geographical location of the site allowed them to maintain their job or to continue vegetable farming also helped.

> Question 2. What are the effects of institutional inputs and social capital on the change in food security among the resettled households?

The institutional inputs and social capital had varying impacts on the resettled households' food security. Despite the sometimes flawed institutional inputs (i.e. basic services and public places), the ability of the households to achieve food security was supported by the decision of the regional government to build the resettlement site at a relatively short distance from the original place of residence. This enabled the households to continue their vegetable gardening.

As explained above, the fast rebuilding of social capital in the resettlement site also strengthened household food security. Households could rely on their new strong ties in securing food. The continued practices of food exchange and food assistance augmented household food security as well. Wives were assured of emergency supplies of cooking oil, sugar, salt, and even small cash through their neighbours and friends. A widow who lived alone admitted that she could get by because of the consistent kindness of her neighbours, who would give her food as well as access to irregular farm jobs.

> Question 3. What are the effects of social capital and institutional inputs on the under-fives' wellbeing after displacement?

The answers to the first and second question already highlight that the households were not less food secure after resettlement. Closer examination regarding the households' children under five revealed that the new bonding social capital (composed of friend and neighbours) was not only the source of food assistance but also critical for important information on food and nutrition for under-fives.

Household food security is a necessary but not sufficient condition for the wellbeing of under-fives, since food wellbeing is just one component of overall wellbeing. Looking at factors other than food that affect under-fives' wellbeing, the findings show that the institutional context proved to be more favourable for the children after resettlement. The physical layout of the site allowed for more supervision (contributing to the children's safety), and health institutions and schools were closer to the new site. Although in each cohort the number of visits of the *posyandu* was lower after resettlement, over time and across cohorts, visits to the *posyandu* increased, presumably because of better accessibility of the *posyandu* in BT compared to the previous location. In case of injuries among under-fives, parents would go to the nearby traditional healer or village midwife. Regarding health insurance (*Jamkesmas*), coverage remained as low after resettlement as before.

To conclude, the resettled households and their young children were doing much better than could be expected based on the literature about potential negative impacts of displacement (Cernea, 2000; Cernea and McDowell, 2000). At the same time, the results confirm the evidence in the literature about the crucial importance of social capital for health and wellbeing in general (Ferguson, 2006; Ferlander, 2007; Helliwell and Putnam, 2004; Mohnen *et al.*, 2015) and for involuntarily resettled families in particular (Curley, 2009; Lamba and Krahn, 2003).

This study has generated insights that can be valuable for resettlement scientists and practitioners. First, bonding social capital of the resettled poor households is indispensable in the process of getting by and getting ahead in their new community. Forging linking ties in an involuntary resettlement context should be initiated and nurtured by the potential linkages. Second, for institutional inputs to be effective for households' wellbeing, they should be human-rights based and introduced in a timely manner. Lastly, in the locational choice of a resettlement site, minimal economic and social dislocation of the resettlers are primary considerations for a successful resettlement.

References

Arif, S., Syukri, M., Holmes, R. and Febriany, V.., 2010. Gendered risks, poverty and vulnerability: case study of the Raskin food subsidy programme in Indonesia. London: Overseas Development Institute. Available at: http://tinyurl.com/ybwqtb73.

Ascher, W. and Mirovitskaya, N., 2016. In: *Development strategies and inter-group violence. politics, economics, and inclusive development.* New York: Palgrave Macmillan, pp. 181-202..

Bang, H.N. and Few, R., 2012. Social risks and challenges in post-disaster resettlement: the case of Lake Nyos, Cameroon. *Journal of Risk Research* 15(9): 1141-1157.

Bhattacharya, J., Currie, J. and Haider, S., 2004. Poverty, food insecurity, and nutritional outcomes in children and adults. *Journal of health economics* 23(4): 839-862.

Bickel, G., Nord, M., Price, C., Hamilton, W. and Cook, J., 2000. *Guide to Measuring Household Food Security.* Alexandria VA: US Department of Agriculture, Food and Nutrition Service.

BPS, 2012. *Statistical glossary.* Jakarta: Badan Pusat Statistik. Available at: http://tinyurl.com/yakrnk5e.

Bradshaw, J., 2011. *The well-being of children in the UK.* Bristol UK: The Policy Press.

Bronfenbrenner, U. and Morris, P.A., 2006. The bioecological model of human development. In: Damon, W. (ed.) *Handbook of child psychology* I. [6th Edition] Wiley, Hoboken, NJ, USA, pp. 793-828.

Cernea, M.M., 1985. Putting people first: sociological variables in rural development. Oxford: Oxford University Press.

Cernea, M., 2000. Risks, safeguards and reconstruction: a model for population displacement and resettlement. *Economic and Political Weekly* 35(41): 3659-3678.

Cernea, M.M. and McDowell, C., 2000. *Risks and reconstruction: experiences of resettlers and refugees.* Washington DC: The World Bank.

COHRE, 2008. *Indonesia.* Geneva: Centre on Housing Rights and Eviction. Available at: http://www.cohre.org/Indonesia.

Contractor, Q., 2008. Understanding the impact of involuntary slum resettlement on women's access to healthcare in Mumbai, India. *Journal of Comparative Social Welfare 24*(2): 153-163.

Curley, A.M., 2009. Draining or gaining? The social networks of public housing movers in Boston. *Journal of Social and Personal Relationships* 26 (2-3): 227-247.

Den Hartog, A.P., Van Staveren, W.A. and Brouwer, I.D., 2006. *Food habits and consumption in developing countries: manual for field studies.* Wageningen: Wageningen Academic Publishers.

FAO, 2013. *The state of food security in the world.* Rome: Food and Agricultural Organization.

FAO, 2016. *Measuring food and nutrition security: an independent technical assessment and user's guide for existing indicators.* Rome: Food and Agricultural Organization. Available at: http://tinyurl.com/yba97aua.

Ferguson, K.M., 2006. Social capital and children's wellbeing: a critical synthesis of the international social capital literature. *International Journal of Social Welfare* 15: 2-18.

Ferlander, S., 2007. The importance of different forms of social capital for health. *Acta Sociologica* 50(2): 115-128.

Green, R., 2000. food security in refugee and return: Some aspects of entitlements, markets, and modalities. In: Cernea, M.M. and McDowell, C. (eds), *Risks and reconstruction: experiences of resettlers and refugees*. Washington DC: World Bank, pp. 253-290.

Guinness, P., 1986. *Harmony and Hierarchy in a Javanese Kampung*. Singapore: Oxford University Press.

Helliwell, J. and Putnam, R.D., 2004. The social context of well-being. *Philosophical Transactions of the Royal Society (London), Series B*: 1435-1446.

Krishna, A. and Shrader, E., 2000. Cross-cultural measures of social capital: a tool and results from India and Panama. In: *Social capital initiative*. Washington DC: The World Bank.

Lamba, N. and Krahn, H., 2003. Social capital and refugee resettlement: the social networks of refugees in Canada. *Journal of International Migration and Integration* 4(3): 335- 360.

Mohnen, S.M., Völker, B., Flap, H., Subramanian, S.V. and Groenewegen, P.P., 2015. The influence of social capital on individual health: is it the neighbourhood or the network? *Social Indicators Research* 121(1 SI): 195-214.

Niehof, A., 2010. *Food, diversity, vulnerability and social change: research findings from insular Southeast Asia*. Wageningen: Wageningen Academic Publishers.

Niehof, A., 2011. Conceptualizing the household as an object of study. *International Journal of Consumer Studies* 35(5): 488-497.

Parasuraman, S. and Cernea, M.M., 1999. *The development dilemma: displacement in India*. Basingstoke: Macmillan.

Pollard, E.L. and Lee, P.D., 2003. Child well-being: a systematic review of the literature. *Social Indicators Research* 61(1): 59-78.

Quetulio-Navarra, M., Niehof, A., Van der Horst, H. and Van der Vaart, W., 2014. Short-term risk experience of involuntary resettled households in the Philippines and Indonesia. *Habitat International* 41: 165-175.

Quetulio-Navarra, M., Van der Vaart, W. and Niehof, A., 2015. Can 'Third-Party Help' improve data quality in research interviews? A natural experiment in a hard to study population. *Field Methods* 27: 426-440.

Westoby, P., 2008. Developing a community-development approach through engaging resettling Southern Sudanese refugees within Australia. *Community Development Journal* 43(4): 483-495.

Woolcock, M. and Narayan, D., 2000. Social capital: Implications for development theory, research, and policy. *World Bank Research Observer* 15(2): 225-249.

World Bank, 2016. *Involuntary resettlement*. Washington DC: The World Bank. Available at: http://tinyurl.com/y84spknb.

Yuan, J., 2010. *The household responsibility system and social change in rural Guizhou, China: applying a cohort approach*. Wageningen: Wageningen University. [PhD Thesis]

Zaman, M., 2002. Resettlement and development in Indonesia. *Journal of Contemporary Asia* 32(2): 255-266.

7. What explains stunting among children living in a rice surplus area in Central Java, Indonesia?

R.C. Purwestri[1,2], Z. Barati[1,3], N.N. Wirawan[2], I. Fahmi[2], J. Lauvai[1] and V. Scherbaum[1,4]*

*[1]Institute for Biological Chemistry and Nutritional Science, University of Hohenheim, Stuttgart, Germany; [2]Faculty of Medicine, Nutrition Department, University of Brawijaya, Malang, Indonesia; [3]Institut for Agriculture Engineering in the Tropics und Subtropics, University of Hohenheim, Germany; [4]Food Security Center, University of Hohenheim, Stuttgart, Germany; *rc.purwestri@uni-hohenheim.de / purwestri@yahoo.com*

Abstract

Demak district is part of the province of Central Java, one of the national rice stock areas in Indonesia. In 2013, in Central Java the districts ranked sixth in rice productivity, but, at the same time, seventh in chronic child malnutrition. This paradox triggered an interest in the correlations between food accessibility and utilisation of households in Demak and the prevalence of stunting among children under five years old. During December 2014-February 2015, this issue was investigated in three sub-districts in Demak that had the highest rice production and poorest nutritional status of children under five. The oldest breastfed children aged 6 to 60 months from farmer families were selected for food intake and nutritional status measurement. Data were obtained on household socio-economic status, general characteristics, anthropometry, and food intake of the children. Nearly one-third (31.9%) of the 335 children in the study was stunted. Remarkably, food accessibility was similar between households with and without stunted children. However, diet quality regarding energy, calcium and iron intake was significantly better in households without stunted children. Unhealthy snacking habits of the children were found to contribute to an inadequate food intake as well. Improvement of complementary foods and family diet, including healthy homemade snacks based on locally available foods from animal and plant sources should be promoted.

Keywords: expenditure, income, stunting, diet

7.1 Introduction

Food and nutrition security indicates a situation in which at all times adequate food is available and accessible in terms of quantity, quality, safety and socio-cultural acceptability, and the food can be satisfactorily utilised by individuals for an active and healthy life (Weingärtner, 2005). This description includes the four pillars of food security, i.e. availability, accessibility and utilisation, as three physical aspects, and stability as the time dimension. The last part of the description refers to nutrition security (cf. Gross *et al.*, 2000).

Food availability is achieved when adequate food is available, from either domestic production or imports. Access to food at the household level, means ensuring that households have sufficient resources for getting adequate nutritious foods through production, purchase or donation. Utilisation of food is accomplished through nutritious and safe diets, an adequate biological and social environment, and proper health care. In the term food utilisation, the aspects of food and non-food items (health) are included. The aspect of stability takes into account the availability and accessibility of food through time, also during lean periods. The four pillars of food and nutrition security are finally manifested in the nutritional status of the individual household members (FAO, 2006; Gartaula *et al.*, 2017; Maxwell and Smith, 1992).

There is an increasing interest among government and non-government organisations in the question whether interventions in agriculture have the potential to improve the nutritional situation of the beneficiaries. Several systematic reviews showed that increased agricultural production does not automatically result in an improved nutritional status of the farmer family members concerned (Berti *et al.*, 2004; Leroy and Frongillo, 2007; Masset *et al.*, 2011, 2012; Ruel, 2001).

In Indonesia, rice is considered the most important staple food. Since more than three decades, the agricultural policies of the national government prioritise increasing rice production through subsidies, infrastructure interventions, monitoring, etc. Following the achievement of rice self-sufficiency in 1984, the national government started to focus on diversifying food crops (Five Year Plan 1989-1994), with the intention to increase employment opportunities and intensify poverty alleviation. However, the progress was slow and up to now there have been no clear guidelines for implementation. Rice continued to be the most important food crop, especially after by the global economic crisis in 1997-98 (Asian Development Bank, 2006; Thomas, 2003). Prioritising rice cultivation was considered to be the best option for the farmers because of a definite market and farmer experiences in producing rice, especially if the farm capital was a major constraint (Purwestri *et al.*, 2017; Siregar and Suryadi, 2006).

In 2013, the Central Bureau of Statistics reported that Central Java province ranked as the second highest national rice producing area (BPS, 2014). However, this high productivity was not reflected in the nutritional status of children living in this area. The district of Demak in Central Java had the sixth highest rice production within this province. However, in the same year, the Indonesian Ministry of Health (MoH, 2013a) reported that the prevalence of stunted children in Demak was above 30%, indicating an alarmingly high prevalence of chronic malnutrition in this area. The 2010 Ministry of Health report also stated that farmers were considered a vulnerable group due to their monthly expenses being in of the lowest quintile of the distribution and having the highest prevalence of stunted children (MoH, 2010).

Recognizing the complex relationship between agricultural production and nutritional status, this study investigated the contribution of household expenditure and dietary intake to the nutritional status of children below five years old from farmer families living in a rice surplus area with a high prevalence of stunted children. Figure 7.1 shows the conceptual framework of the study.

Figure 7.1 was developed based on UNICEF (1998), Gross *et al.* (2000) and Masset *et al.* (2011). The framework shows agricultural production's impact on food accessibility, as indicated by the economic situation and dietary diversity of the family. The provided and consumed foods and breastmilk intake can be converted into macro- and micronutrient intake, which determines the nutritional status of the children below five years of age. Care plays an important role in the nutritional status of children through feeding practices as well as hygiene behaviour, sanitation and health care utilisation. The figure shows the interrelationships.

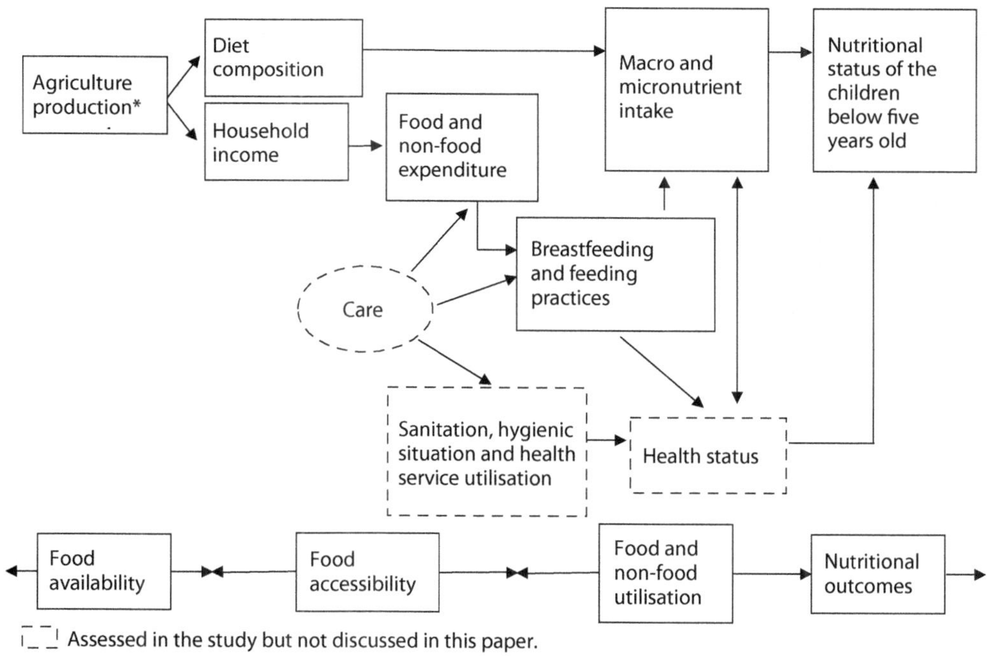

*Figure 7.1. Conceptual framework of the study. * = discussed in detail by Purwestri et al. (2017).*

7.2 Methodology

A cross sectional survey was conducted by the first author in the period December 2014-February 2015, in Demak district, Central Java, Indonesia. Of the 35 districts in Central Java, Demak had the sixth highest rice productivity in 2013 of 6,117 kg per hectare and in 2012 the seventh highest percentage of child underweight (indicated by weight-for-age Z-score <-2SD) of 9.4% (Purwestri *et al.*, 2017). In the same period, the prevalence of stunted children (height-for-age z-score <-2 SD of the reference population) in Central Java was above 30% (MoH, 2013a). Nutritional status based on height-for-age (HAZ) scores of Indonesian children at province level is usually reported by the Ministry of Health as part of a regular national health and nutrition survey. Since children's height is not commonly measured during the monthly growth monitoring carried out by voluntary workers at the integrated health services posts, called *posyandu*, at the village level, Central Java province only reported incidence of underweight based on the WAZ scores of the children. In Demak, the three sub-districts of Karanganyar, Dempet, and Gajah, were selected for the research because they had a high rice productivity and poor nutritional status of children under five. In total, eight villages in Karanganyar, seven villages in Dempet and one village in Gajah were identified as rice surplus areas and were then chosen as sites for data collection.

The oldest non-exclusively breastfed children under five years (aged between 6 and 60 months) living in farmer households were selected to capture as much as possible stunting in children. Stunting reaches its highest level between 18 and 24 months, often as a consequence of inadequate complementary feeding (Gross *et al.*, 1996; Seth *et al.*, 2011). Farmers in the study included farmers with land, tenant farmers, farm workers, sharecroppers and livestock raisers. Lists of eligible children in the selected villages in Karanganyar, Dempet and Gajah were obtained from village midwives or *posyandu* workers to establish the proportions of children, and then draw up a random sample of eligible children (Purwestri *et al.*, 2017).

A structured questionnaire was used for data collection. Mothers of the eligible children were interviewed at home and were asked about the child's general characteristics and household expenditure. Data on the dietary intake of the children was collected by a 24-hour food intake recall, including frequency and estimated duration of breastfeeding. The children's height was measured using a height/length board depending on the age of the children, and their weight was determined using a SECA 201 scale. Two focus group discussions (FGDs) were conducted in Karangwaru Kidul village and Dempet village in the sub-districts of Karanganyar and Dempet, respectively, to gather information regarding food habits of children under five years of age. FGD respondents (about six to maximum ten women per FGD) were selected based on lists provided by village midwives or voluntary health workers of the *posyandu*. In addition, in-depth interviews with two midwives and one nutritionist at the local health centre were carried out.

7.2.1 Sample size

A minimum sample size of 330 children was deemed necessary for a cross-sectional study in the area having a stunting prevalence of 31.1% in Central Java (MoH, 2013a) with a confidence level of 95% and a power of 0.8. In addition, for detecting differences between the group of stunted versus non-stunted children, a minimum of 95 children per group was calculated (Lwanga and Lemeshow, 1991). The final sample size needed for this study was 335 eligible children.

7.2.2 Data analysis

An analysis was performed, testing the relationship of agricultural production and revenue, dietary intake and HAZ-score of the 335 respondents. Data was checked for compliance with a normal distribution using QQ-plots of normality. For normally distributed data, differences in means were tested with the Independent T-test, while for the non-normally distributed data the Mann-Whitney U-test was used. Statistical analysis was performed using IBM SPSS Statistics Version 22. Anthropometric data were transformed to Z-scores using Emergency Nutrition Assessment version 2011 (ENA for SMART). Two groups of children were categorized based on HAZ-score; below -2SD was defined as stunted, while a HAZ-score ≥ -2 SD was considered non-stunted. Data from one day 24-hour recall food intake of the children was converted into macro- and micronutrient intake using Nutrisurvey. Breastmilk intake was estimated using Butte *et al.* (2002) reference data. In the analysis, snacks, either homemade or bought from nearby shops or street vendors, which were consumed by the children were separately assessed for their contribution to the children's diet. Nutrition intake was compared to the guidelines of nutrient intake for Indonesian well-nourished children (MoH, 2013b). FGDs and in-depth interviews data on dietary habits were transcribed, evaluated and coded. The important keywords were listed and grouped in Microsoft Excel. Total household expenditure was calculated based on monthly (routine) and yearly (non-routine) expenditures, and used as the indicator of food accessibility per month. Household expenditure per capita was divided into food- and non-food items expenses per month. Cost of snacks as part of food expenditures was reported separately. For more details on site selection, sampling and quality control of the study the reader is referred to Purwestri *et al.* (2017).

7.2.3 Ethical approval and informed consent

All procedures performed in studies involving human participants were in accordance with the ethical standards of the institutional and/or national research committee and with the 1964 Helsinki declaration and its later amendments or comparable ethical standards. The study protocol was approved by the Ethics Committee of the Faculty of Medicine, University of Brawijaya, Malang, Indonesia. Informed consent was obtained from all individual participants included in the study who volunteered to join the research. The informants in the in-depth interviews also gave the permission to record their statements.

7.3 Results

The total sample of this study comprised of 335 children aged above six months and below five years. Of the total sample, 107 children were stunted (31.9%), the other 228 (68.1%) were categorized as non-stunted children, and 62% of the children were still breastfed on the day of the survey (59.8% and 62.3% of stunted and non-stunted children, respectively). As already reported by Purwestri *et al.* (2017), stunted children were significantly older, lighter and shorter (34.3±13.0 months, 10.6±2.1 kg, 83.4±7.9 cm, respectively) than non-stunted children (29.9±15.3 months, 11.3±2.9 kg, 86.5±10.7 cm, respectively), irrespective of their sex.

No stunted children were found in the youngest age group of 6 to 12 months. However, the prevalence of HAZ-score below -2SD increased strikingly after the age of one year (35.8% for children aged 12 to 36 months and by 36.1% for children aged 36 to 60 months). Of the 50 stunted children aged between 12 and 36 months, 18 (10.9%) suffered from severe stunting (HAZ-score below -3SD). A similar proportion (9.8%) of severe stunting was found among the group of children above three years of age. Table 7.1 shows the characteristics of the children per age group. It is striking to see that nearly 29% of the children were still breastfed at the age of three to five years, indicating that prolonged breastfeeding is a well-established practice in the study area. Additionally, 10.1% of the children in the sample were wasted, 3.3% of them severely so. When the children were grouped according to

Table 7.1. Characteristics of the study children according to age groups.[1]

Characteristics	Age groups		
	≥6 to <12 months n=37	≥12 to <36 months n=165	≥36 to <60 months n=133
Child's sex (girls)	43.2 (16)	46.1 (76)	48.9 (65)
Breastfeeding (yes)	91.9 (34)^{c}	81.2 (134)^{c}	28.6 (38)^{a,b}
Child's age (months)	9.4±1.7^{b,c}	23.8±6.9^{a,c}	46.7±6.8^{a,b}
Child's height (cm)	71.1±2.9^{b,c}	81.2±5.9^{a,c}	95.1±5.5^{a,b}
Child's weight (kg)	7.8±1.0^{,c}	10.0±1.5^{a,c}	13.4±2.3^{a,b}

[1] Data is presented as % (n) or mean ± sd.

[2] Group comparisons were analysed using Chi's square test (categorical data) or ANOVA continued by LSD test (continuous and normally distributed data). ^{a,b,c} Values followed by the same letters are not significantly different at $P<0.001$.

breastfeeding status, the prevalence of the children having HAZ and WHZ-score below <-2SD was similar among breastfed and non-breastfed children.

Both groups of children were from similar socio-demographic backgrounds as indicated by housing type, house ownership, parental occupation, and educational level (Purwestri *et al.*, 2017). They also had similar household size (median: 4, minimum: 3 and maximum: 9). Table 7.2 shows the economic indicators of stunted and non-stunted children in the study area. It was found that the prevalence of stunting of children was similar across the household's total yearly farm income (wealth) quintiles per capita. Likewise, the indicators of monthly household expenditure (total, food- and non-food items expenses) of the stunted and non-stunted children were not statistically different. In addition, snacks cost contributed to 68.3 and 57.6% of the total monthly food cost for stunted and non-stunted children.

The contribution of breastmilk, complementary foods/home-based meals, and snacks in the dietary intake of stunted and non-stunted children is presented in Figure 7.2. Overall, dietary intake fulfilment of non-stunted children was better than that of the stunted ones. Fulfilment of energy, calcium and iron intake from home-based family diet of non-stunted children was significantly higher than stunted children ($P<0.05$), and the other fulfilment of nutrients (protein, vitamin B6 and vitamin A intake) of non-stunted children was

Table 7.2. Economic indicators of stunted and non-stunted children.[1]

Socio-economic indicators	Non-stunted children n=228	Stunted children n=107	*P* value[3]
Wealth quintiles per capita per year			0.853
Lowest	20.6 (47)	19.6 (21)	
Second	19.3 (44)	21.5 (23)	
Middle	18.9 (43)	22.4 (24)	
Forth	20.2 (46)	19.6 (21)	
Highest	21.1 (48)	16.8 (18)	
Monthly household expenditure per capita per month (in €)[2]			
Non-food items	24.9 (16.0; 37.8)	23.7 (13.8; 34.6)	0.220
Food items	10.6 (8.1; 14.5)	10.6 (7.4; 14.7)	0.529
Snacks	4.9 (3.2; 6.6)	4.9 (3.2; 7.9)	0.372
Total	37.7 (25.8; 55.3)	33.6 (26.5; 49.3)	0.196

[1] Data is presented as % (n) or median (25th, 75th percentiles).
[2] Exchange rate during data collection: €1 = Rp15,240.
[3] Group comparisons was analysed using chi-square test (categorical data) or Mann-Whitney test (continuous and not normally distributed data).

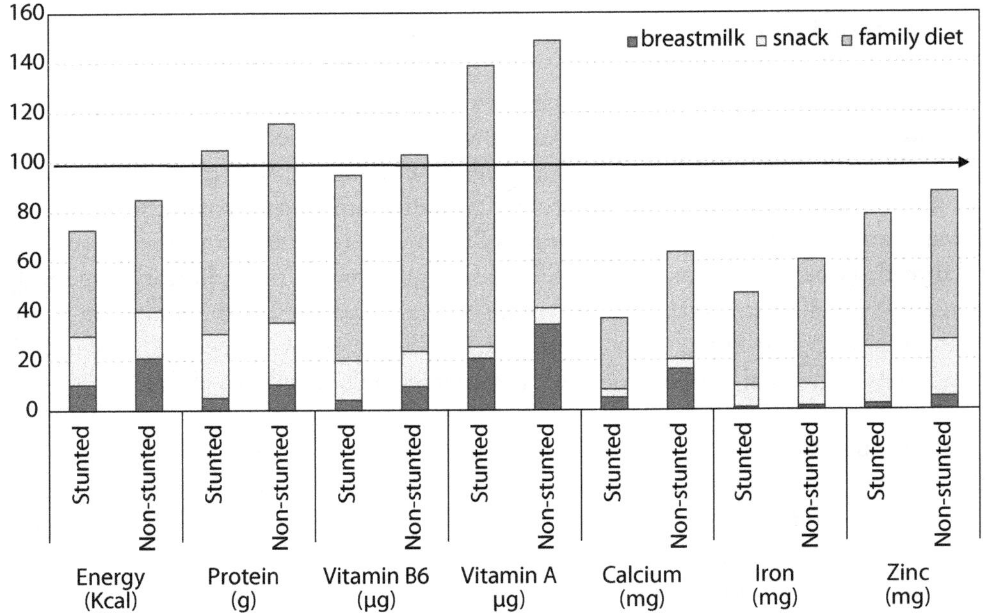

Figure 7.2. Contribution of breastmilk, snacks and complementary feeding/family diet to the total daily habitual intake of stunted and non-stunted children.

slightly higher compared to the stunted counterparts. Protein and vitamin A intake of both categories of children exceeded the Indonesian dietary recommendation for well-nourished children; however, intake of selected minerals (calcium, iron and zinc) did not reach the 100% dietary intake fulfilment. In addition, vitamin B6 intake of non-stunted children also surpassed the dietary guidelines for Indonesian children. The contribution from breastmilk to dietary intake fulfilment of selected macro- and micronutrients intake of non-stunted children was significantly higher than that of stunted children ($P<0.001$). On the other hand, contribution of snacks to the fulfilment of selected macro- and micronutrient intake of stunted and non-stunted children did not differ significantly.

The contribution of snacks as a source of nutrients also increased with increasing age. Snacks contributed 9.7, 21.0, and 19.4% to the total daily energy intake of children in the age groups ≥6 to <12 months, ≥12 to <36 months, and ≥36 to <60 months, respectively, based on the recommended intake of well-nourished Indonesian children. In total, 94.4% of stunted children consumed snacks at the time of the survey, a significantly higher proportion than that of 88.7% of the non-stunted ones ($P=0.01$).

In total, 15 mothers (7 in Karanganyar and 8 in Dempet sub-district) participated in the focus group discussions. The mothers reported that rice was the main staple food for the child's meal. Meal frequency per day was two to three times depending on age of the child. Based on the family financial situation, the most commonly side dishes offered were fried-egg, tofu, *tempeh* (fermented soya), and catfish (fresh water fish) with spinach clear (*'sayur bayam bening'*) or carrot-cabbage soup. The mothers also reported a high consumption of snacks bought from the nearby shop or street vendors, which made their children consume these snacks at the expense of eating the offered family meals, saying that if their children cried for the snacks they would try to buy and give the foods, even during the lean season.

7.4 Discussion and conclusion

The contradictive situation of child malnutrition in a rice-surplus area motivated us to investigate specific factors that could contribute to the high prevalence of stunting among children in the study area of Demak, Central Java. In Indonesia, rice is the staple in the typical local diet. A hundred grams of white rice contains 366 kcal energy, but provides only 1 mg of iron, 0.63 mg of zinc and no vitamin A (Jati *et al.*, 2012). Additionally, Sekiyama *et al.* (2012) reported that high consumption of snacks among Indonesian children in another Java province limited the intake of adequate nutritious food among children. These findings triggered our interest to analyse the relationships between food accessibility, dietary intake and stunting among children in Demak. Purwestri *et al.* (2017) reported that both groups of non-stunted and stunted children in the study area were from similar socio-economic background. The stunting prevalence in the whole sample of children was high (31.9%). From 335 households, 66 did not have any agricultural production in the previous year. However, at the time of the survey, all of the respondents had at least one family member who earned some revenue from on-farm activities as a farm worker. The other 269 households (80.3%) had between one to four types of agricultural production (crops and/or livestock) (Purwestri *et al.*, 2017).

Masset *et al.* (2011) presented a trajectory of household income from agricultural production, through household expenditure (including food expenses), to food intake and nutritional status. However, for our study area, Purwestri *et al.* (2017) reported that the income of households with non-stunted children was only slightly higher than that of households with stunted children. When the family per capita income was categorised into wealth quintiles, the proportions of stunted children were similar across all quintiles. In the same vein, monthly expenditure per capita on food and non-food items of households with non-stunted children was slightly, but not significantly, higher than that of households with stunted children. Hence, it can be concluded that the conditions in terms of food availability and accessibility are more or less the same for stunted and non-stunted children. The only difference was that households with non-stunted children significantly spent more money on yearly non-routine health care activities and improved environmental living conditions than households with stunted children (Purwestri *et al.*, 2017).

The first 1000 days of an infant's life (from the first day of pregnancy until 2 years of age) constitute a critical window of opportunity for growth and development. During this period, inadequate dietary intake, frequent infections, and unfavourable environmental conditions may lead to short- and long-term health consequences and poor nutritional status among children (Black *et al.,* 2008; Rosado, 1999; Semba *et al.,* 2008; Stephensen, 1999; Victora *et al,.* 2008, 2010). Stunting starts mostly at the age of six months, when the first complementary food is commonly introduced, and can continue up to older ages (Gross *et al.,* 1998; Seth *et al.,* 2011). In the study area, there were no stunted children in the age group of 6 to 12 months, but the prevalence of children with HAZ-score below -2SD increased significantly after the age of 12 months by 35.8% for children aged 12 to 36 months and by 36.1% for children aged 36 to 60 months.

Figure 7.2 shows that breastmilk contributed to dietary quality. It protects the children from being stunted by decreasing the fulfilment gap of energy, vitamin B6, calcium, zinc and iron intake. Even, the bioavailability of lactoferrin in breastmilk is the highest (>50%) compared to heme iron (15-30%) and non-heme iron (<5%) (Halberg, 1981; Saarinen *et al.,* 1977). The composition of human milk is unique, dynamically and actively adapting to the individual infant's nutritional needs, and modifying with prematurity and postnatal age (Scherbaum and Srour, 2016). Breastmilk provides all the needed macro- and micronutrients and fulfils the recommended dietary allowance for infants under six months of age (Lubetzky *et al.,* 2013). After the age of six months, however, complementary food with adequate macro- and micronutrients should be introduced (WHO, 2003), while breastmilk remains the main source of energy and nutrients for infants and young children. Approximately, 50% and 30% of an infant's energy intake is provided by breast milk within the first year and the second year of life, respectively. Moreover, breast milk has long been recognized to provide high-quality nutrients and protection against infectious diseases (Sankar *et al.,* 2015; WHO, 2003). Several studies pointed out that mothers are more likely to breastfeed stunted, wasted or underweight children for a longer period of time compared to healthy children. This so-called inverse infant feeding causality leads to the wrong assumption that prolonged breastfeeding impairs growth (Simondon and Simondon, 1998). In this study, like Giugliani *et al.* (2015) in their study, we found no significant relationship between breastfeeding practices and stunting prevalence.

In the study area, the family diet consumed by the non-stunted children was significantly better in percent fulfilment of energy, calcium and iron intake than that of stunted children, although calcium, iron and zinc intake were still below the level of the Indonesian dietary guidelines for both stunted and non-stunted children. It is likely that better food choices for the children were found among families with non-stunted children, however, the fulfilment gap of the selected mineral intake from both categories of households indicate the need of improving the diet quality of home-based foods in general.

Children's diets should consist of a variety of foods from animal and plant sources. Iron, zinc, and vitamin A are more readily available in animal sources, while several vitamins like B and C are present in foods that are plant-based. A balanced diet is an important health indicator especially for young children (Gibson, 2005; Murphy and Allen, 2003). In Indonesia, home-based meals are characterized as a rice-based diet with mainly plant sources rather than animal sources of protein and a moderate intake of fruits (Jati *et al.*, 2012). Furthermore, Santika *et al.* (2009) and Fahmida (2013) reported an increased consumption of fortified factory-produced, ready-to-use complementary foods among Indonesian children. By use of these products the micronutrient content of complementary food offered to children could be enhanced (Santika *et al.*, 2009). However, these ready-to-use baby foods often increase the risk of childhood obesity and other related chronic degenerative diseases mainly because of preservative agents and a high sugar content (Barness, 1993; Bentley, 2014).

We also found a high snack consumption, especially by older children. Snacks that are either homemade or bought from nearby shops or street vendors were introduced after the child was able to consume something other than breastmilk. Typically, different types of snacks that can be bought outside of home are offered to children such as ready-to-eat snack foods, fried snack foods (e.g. tofu, *tempeh*, and bananas), meatballs (*bakso*), and soft drinks, including homogenised milk and fruit drinks. Unfortunately, we discovered that the children often consume these snacks instead of taking their meals. The monotonous rice-based diet combined with high consumption of snacks seemed to contribute to low HAZ-scores of children. Snacks have a high energy content, but are often low in micronutrient content (Sekiyama *et al.*, 2012). During the focus group discussions, it transpired that the mothers did not fully understand the consequences of frequent consumption of these snacks for their children. They would often give in to the children's demand for snacks, instead of persuading the children to consume homemade food. Snacks could be bought for a relatively low price ranging from Rp 500 to Rp 10,000 (€0.03 to €0.66). A significantly higher proportion of stunted children (94.4%) consumed more snacks per day than non-stunted children (88.7%).

The households appeared to spend more than half of their monthly food budget on snacks. This food choice in food expenditures is problematic, also because the hygienic conditions of food sold by the street vendors is questionable. Previous studies in Indonesia regarding food purchase from street vendors found poor hand-washing practices and direct hand contact with foods, thus increasing the risk of food-borne diseases. (Gasem *et al.*, 2001; Vollaard *et al.*, 2004).

In this study, the food availability and accessibility (as indicated by farm income and expenditure per capita) of households with stunted children were similar to those of non-stunted children, indicating that these factors do not explain the high prevalence of stunting. However, non-stunted children had a lower snack intake and were provided with a significantly better home-based diet quality than stunted children. Furthermore, the study showed that breastmilk intake protected non-stunted children by decreasing the gap between selected daily nutrient intake and recommended intake. Thus, improving health and nutrition knowledge and practices of mothers regarding the benefits of breastfeeding and appropriate complementary feeding, as well as the promotion of healthy homemade diets and discouragement of snacks, seem to be crucial in supporting families to making healthier food and feeding choices.

Acknowledgement

The authors would like to thank the Neys-Van Hoogstraten Foundation for its financial support. Dr. R.C. Purwestri was also funded by the Federal Ministry of Education and Research within the project Humboldt reloaded (01PL11003) at the University of Hohenheim, Germany. We greatly appreciate the assistance of Daniel Pfaff, Loretta Renz, Mief Qur'anin Setyohadi, enumerators, voluntary workers, and health workers, village heads, and the mothers and their children in the district of Demak. We are also thankful to Prof. Dr. Hans-Peter Piepho for the statistical advice.

References

Asian Development Bank, 2006. Indonesia: strategic vision for agriculture and rural development. Manila: Banco de Desarrollo Asiático. Available at: http://tinyurl.com/yb9s3g3h.

Barness, L.A. (ed.), 1993. *Pediatric nutrition handbook.* [3rd ed.]. Elk Grove Village IL: American Academy of Pediatrics, p. 23.

Bentley, A., 2014. *Inventing baby food: taste, health, and the industrialization of the American diet.* New York: University of California Press.

Berti, P.R., Krasevec, J. and Fitzgerald, S., 2004. A review of the effectiveness of agriculture interventions in improving nutrition outcomes. *Public Health Nutrition* 7(05): 599-609.

Black, R.E., Allen, L.H., Bhutta, Z.A., Caulfield, L.E., De Onis, M., Ezzati, M., Mathers, C. and Rivera, J., 2008. Maternal and child undernutrition: global and regional exposures and health consequences. *The Lancet* 371(9608): 243-260.

BPS, 2014. *Jawa Tengah dalam angka 2014* [Central Java in figures 2014]. Semarang, Indonesia: Province and Regional Development Planning Bureau.

Butte, N.F., Lopez-Alarcon, M.G. and Garza, C., 2002.Nutrient adequacy of exclusive breastfeeding for the term infant during the first six months of life. Geneva: World Health Organization.

FAO, 2006. Food security. FAO Policy Brief. June 2006, Issue 2. Available at: http://tinyurl.com/q4k5rn8.

Fahmida, U., 2013. Use of fortified foods for Indonesian infants. In: Preedy, V., Srirajaskanthan, R. and Patel, V. (eds), *Handbook of food fortification and health: from concepts to public health applications.* Vol. 2. *Nutrition and health.* New York: Humana Press, pp 383-393.

Gartaula, H., Patel, K., Johnson, D., Devkota, R., Khadka, K. and Chaudhary, P., 2017. From food security to food wellbeing: examining food security through the lens of food wellbeing in Nepal's rapidly changing agrarian landscape. *Agriculture and Human Values* 34(3): 573-589.

Gasem, M.H., Dolmans, W., Keuter, M. and Djokomoeljanto, R., 2001. Poor food hygiene and housing as risk factors for typhoid fever in Semarang, Indonesia. *Tropical Medicine and International Health* 6(6): 484-490.

Gibson, R.S., 2005. *Principles of nutritional assessment.* New York: Oxford University Press.

Giugliani, E.R., Horta, B.L., Loret de Mola, C., Lisboa, B.O. and Victora, C.G., 2015. Effect of breastfeeding promotion interventions on child growth: a systematic review and meta-analysis. *Acta Paediatrica* 104(S467): 20-29.

Gross, R., Karyadi, D., Sastroamidjojo, S. and Schultink, W., 1998. Guidelines for the development of research proposals following a structured, holistic approach for a research proposal (SHARP). *Food and Nutrition Bulletin* 19(3): 268-282.

Gross, R., Schoeneberger, H., Pfeifer, H. and Preuss, H.J., 2000. The four dimensions of food and nutrition security: definitions and concepts. *SCN News* 20: 20-25.

Gross, R., Schultink, W. and Sastroamidjojo, S., 1996. Stunting as an indicator for health and wealth: an Indonesian application. *Nutrition Research* 16(11-12): 1829-1837.

Hallberg, L., 1981. Bioavailability of dietary iron in man. *Annual Review of Nutrition* 1: 123-147.

Jati, I.R.A.P., Vadivel, V., Nohr, D. and Biesalski, H.K., 2012. Dietary formulation to overcome micronutrient deficiency status in Indonesia. *Nutrition and Food Science* 42(5): 362-370.

Leroy, J.L. and Frongillo, E.A., 2007. Can interventions to promote animal production ameliorate undernutrition? *The Journal of nutrition* 137(10): 2311-2316.

Lubetzky, R., Mandel, D. and Mimouni, F., 2013. Vitamin and mineral supplementation of term infants: are they necessary? In: Szajewska, H. and Shamir, R. (eds), *Evidence-based research in pediatric nutrition. World Review of Nutrition Dietetics.* Basel, Switzerland: Karger Publishers, pp 79-85.

Lwanga, S.K. and Lemeshow, S., 1991. *Sample size determination in health studies: a practical manual.* Geneva: World Health Organization.

Masset, E., Haddad, L., Cornelius, A. and Isaza-Castro, J., 2011. *A systematic review of agricultural interventions that aim to improve nutritional status of children.* London: EPPI-Centre, Social Science Research Unit, Institute of Education, University of London.

Masset, E., Haddad, L., Cornelius, A. and Isaza-Castro, J., 2012. Effectiveness of agricultural interventions that aim to improve nutritional status of children: systematic review. *British Medical Journal* 344: d8222.

Maxwell, S. and Smith, M., 1992. Household food security: a conceptual review. In Maxwell, S. and Frankenberger, T. (eds), *Household food security: concepts, indicators, measurements.* Rome and New York: IFAD and UNICEF, pp 1-72.

MoH, 2010. *Riset Kesehatan Dasar 2010.* [Primary health research 2010]. Jakarta: Ministry of Health Republic of Indonesia. Available at: http://tinyurl.com/y9egn7xz.

MoH, 2013a. *Riset Kesehatan Dasar 2013.* [Primary health research 2013]. Jakarta: Ministry of Health Republic of Indonesia. Available at: http://tinyurl.com/yau4gjmz.

MoH, 2013b. Angka kecukupan gizi. Peraturan Menteri Kesehatan Republik Indonesia Nomor 75 Tahun 2013.[Nutrition figures, MoH Regulation 75]. Jakarta: Ministry of Health Republic of Indonesia. Available at: http://tinyurl.com/ya2cydsl.

Murphy, S.P. and Allen, L.H., 2003. Nutritional importance of animal source foods. *The Journal of Nutrition* 133(11): 3932S-3935S.

Purwestri, R.C., Renz, L., Wirawan, N.N., Jati, I.R.A., Fahmi, I. and Biesalski, H.K., 2017. Is agriculture connected with stunting in Indonesian children living in a rice surplus area? A case study in Demak regency, Central Java. *Food Security* 9(1): 89-98.

Rosado, J.L., 1999. Separate and joint effects of micronutrient deficiencies on linear growth. *The Journal of Nutrition* 129(2): 531S-533S.

Ruel, M.T., 2001. *Can food-based strategies help reduce vitamin A and iron deficiencies?* Washington DC: International Food Policy Research Institute (IFPRI). Available at: http://tinyurl.com/y8w9ujps.

Saarinen, U.M., Siimes, M.A. and Dallman P.R., 1997. Iron absorption in infants: high bioavailability of breast milk iron as indicated by the extrinsic tag method of iron absorption and by the concentration of serum ferritin. *The Journal of Pediatrics* 91(1): 36-39.

Santika, O., Fahmida, U. and Ferguson, E.L, 2009. Development of food-based complementary feeding recommendations for 9- to 11-month-old peri-urban Indonesian infants using linear programming. *The Journal of Nutrition* 139(1): 135-141.

Sankar, M.J., Sinha, B., Chowdhury, R., Bhandari, N., Taneja, S., Martines, J. and Bahl, R., 2015. Optimal breastfeeding practices and infant and child mortality: a systematic review and meta-analysis. *Acta Paediatrica* 104(S467): 3-13.

Scherbaum, V. and Srour, M.L., 2016. The role of breastfeeding in the prevention of childhood malnutrition. In Biesalski, H.K. and Black, R.E. (eds) *Hidden hunger: malnutrition and the first 1,000 days of life: causes, consequences and solutions*. World Review of Nutrition Dietetics, Basel: Karger Publishers, pp 82-97.

Sekiyama, M., Roosita, K. and Ohtsuka, R., 2012. Snack foods consumption contributes to poor nutrition of rural children in West Java, Indonesia. *Asia Pacific Journal of Clinical Nutrition* 21(4): 558-567.

Semba, R. D., de Pee, S., Sun, K., Sari, M., Akhter, N. and Bloem, M.W., 2008. Effect of parental formal education on risk of child stunting in Indonesia and Bangladesh: a cross-sectional study. *The Lancet* 371(9609): 322-328.

Seth, V., Garg, A. and Vir, S., 2011. Complementary feeding of infants and young children. *Public Health and Nutrition in Developing Countries (Part I and II).* New Dehli: Woodhead Publishing India.

Simondon, K.B. and Simondon, F., 1998. Mothers prolong breastfeeding of undernourished children in rural Senegal. *International Journal of Epidemiology* 27(3): 490-494.

Siregar, M. and Suryadi, M., 2006. *Changes in food crop diversification in Indonesia*. Jakarta: Indonesian Centre for Agricultural and Socio Economic Research and Development.

Stephensen, C.B., 1999. Burden of infection on growth failure. *The Journal of Nutrition* 129(2): 534S-538S.

Thomas, H., 2003. *WTO agreement on agriculture. The implementation experience – developing country case studies*. Food and Agriculture Organization of the United Nations, Rome. Available at: http://tinyurl.com/y8e96kr3.

UNICEF, 1998. The state of the world's children 1998: a UNICEF report. Malnutrition: causes, consequences, and solutions. *Nutrition Review* 56(4 part I): 115-123.

Victora, C.G., Adair, L., Fall, C., Hallal, P.C., Martorell, R., Richter, L. and Sachdev, H.S., 2008. Maternal and child undernutrition: consequences for adult health and human capital. *The Lancet* 371(9609): 340-357.

Victora, C.G., de Onis, M., Hallal, P.C., Blössner, M. and Shrimpton, R., 2010. Worldwide timing of growth faltering: Revisiting implications for interventions. *Pediatrics* 125(3): e473-80.

Vollaard, A., Ali, S., Van Asten, H., Ismid, I. S., Widjaja, S., Visser, L., Surjadi, C.H. and Van Dissel, J.T., 2004. Risk factors for transmission of foodborne illness in restaurants and street vendors in Jakarta, Indonesia. *Epidemiology and Infection* 132(5): 863-872.

Weingärtner, L., 2005. The concept of food and nutrition security. *Achieving Food and Nutrition Security* 3: 21-52.

WHO, 2003. *Guiding principles for complementary feeding of the breastfed child.* Washington D.C., Pan American Health Organization/World Health Organization. Available at: http://tinyurl.com/jz7yuaz.

8. Gender, intra-household food allocation and social change in two Himalayan communities in Nepal

D.S. Madjdian

Department of Social Sciences, Wageningen University and Research, P.O. Box 8130, 6700 EW Wageningen, the Netherlands; donya.madjdian@wur.nl

Abstract

In large parts of South Asia, gendered inequalities in intra-household food allocation cause a gender gap in food and nutrition security. Such inequalities are revealed in customary practices relating to food, health and care. This study is about the relationship between intra-household food allocation and women's vulnerability to food and nutrition insecurity in Nepal. It highlights the differential effects of family type, household composition and ethnicity on this relationship and on the impact of social change on women's vulnerability to food and nutrition insecurity. The study uses the moral household economy framework and applied a mixed-methods approach with an emphasis on qualitative methods. Thirty women from a Hindu and a Buddhist community in Humla, one of the poorest districts in Nepal, were interviewed about their perspectives on food and nutrition security. Their nutritional status was assessed as well. While in both communities, practices of food allocation and consumption were highly gendered, only in the Hindu households gender inequality was reflected in women's vulnerability to food and nutrition insecurity. Buddhist women had more decision-making power than the Hindu women had, were less affected by deeply rooted cultural practices that disfavour pregnant and lactating women, and seemed to benefit more from social change.

Keywords: food and nutrition security, moral household economy, women, family, ethnicity

8.1 Introduction

Food scarcity and malnutrition is a persisting problem in South Asia that is affecting mostly women and children (De Schutter, 2013; Ramachandran, 2007). Nepal still belongs to the ten countries in the world with the highest prevalence of stunting, an indicator of chronic undernutrition, even though – at the same time – Nepal ranks high in terms of food availability (MOAD *et al.*, 2016). Despite considerable progress over the last decades in economic and human development (Samir *et al.*, 2016; Sharma *et al.*, 2014), undernutrition is a persisting problem in many parts of the country. The problem is more severe in rural than in urban areas and stunting is more pronounced in the mountainous areas than in the hills and lowlands (MOAD *et al.*, 2016). In addition, regions where food availability is sufficient may have high undernutrition rates. Moreover, malnutrition is more frequent among females than males (MOAD *et al.*, 2016), and even in food secure households, women and girls

DOI 10.3920/978-90-8686-864-3_8,

may suffer from malnutrition (Gittelsohn *et al.*, 1998; Maharjan and Khatri-Chhetri, 2006; Messer, 1997). Such a gender gap is not uncommon in South Asia (DeRose *et al.*, 2000).

Food security is defined as a situation in which all people, at all times, have physical, social and economic access to sufficient, safe and nutritious food that meets their dietary needs and food preferences for an active and healthy life (FAO, 1996). Food security also requires that 'individuals do not face social barriers in feeding themselves' (Narayanan, 2015: 198). Unequal distribution of food at the household level plays a crucial role in the gender gap in nutrition (Gittelsohn, 1998; Maharjan and Khatri-Chhetri, 2006). Such inequalities are caused by deeply rooted cultural practices and norms and values related to food, health and care. These vary between social groups and according to ethnicity, caste, religion, kinship systems and household structures. Gender inequalities in society affect the food and nutrition security of women and girls by disfavouring women in terms of food allocation, especially during their reproductive life. Because of social norms surrounding early marriage and family formation (e.g. son preference), often young, pregnant and lactating women are disadvantaged most.

The gendered cultural practices in intra-household food allocation that disfavour women, combined with poor food availability at the household level, create for women a vicious cycle of nutritional deprivation (DeRose *et al.*, 2000). Undernutrition of girls and women will be passed on to the next generation, hindering girls and women to reach their full potential, and, thereby negatively affecting society's development. If today's women are indeed the 'powerful source for improving health, longevity, mental and physical capacity and productivity of the next generation of adults' (De Schutter, 2013), we need to understand the mechanisms of unequal food allocation within households, taking into account that those patterns are context-specific and time-bound. In Nepal, with its socially and culturally diverse population, different factors may affect such patterns. Moreover, processes of social and life style changes add to shaping such patterns over time.

Many studies investigated gender gaps in food security and nutrition at the household level. Evidence points at a pro-male bias in South Asia, and a pro-female bias in the African context (Backstrand *et al.*, 1997; Bouis, 2003; Darnton-Hill *et al.*, 2005; Haddad, 1999; Hadley *et al.*, 2008). However, there are notable differences between and within regions and contexts (DeRose *et al.*, 2000; Fledderjohann *et al.*, 2014; Haddad *et al.*, 1996). The key questions still are: who gets what and why, and what impact does social change have on this? To improve women's food and nutrition security at the individual and household level, micro-studies are needed that focus on intra-household dynamics, the role of gender, and household characteristics (DeRose *et al.*, 2000; Gittelsohn, 1991; Gittelsohn *et al.*, 1997, 1998; Griffiths *et al.*, 2002; Messer, 1997; Miller, 1997). For adequate nutritional outcomes for women and girls, uncovering the role of cultural and social variables in women's food and nutrition insecurity is essential (Ramachandran, 2007). Equally essential is the need for understanding intra-household food allocation as well as food-related norms and values, practices and decision-making processes, in order to resolve local food challenges in a context of social change. Though acknowledging social causes, studies on women's food and nutrition insecurity at the household level rarely incorporate the notion of social change (Noack and Pouw, 2014). This study advances on previous research by investigating intra-household food allocation as a factor in women's food and nutrition (in)security in two ethnically different communities in the same geographical region, emphasizing the influence of the immediate contexts of household, kinship and culture in a situation of social change.

8.2 Conceptual framework

This study aims at gaining insights into the relationship between practices of intra-household food allocation and women's vulnerability to food and nutrition insecurity in Nepal by highlighting socio-cultural differentiation related to family type, household composition and ethnicity. Furthermore, the actual and possible impact of social change on women's vulnerability to food and nutrition insecurity is investigated. In this section, the concepts used in the study are introduced and elaborated. Conceptually, this study uses the framework of the moral household economy.

8.2.1 Household food security and intra-household food allocation

In studying intra-household food allocation, the household is defined as 'a co-residential unit, usually family-based in some way, which takes care of resource management and primary needs of its members' (Rudie, 1995: 228). Using this definition allows for revealing 'the ways in which people structure their daily lives in a certain setting to provide for their needs and well-being, assuming responsibilities for dependents and family members' (Niehof, 2011: 494-495). The household is the site where on a daily basis people negotiate and perform the distribution of, and access to food and care (Niehof and Wahlen, 2016).

A household's food security situation is embedded in a broader context that includes governmental policies, food supply, seasonality, cultural and social aspects, and community-level factors such as health services, sanitation, and water supply (Gittelsohn *et al.*, 1998). Within households, an individual's nutritional status is the outcome of intra-household food allocation patterns, feeding and care practices and health status, which are affected by factors such as socio-economic status, caste group, household size and composition, norms and values (Gittelsohn *et al.*, 1998; UNICEF, 1991). Decision-making and negotiations on food at the household level are also based on norms and practices relating to family and kin relations, referred to as family systems (Skinner, 1997). These exert an influence on an individual's life outcomes and opportunities (Das Gupta, 1999), hence can be expected to have an effect on an individual's nutritional status as well.

Among the four dimensions of household food security (availability, access, utilisation and sustainability), this study focuses mainly on the utilisation dimension, i.e. nutritional intake, which depends on care and feeding practices, dietary diversity and intra-household food allocation (Den Hartog *et al.*, 2006; Gross *et al.*, 2000). Gittelsohn (1991) framed intra-household food behaviour as comprising three stages: food selection, preparation and serving. Food selection refers to who is in charge of buying or supplying food. Food preparation is about the actual cooking, and food serving includes who serves what (type of food), how much (amount of food), in what way and to whom (serving order). These stages determine food allocation outcomes, which can be divided into four categories: egalitarian, favouritism in terms of quantity, favouritism in terms of quality, and food allocation during transitory stages. The latter refers to temporary physical conditions, such as illness and – in the case of women – menstruation, pregnancy, and lactation (Gittelsohn, 1991). Behrman (1997) defined another set of outcomes. The first is equity, in which everyone gets a fair share, the second equal outcome, in which food is allocated to equalize the welfare of household members. Third is allocation according to the household maximisation rule (fair return), where food allocation follows the actual or expected contribution of household members to the household economy. Finally, there is allocation according to the individual maximisation rule, where those expected to benefit most from a certain food product, get the largest share. Particularly this latter outcome is perceived as resulting in more inequality (Behrman, 1997).

8.2.2 The gender-nutrition gap

Women's food and nutrition security status is shaped in the contexts of household, kinship and culture. Household power structures and decision-making may disfavour women in terms of food intake. Gittelsohn (1991) and Gittelsohn *et al.* (1997) found for Nepal that when women have a low social status in the household and weak decision-making power, they are less likely to meet their dietary requirements than men. This can occur even when women are responsible for food preparation (as they often are), because men are in charge of food distribution (De Schutter, 2013; Griffiths *et al.*, 2002; DeRose *et al.*, 2000). Women's social position in the household is defined by culture and kinship (Das Gupta, 1999; Gittelsohn, 1991) and reflected in intra-household food allocation as well as in access to health care services (Ramachandran, 2007). Gender bias may also result from the perception of women as being an 'economic drain on the household' (Gittelsohn *et al.*, 1997). Those who contribute most to the household income, have the (bargaining) power to decide upon food allocation. Where men are the breadwinners and women are kept subservient, women's nutritional status and, indirectly, that of their children, is negatively affected (Ramachandran, 2007; De Schutter; 2013). Where women are in charge of household expenditures, they direct these more towards nutrition and health than when men are in charge (Alam, 2012; Haddad *et al.,* 1996). Gendered intra-household power inequalities thus directly affect food and nutrition security status through decision-making and gendered responsibilities for producing, purchasing, preparing and consuming food (Ramachandran, 2007). DeRose *et al.* (2000) point out that also when there is no systematic gender bias that disfavours women in terms of food quantity, there could still be one in terms of food quality.

Furthermore, pregnant and lactating women do often not have the food they need in terms of quality and quantity, which is problematic in high-fertility settings where women are often pregnant or lactating.

The evidence on a gender bias before reaching adulthood is less consistent, although in South Asia a pro-male bias is often found because of the economic potential of boys (Gittelsohn, 1991). According to Miller (1997), this economic-benefit hypothesis could explain preferential food allocation to boys, especially in poor households where gender was found to be the most significant determinant of malnutrition among young children (Rahman, 2002). For Nepal, Gittelsohn (1991) found no differences between boys and girls until the age of eight years or until they moved into adolescence. Then, preferential food allocation became visible in the serving order and in the amounts and quality of food served. Gittelsohn concludes that cultural values such as a son preference play a role in intra-household resource allocation.

Kinship relations may also play a role in food allocation (Ramachandran, 2004). Kinship and family systems influence the way in which households manage resources, thereby influencing health behaviour and outcomes (Das Gupta, 1999). In nuclear family systems that are based on household units of couples with their (unmarried) children, women's position is relatively good. In stem, extended and joint family systems, households comprise more than one nuclear unit. Here, women are often in a weaker position, especially when upon marriage they leave their natal home to live with their family-in-law. Generally, gender bias is most pronounced in patrilineal and virilocal joint family systems and least in nuclear family systems, although the extent of gender bias varies with the local context (Therborn, 2004).

8.2.3 The moral household economy and social change

As Niehof and Wahlen (2016: 152) put it in their conceptual framework on the relationship between gender and food at the household level, households are gendered moral economies, describing these as 'social constructions shaped by social practices, in which morality and gender are performed.' Women's normative responsibilities for food preparation and for properly 'feeding the family' are morally loaded. Hence, household food distribution and consumption practices are morally grounded and gendered. In the sharing and caring related to food distribution, morally justified gendered power inequalities may increase women's vulnerability to food and nutrition insecurity.

Gender bias can be more pronounced in times of food scarcity, when women have to adjust their food allocation practices to be able to deal with food shortages when feeding their families, particularly their children (Ramachandran, 2004). Niehof and Wahlen (2016) see the moral grounding of gendered food allocation practices as related to the moral principles of lay normativity, which refer to our sense of what is good or bad and to what we or others should do, such as being a good mother, wife, or daughter-in-law. Such lay concepts of gender are embedded in culture and traditions (Sayer, 2005). Views on appropriate gender behaviour and beliefs on what are good and bad foods, for whom and when – permanently or only within certain life stages – are deeply rooted in cultural notions about best practices (Den Hartog *et al.*, 2006). For example, Gittelsohn *et al.* (1997) found pregnant and lactating women in Nepal adhering to certain food proscriptions and prescriptions that in fact reduced their micronutrient intake, because of Hindu beliefs on women's impurity during specific life stages and norms of being a good mother, i.e. not passing on illness to the (un)born child by eating certain foods. In their framework, Niehof and Wahlen (2016) give a prominent place to agency, capabilities, resources and entitlements, which are all needed to perform the desired practices and are embedded in family norms and systems that vary according to ethnicity, social class, and ideology (patriarchy).

Processes of social change comprise changes in social structures and cultural patterns. Social structures include family norms and systems, in which established normatively underpinned routines of social interaction might be uprooted. Similarly, there can be changes in the cultural patterns of shared ways of living and thinking, and in the norms and values that make up lay normativity. Social change can also result from local or even individual changes, such as in the domains of education or work. It affects the answer to the question of 'who gets what and why' by changing gendered and moral food allocation practices.

In this study, the relationship between women's vulnerability to food and nutrition insecurity and cultural and social factors and the influence of social change on this relationship were addressed through two main questions:

- How is women's vulnerability to food and nutrition insecurity mediated by cultural and social factors?
- What is the actual and possible impact of social change on women's vulnerability to food and nutrition security?

8.3 Research area and methodology

8.3.1 Research area

The study was conducted in two Himalayan communities in Humla district, situated in the far-west corner of Nepal, at an altitude of up to 7,300 meters. Humla is considered one of the most isolated areas of Nepal. The district is not connected to Nepal's road network, and is only accessible by foot or small aircraft that – when weather conditions are favourable – can land at Simikot district headquarters. The area is prone to food insecurity, because of poor soils and a climate that leaves Humla covered with snow for at least five months each year.

Humla is one of the most deprived districts in terms of health, education, and sanitation (UN, 2014). It is the second-least developed district of Nepal with a Human Development Index of 0.367 and has the highest poverty index (UN, 2014). Humla's food production cannot meet the total food need of its population, resulting in food scarcity (Adhikari, 2008). The living and health conditions of the people are compromised by poor basic health care services and educational facilities, lack of electricity and shortage of safe drinking water. As a result, malnutrition, epidemics, and maternal diseases are common and child mortality is over 30%. Humla has the highest prevalence of child malnutrition (66%) in Nepal, life expectancy averages only at 64 years, and adult illiteracy is 63% (UN, 2014). The most disadvantaged groups in Humla are women, adolescent girls, and the lower caste groups.

Humla has a population of about 50.000 persons. Most villages, in particular those in the South, are populated by Hindus (mainly Chhetri, Brahmin, Thakuri, and Dalit). The Chhetri and Thakuri groups (about 50% of the Hindu population) belong to the top of the caste hierarchy. Traditionally they have a relatively high social and political status (Gautam and Andersen, 2016). While farming is their main way to generate income, they also have non-farm secondary jobs. The Dalit (about 14%) are lowest in the caste hierarchy. They are the most disadvantaged group in Humla. To date, the Dalit face social exclusion, discriminatory practices, and violence. They are denied access to certain foods, water taps, and temples (McKay, 2002). They are worst off in terms of socio-economic wellbeing, income and landholdings, which makes them especially vulnerable to food insecurity (Gautam and Andersen, 2016).

At a higher altitude, the Tibetan Lama or Nyinba (16% of Humla's total population) mainly populate northern Humla. They have always practiced fraternal polyandry, with two or more brothers sharing one wife, a system that kept the land in the family and regulated the amount of household labour (Ross, 1981; Gautam and Andersen, 2016). However, nowadays things are changing, as reflected in divorce, acceptance of monogamy, and conflicts between brothers. The Lama people exploit a different environment and do better in terms of food security than the Hindu groups (Ross, 1981; Gautam and Andersen, 2016).

Humla consists of 27 Village Development Committees (VCDs), which are villages or clusters of villages. For the study, the communities of Bargaun and Torpa in Bargaun VDC and Thehe in Thehe VDC were selected. Bargaun has a population of 1023 persons, 53% male and 47% female. The Lama are the dominant group (Knoema, 2014). Most families live in joint-family households, some of them practicing polyandry. A few hours walk downhill lies Thehe, which has 2165 inhabitants, 52% male and 48% female (Knoema, 2014). This is a VDC where people have extremely poor living standards (McKay *et al.*, 2007). The main caste groups are the Chhetri and the Dalit, who live in separate wards. Although historically households were joint-family based, nowadays households based on nuclear families are found and, especially in the higher caste groups, more often preferred. In both communities sons are highly valued, because of traditional practices such as the role of sons in funerals and because daughters usually leave their natal home upon marriage.

8.3.2 Methodology

The study used a comparative mixed-methods approach, based on qualitative ethnography. The main method used was in-depth interviews with women. In these interviews, the focus was on women's personal stories and subjective experiences regarding food and nutrition. Interviews were two-on-one, with an interpreter asking the questions in the local languages, Lama in the Buddhist community and Nepali in the Hindu community, and then translating the answer into English. Depending on the participants' time availability and the interview process, the interviews lasted between 45 and 90 minutes, and were conducted at the interviewee's house or in front of it. The interviews were recorded, and notes were taken. Prior to the in-depth interview, women were asked to name all household members (present and absent) and provide an overview of their birth history (pregnancies, live births, stillbirths). Documenting these data was done by adapting the questionnaires used in UNICEF's Multiple Indicator Cluster Surveys, MICS (UNICEF, 2012).

The interview themes were derived from the conceptual framework discussed above. An interview guide was developed around the theoretical concepts, with questions adapted from the MICS questionnaires. It allowed space for topics that could emerge during the interview. The interview consisted of four parts. The first included introductory questions to get background information, e.g. on age, marital status, age and (subjective) health status. Women were asked to rate their health as good, neutral or poor. The second part focused on reproductive issues, e.g. pregnancy, the post-partum period, birth spacing, contraception, and maternal care. The third part covered intra-household food allocation, both in times of food abundance and food scarcity. These questions were adapted from the manual by Den Hartog *et al.* (2006). The last part focused on food, culture and gender. It included questions on general food beliefs and food beliefs during women's vulnerable stages such as menstruation, pregnancy, immediate post-partum and lactation. A pilot interview was conducted in Bargaun, on the basis of which some minor adjustments were made.

An anthropometric assessment of women's nutritional status was done by measuring women's weight and height and calculating their Body Mass Index (BMI). For pregnant or lactating women the Mid-Upper Arm Circumference (MUAC) was used. The fieldwork was conducted between September and December 2014.

Participants were purposively sampled and on site invited to participate in the study. The snowball method was used since no up-to-date registration of households and villages was available. Participants included married women in the reproductive ages of 15 to 49 years old, who were either pregnant at the time of the survey or had been pregnant at least once. When households consisted of more women, the youngest woman matching the inclusion criteria was invited. The total sample comprised 30 women. Table 8.1 provides information on their residence and ethnic background.

Table 8.1. Composition of the sample (n=30) of the in-depth interviews.

Community	Nyinba (Buddhist Lama)	Hindu	
		Chhetri	Dalit
Bargaun	7		
Torpa	2		
Simikot & Bargaun	6		
Thehe		7	8
Total	15	15	

Interviews were recorded and transcribed verbatim with F5 Audio Transcription software. The transcripts were checked by the interpreter for accuracy and compared with the field notes. The conceptual framework and research questions guided coding the transcripts, using Atlas Ti. Although codes were predefined based on the conceptual framework, new concepts were allowed to emerge. With this new list of codes in mind, interviews were read again to check for consistency, completeness, and codes. Some codes were merged. Finally, the codes and themes were linked to the research questions and the findings were described.

BMI values were used to assess adult malnutrition based on the, for use in Asia validated, WHO International BMI classification (WHO, 2004). For pregnant and lactating women the MUAC was used as an indicator for maternal malnutrition. A MUAC below 23 centimetres indicates maternal malnutrition, also in Asia (cf. Ververs *et al.*, 2013).

Informed consent was obtained from the participants before recording the interviews. Given the high illiteracy levels of women, the interpreter informed the participants orally about the purpose of the study and asked permission to use a tape recorder. The women were informed that participation was voluntary, that they could withdraw at any time during the interview, and that names and other identifying information would not be linked to the data in order to ensure privacy and confidentiality.

8.4 Results

This section starts with a brief profile of the participants and then moves on to present the findings related to women's subjective experience of food and nutrition insecurity, mediating cultural and social factors, and the actual and possible impact of social change. Cultural and social factors are those in the women's immediate household and kinship environment, and include family norms and systems, lay normativity, and resources and capabilities.

8.4.1 General characteristics

Table 8.2 presents an overview of the women's characteristics in terms of household structures, educational attainment and average ages at marriage and first pregnancy.

The women's ages ranged from 18 to 48 years old. The average age was 29. The majority of the Nyinba women lived in joint families, of which 40% practiced polyandry. In the Hindu community, most women lived in nuclear-family households, though often on the compound of their husband's family. The Nyinba women's average at marriage is higher than that of the Hindu women, and they score higher levels of educational attainment. In both groups, women got pregnant of their first child within one or two years after marriage.

8.4.2 Intra-household food allocation

Intra-household food allocation patterns are presented according to the three stages of food selection, preparation and serving (cf. Gittelsohn, 1991).

Food selection

Food selection includes buying food and bringing food into the household. In general, in most Nyinba families, decision-making regarding buying food was seen as the responsibility of either the eldest woman in the household, in extended families the mother-in-law, or women's husband(s). Husbands had the final say in buying food when they were the breadwinner. In many nuclear families, women bought and selected the food as part of

Table 8.2. Participants' characteristics.

Participants (n=30)	Nyinba (n=15)	Chhetri (n=7)	Dalit (n=8)
Average age	29	28	29
Family type (%)			
Nuclear	40	71	100
Extended	60	29	0
Fraternal polyandry (%)	140	0	0
Average age at marriage	19.5	16.5	14.5
Average age 1st pregnancy	21	17.5	16.5
Education (%)			
None	13	29	50
Primary	67	71	50
Secondary/higher	20	0	0

their responsibility for cooking. In nuclear-family households where husbands were labour migrants (domestic or international), hence were absent for longer periods, women were responsible for buying and selecting food. Interestingly, women from nuclear families living alternately in Simikot and Bargaun were either solely responsible for buying and selecting food, or shared this responsibility with their husbands. Reasons mentioned were that their husbands were often away for work or that women generated income too. In Hindu nuclear-family households, however, husbands remained responsible for food selection, although women decided what to cook. In some cases, the women's mother-in-law decided on this. Many women said that deciding on what to cook was not so interesting, since meals were the same every day. 'We only eat *chapatti* [flatbread], so we simply don't have to decide upon anything', said a Chhetri woman (20 years old). Women had little or no choice regarding side dishes, as these were either unavailable or the same every day (chili's, radish).

Food preparation

Food preparation includes the actual cooking. In both communities, this was done by wives, in nuclear families sometimes with help of the eldest daughter. In joint- or extended-family households, the daughter-in-law was responsible for cooking. In the Hindu community, men or the eldest daughter took over cooking when the woman responsible was 'impure' because of menstruation or immediate-post-partum.

> [After delivery] we cook for ourselves. We stay in a different room. There is a separate room made for after delivery. We have to stay there, sometimes for 20 days. [...] Because of our god [read: religious beliefs]; if we don't cook for ourselves, or if we touch or enter our kitchen, wrong things may happen to our family or to ourselves.
> (Chhetri woman, 35 years old)

Food serving

Food serving includes the distribution of food within the household. In both communities, families eat together and from separate plates. As a rule, the one who cooks also serves the food. Therefore, when because of 'impurity' of the regular cook (Hindu community) the mother-in-law or the woman's husband cooked, they would also serve the food. The order of food serving followed the household hierarchy and was similar in both communities. The male head of the household, which could be a woman's husband or father-in-law, is always served first. In case of more husbands, the eldest is served first. Also a guest (own experience) is served after the eldest man in the household. Next, children are served, usually starting with the eldest.

> First I serve my father-in-law, then my small father-in-law [read: second husband of mother-in-law], my mother-in-law, and then my sister-in-law, [then] my son, and I eat last [...] Sometimes I eat with the family, sometimes I wait until they're finished. [...] I just wait, there is no reason... (Nyinba woman, 20 years old)

In Nyinba families, food is divided according to appetite, and upon request (more servings). There were no food pre- or proscriptions for women, men, boys or girls. Everyone gets the same food (quality). In Hindu families, often male children or the child who is most hungry (also when a girl) are served first, sometimes even before their father. The one who serves always eats last, and adolescent girls end up second last in the serving order. The amount of food served differs between men, women and children. In nuclear-family households, men and children are served first and until satisfied. What is left (which can be nothing) is for the one who served and for adolescent girls. This seemed normal practice to all Dalit women and many Chhetri women, also in times of food scarcity. The one exception was of a woman whose husband insisted on her eating first, and until she was satisfied. The reason was that she was pregnant and that previous pregnancies had ended in miscarriages and stillbirths. The following quote presents the reflections of a Dalit woman on the gendered serving order.

> Lastly, I serve to my daughter (10 years old). Girls are suppressed everywhere. [...] Nowadays she [daughter] works all day, so I serve these small kids first. I do not know why, I do not know how, but girls are always suppressed. Even if I try to serve her first, she ends up last. [...] When they were small, they [boys and girls] were treated equally, but now she is grown up, she works. (Dalit woman, 35 years old)

This quote suggests that the woman feels her daughter should have been served first. However, even though she works for the family, she ends up last in the serving order.

8.4.3 Women's perspectives on food, nutrition and maternity

Most women from the Buddhist community did not experience food scarcity. These households all owned some livestock and land on which they cultivated crops such as local rice, pulses, buckwheat, wheat, millet, potatoes and vegetables such as pumpkin, radish, chilies, spinach and mustard greens. However, farming is usually not sufficient to feed the family, so families have to buy food, some of which is flown into Simikot (e.g. white rice). Families eat four times a day, and diets are varied (Madjdian and Bras, 2016). When women mentioned food scarcity, they referred to a shortage of white rice. White rice is seen as better than locally grown rice, which is considered poor men's food. During my stay in the villages, I was denied local rice and ate white rice at all times. Having access to white rice is seen as luxury by those who cannot afford it, hence can be considered a proxy of household wealth.

> We don't face any [food] scarcity, because of the farm. We buy [white] rice and wheat flower that comes from upper China, we have lots of horses and mules, so we do not

> face problems. We get food from China. If we didn't have that, if we had to depend on our land, we would have problems. (Nyinba woman, 34 years old)

In the Hindu community, only the Chhetri households owned some small plots of land, where they cultivated mainly local rice, millet and buckwheat. Women said that the land was not sufficient to grow enough food for their household. Therefore, they also worked on the land of Buddhist families in return for a certain percentage of the yield. Livestock is not for consumption but to support farming. Families eat less than three times a day, often skipped breakfast, and have a less varied diet. Vegetables and particularly meat (chicken or mutton) are rarely part of everyday meals. Food scarcity was explained in terms of shortages of rice, meat and vegetables.

> We don't have much food varieties to choose from. Whatever we get – when we work in the fields of other people – whatever we get, we bring home. We cook and eat it. (Dalit woman, 35 years old)

Most Dalit families experienced chronic food scarcity, especially in the pre-monsoon period and when winter harvests are depleted and new crops yet have to grow. During winter, fields are covered with snow, hence no work and food problems. In these times, women mentioned it was important to feed their family first, even during pregnancy or lactation.

> Sometimes when the food is not sufficient, after feeding my husband and kids, I still have an empty stomach. I try to feed my husband nicely [...] Even during pregnancy, I have to serve my husband first. No, they don't say 'you take more', they don't care about that. (Dalit woman, 23 years old).

There was one exception. A young woman from a nuclear-family household said she equally divided the food among the household members during food scarcity. She would serve her children first, and then divide the rest of the food between her and her husband. This family had also decided to use contraceptives in order to keep the family small, to enable them to better cope with food scarcity.

Some women linked food scarcity to family size. Because women should have their first child – preferably a son – soon after marriage, contraception is not an option. In the Buddhist community, the ages at marriage and first pregnancy are higher (Table 8.2) and increasing, and no longer are marriages always arranged by families. However, some women mentioned the tensions between the tradition of having large families and social change (e.g. importance of education).

> These days we are told that we shouldn't get married until we've finished our education. But our tradition is different, right? So we are stuck in the middle. We are like a pendulum. We learn something new, but we have to follow the old people. (Nyinba woman, 20 years old)

> At that time I was only 15 years old. I felt like a child. I thought I would not be able to give birth to this child. I felt I would die during birth. There are a lot of mothers who die, right? Back then, it was common that mothers died. There was no hospital and they did not go for check-ups. So I thought I would die, I would not live long. Even after I gave birth to my child, I thought I would not be able to take care of this child. Mentally and physically, I felt not ready. (Dalit woman, 33 years old)

> I don't want to get more babies, because it's very difficult to raise them. But my family puts me under pressure to increase our family size. Even my father-in-law tells me to increase my family size. You have only one son, he says. You should have at least three children. (Dalit woman, 33 years old)

During periods such as menstruation, pregnancy, after childbirth and when lactating, women adopt specific food habits based on cultural rules. Hindu women are regarded unclean during menstruation and immediate post-partum. They cannot cook, enter the kitchen, or touch men. Chhetri women are banned from certain areas of the house. Dalit women are confined to animal sheds during menstruation and the first two weeks after delivery (with their newborn child), a social practice called *chhaupadi*. By now, the Nepal government officially declared it a criminal and punishable custom.[1] Especially during winter (December-March), the cold and unhygienic circumstances of their confinement are hard for the Dalit women:

> We do not have sufficient clothes to wear, we do not have enough sleeping mats, and we do not have soap. After our stay at the cowshed, we have to wash all cloths. We have to sit there the whole time. It stinks, there is no light. People say now we have to take care of menstruating girls, because they become weak and our blood pressure becomes low. Back in the old days, we did not know anything about this. These days we know that, but what to do? Even though we know that we have to treat girls better, we do not have any option. (Dalit woman, 35 years old)

During pregnancy, Buddhist women do not change their diet. Some women avoid certain food, such as black millet and chapatti, but this is mainly because of loss of appetite. Some eat extra rich foods, such as eggs, meat, milk, fruit and vegetables. Some Hindu women were

[1] Criminal Code 2074 into effect as from august 17, 2017. See e.g. http://kathmandupost.ekantipur.com/news/2017-08-10/new-law-criminalises-chhaupadi-custom.html

denied certain foods, including millet, buckwheat and radish. Other women said they did not avoid food, because of food scarcity. During pregnancy, women are not given more food:

> We face many shortages. Whatever we can get during pregnancy or menstruation, we take. We have to work hard. When we don't eat during pregnancy and we have to work hard, blood will come out. (Dalit woman, 18 years old)

After delivery and during the post-partum period, Buddhist women go to their natal home or a relative's home for some time (two weeks to three months), also depending on household composition, where they are taken care of by their mothers. Being the only woman in the household, women from nuclear-family households often face more difficulties. They would have no time (or were not permitted) to rest during pregnancy and after delivery:

> After delivery I only rested for eight days, and after that I had to do all the household chores. My mother-in-law was not with me and I was the only woman in my family and had to do all the work myself: cutting grass, cooking, weaving, I didn't rest properly during and after my pregnancy, since I had to work until late at night and wake up early again to work. (Nyinba woman, 48 years old).

During the first month of breastfeeding, Nyinba women change their diet in order to gain strength. They consume more butter, honey, meat, and eggs, and eat the local red rice instead of white rice, even though local rice is looked down upon at other times. Buckwheat pancakes, white rice, black pulses, flatbread, turnips, spinach, radish, chili, pumpkin and potato are taboo. While women could not explain why, some said these foods would cause illness.

Hindu women are considered impure during this time and cannot touch stoves, men, and the food of other family members. For a period of 10 days to a month, they stay in separate rooms or animal sheds, where they take care of themselves and their child. They have to collect wood and water themselves and cook their own meals. Some mothers received support from other women or friends. When breastfeeding, they have to avoid sour, spicy and oily food, almost all available fruits, radish and flatbread made of buckwheat, but they should eat extra local rice, eggs and green vegetables. However, food scarcity induces non-compliance.

8.4.4 Nutritional status and perceived health status

The nutritional status of the interviewed women was assessed by calculating their Body Mass Index or MUAC in case of pregnancy or lactation. At the time of the research, one woman was pregnant and eight women were breastfeeding. In the Nyinba community, most women had a good nutritional status according to their BMI or MUAC. Two women could be classified as being overweight, two other women as thin (grade 1). In the Hindu community, more women were undernourished. Dalit women had the poorest nutritional status (Table 8.3).

When asked about perceived health status, 67% of Nyinba women rated their health as good, 33% as neutral. More or less in line with their nutritional status (Table 8.3), only 43% of Chhetri women and 12.5% of Dalit women rated their health as good. More women perceived their health as poor or neutral, with Chhetri women reporting more often a poor health status than Dalit women. Half of the Dalit women reported their health as neutral (Table 8.4).

Table 8.3. Nutritional status indicators of the interviewed women.

	Body mass index / mid-upper arm circumference			Total
	Low (<18.5/ <23 cm)	Normal (18.5-24.9/≥23 cm)	High (≥25)	
Nyinba	2	11	2	15
Chhetri	4	3		7
Dalit	6	2		8
Total	12	16	2	30

Table 8.4. Self-perceived health status of the interviewed women.

	Poor	Neutral	Good	Total
Nyinba	0 (0%)	5 (33%)	10 (67%)	15
Chhetri	3 (43%)	1 (14%)	3 (43%)	7
Dalit	3 (37,5%)	4 (50%)	1 (12,5%)	8
Total	6	10	14	30 (100%)

8.5 Conclusion and discussion

This study investigated the following questions: (1) How is women's vulnerability to food and nutrition insecurity mediated by cultural and social factors; and (2) What is the actual and possible impact of social change on women's vulnerability to food and nutrition insecurity. Women from two ethnic communities within the same geographical region were compared. This section presents and discusses the answers to the research questions.

In both communities, decision-making on food selection and serving is primarily the responsibility of the male household head or, in the case of extended-family households, his mother (the eldest woman in the household). Daughters-in-law are usually responsible for cooking but are lowest in the household hierarchy and last in the serving order. Only in the Hindu households food allocation clearly disfavoured women and adolescent girls in terms of quantity and quality, which tallies with the household maximization rule. Especially in times of food scarcity, these women said to be dependent on food leftovers. Compared to the Buddhist women, Chhetri and Dalit women tended to be undernourished, and more often reported a poor health status. Notably, whereas the Buddhist women in nuclear-family households have more decision-making power in food selection and preparation than women in the joint-family households, in the Hindu nuclear-family households the husband (or his parents) keeps this decisive power.

During the vulnerable periods of pregnancy, immediate post-partum and lactation, some women from the Buddhist community received or took more nutritious foods. Hindu women, however, reported they were denied certain nutritious foods in addition to being seen as unclean, and the Dalit women reported confinement in animal sheds for some time.

Although food availability and poverty seem to play a major role in the relationship between food allocation at the household level and women's nutritional status, the findings also show that cultural and social factors strongly influence women's vulnerability to malnutrition. Gender, ethnicity, family systems and deeply rooted traditional food beliefs drive intra-household food allocation practices within the different types of households. While in both communities, household hierarchies were reflected in the food serving order, only in the Hindu households the negative impact of this on women's vulnerability to food and nutrition insecurity could be observed.

Using the lens of the moral household economy, this study shows that in both groups the everyday practices of food allocation and consumption are highly gendered, often to the disadvantage of women. This is perceived as normal. In fact, women themselves or their mothers-in-law may perpetuate the gender bias. As was noted in a review on the determinants of intrahousehold food allocation in South Asia, women's self-identity (a low and subservient position in the household) is perpetuating pro-male food allocation (Harris-Fry *et al.*, 2017). However, with social change, i.e. a shift towards nuclear-family households and an increase of educational and working opportunities for women, the practices seem to slowly change in favour of women and girls. However, it could still be observed that even a working daughter was served last. In Bangladesh it was found that mothers controlled their daughters' food intake, not only because of food scarcity, but also in anticipation of the girls' future with their in-laws where they could be denied certain foods as well (Ali, 2005).

Practices are shaped by household resources and women's capabilities, which require agency and entitlements (Niehof and Wahlen, 2016), but also by lay normativity (Sayer, 2005) and the family system. This study shows how women's resources and capabilities clearly differ between the two communities, although in both groups women are morally responsible for feeding the family, in particular the children. However, the Hindu households, and especially the Dalit households, experience more food scarcity than the Buddhist households. When food is scarce, Hindu women perform their responsibility to feed the family even if at the expense of their own food intake and health. By doing so, they comply with the moral expectations of being a 'good' wife, mother, or daughter-in-law.

Inequalities in land ownership and access to food also explain the differences between the groups. A study on livelihood and wellbeing in Humla by Gautam and Andersen (2016), found that non-farm activities including trade and salaried jobs are instrumental for wellbeing. The Tibetan-Buddhist groups are more engaged in such activities than other groups. This is because historically they inhabited the higher regions of the district where land is hard to till, and because their polyandrous household structure provided them with more working-age members than nuclear-family households. In this study, Buddhist families owned more land and livestock than Hindu families. Buddhist men were more involved in other economic activities than farming, a tradition that was already observed by Levine (1987) who found that Nyinba women usually concentrated on agriculture and men engaged in off-farm activities, including trading. In Humla, as a consequence of male labour out-migration, women had more autonomy and decision-making power. Gartaula (2011) found similar patterns in the eastern plains of Nepal where left-behind women were empowered through their enhanced role in agricultural decision-making, although more so among women in (de facto) women-headed households than among women living with their in-laws. To date, the polyandrous family system may still explain women's agency, because it puts women at the virtual centre of the household and kinship relations and renders them a certain control over economic activity and the power to influence decisions. As Luintel (2004: 70) noted: 'Nyinba women are not just subservient 'housewives', they are also active agents in exercising power and agency with their male counterparts.' The Chhetri own some plots of land and have a higher social and political status than de Dalit, who are from origin landless and disadvantaged in terms of socio-economic status. The caste system still poses 'strict sanctions on mobility, interaction, and dietary conduct for all caste groups' (Nagoda and Erikson, 2014: 247). Many differences between the Buddhist and Hindu communities persist up to the present day and are visible in the gendered normativity of household food practices.

In both communities, dietary rules surrounding pregnancy and lactation are rooted in tradition. However, in particular the women from nuclear-family Buddhist households did not always adhere to food restrictions, for which the lack of social control by the mother-in-law was one of the reasons given. These women did also attain higher educational status than the Hindu women, and education can provide women with the knowledge and agency to decide upon their own health. Also some Hindu women said not to follow these restrictions because food scarcity makes them eat what is available. Even so, unequal food allocation within households and being deprived of certain foods or sufficient food during the vulnerable life stages of pregnancy and lactation, cause many Hindu women to suffer from a double burden of food deprivation.

At the time of the research, social change seemed to benefit especially the Buddhist women. With the shift towards nuclear-family households, they gained more power within their households. To the Hindu women, this did not seem to apply. One could argue that for such a change to be beneficial, it has to be accompanied by changes and opportunities in other domains such as education, employment and economic development. Women's education and working opportunities, more pronounced within the Nyinba community, can contribute to their increased agency. Tiwari (2007) found that educated Nepali Gurung women exercised considerable agency in terms of fertility choices and earning money. Like the Nyinba women in this study, they have more freedom of movement than Hindu women, also those from the Chhetri group. Where women contribute to the household income, their decision-making power tends to increase (Boserup, 1990). Based on their review of food allocation practices in South Asia, Harris-Fry *et al.* (2017: 18) suggest a 'non-linear relationship between socioeconomic status and inequity', noting that the most equitable households were low-caste households where women earned an income and the educational status of the household head was low. In the Hindu group, traditional family norms and systems and men's patriarchal control over household resources and food distribution seem to persist, which perpetuates women's disadvantaged position and the gender bias in food allocation.

In their study on perceptions of social change in Nepal, Aaslung and Haug (2011) posit that all people, regardless of background, experience social change. The Dalit, however, seem to do so to a lesser extent than the other ethnic groups. The authors argue that this is because of their low socio-economic status, lack of land ownership, poor access to facilities, and the social barriers posed by ethnicity, caste and religion. In the present case, the remoteness of Humla district might play a role as well. Interestingly, Aaslung and Haug (2011: 199) found that people in Nepal 'ascribe the positive change to personal agency.' It could be that if these disadvantaged groups receive more opportunities, also intra-household resource allocation could be more equal through women's increased agency.

Ironically, while women are often the first to suffer from malnutrition within families, their role in mitigating hunger is indispensable (Gillespie, 2001). To play this role, they need access to resources, and an 'increase of resources at the household level does not necessarily result in a more equitable allocation' (Niehof and Wahlen, 2016: 14). Access to resources should be boosted by investments in women's human and social capital. That would increase their power to exercise their agency, and could make it their moral responsibility to not only feed their family, but also themselves. Finally, a gender-sensitive approach should engage both men and women in discussions on women's vulnerability to food insecurity.

Acknowledgements

I would like to express my gratitude to the Neys-Van Hoogstraten Foundation for funding this study. Sincere thanks go to the Nepal Trust for their logistic and practical support during the fieldwork of this study and in particular to Yeshi Wangmo for her translations and data collection assistance. I owe special thanks to my supervisor at the university, Prof. Dr Hilde Bras, for her guidance, continued support and encouragement. Lastly, I am grateful to the women who participated in this study.

References

Aaslung, A. and Haug, M., 2011. Perceptions of social change in Nepal: are caste, ethnicity, and region of relevance? *Journal of Asian and African Studies* 46(2): 184-201.

Adhikari, J., 2008. *Food crisis in Karnali: a historical and politico-economic perspective.* Kathmandu: Martin Chautari.

Alam, S., 2012. The effect of gender-based returns to borrowing on intra-household resource allocation in rural Bangladesh. *World Development* 40(96): 1164-1180.

Ali, A., 2005. *Livelihood and food security in rural Bangladesh-the role of social capital.* PhD thesis. Wageningen: Wageningen University and Research.

Backstrand, J.R., Allen, L.H., Pelto, G.H. and Chavez, A., 1997. Examining the gender gap in nutrition: an example from rural Mexico. *Social Science & Medicine* 44(11): 1751-1759.

Behrman, J.R., 1997. Intrahousehold distribution and the family. In: Rosenzweig, M.R. and Stark, O. (eds) *Handbook of population and family economics*, Volume 1, part A. Amsterdam: Elsevier, pp. 125-187.

Boserup, E., 1990. *Economic change and the role of women.* In: Tinker, I. (ed.) *Persistent inequalities: women and world development.* New York and Oxford: Oxford University Press, pp. 14-24.

Bouis, H., 2003. Commercial vegetable and polyculture fish production in Bangladesh: Impacts on income, food consumption, and nutrition. In: Quisumbing, A.R. (ed.) *Household and decisions, gender and development: a synthesis of recent research.* Baltimore, USA: The John Hopkins University Press for IFPRI, pp. 73-77.

Darnton-Hill, I., Webb, P., Harvey, P.W.J., Hunt, J.M., Dalmiya, N., Chopra, N., Ball, M.J., Bloem, M.W. and De Benoist, B., 2005. Micronutrient deficiencies and gender: Social and economic costs. *The American Journal of Clinical Nutrition* 8(suppl.): 1198-1205.

Das Gupta, M., 1999. *Lifeboat ethics versus corporate ethics – social and demographic implications of stem and joint families* (Policy Research Working Paper Series 2127). Washington DC: Development Economics Group, World Bank.

De Schutter, O., 2013. *Gender equality and food security. Women's empowerment as a tool against hunger.* Mandaluyong City, Philippines: Asian Development Bank.

Den Hartog, A.P., Van Staveren, W.A. and Brouwer, I.D., 2006. *Food habits and consumption in developing countries. Manual for field studies.* Wageningen: Wageningen Academic Publishers.

DeRose, L.F., Das, M. and Millman, S.R., 2000. Does female disadvantage mean lower access to food? *Population and Development Review* 26(3): 517-547.

FAO, 1996. *Rome declaration on world food security*. World Food Summit, 13-17 November 1996. Rome: Food and Agriculture Organisation of the United Nations.

Fledderjohann, J., Agrawal, S., Vellakkal, S., Basu, S., Campbell, O., Doyle, P., Ebrahim, S. and Stuckler, D., 2014. Do girls have a nutritional disadvantage compared with boys? Statistical models of breastfeeding and food consumption inequalities among Indian siblings. *PloS ONE* 9(9): 107-172.

Gartaula, H.N., 2011. *The other side of migration in rural Nepal: sociocultural transformation and the women left behind*. PhD thesis. Wageningen: Wageningen University and Research.

Gautam, Y. and Andersen, P., 2016. Rural livelihood diversification and household well-being: insights from Humla, Nepal. *Journal of Rural Studies* 44: 239-249.

Gillespie, S.R., 2001. *Strengthening capacity to improve nutrition*. Washington DC: IFPRI.

Gittelsohn, J. Mookherji, J. and Pelto, G., 1998. Operationalizing household food security in rural Nepal. *Food & Nutrition Bulletin* 19(3): 210-222.

Gittelsohn, J., 1991. Opening the box: intrahousehold food allocation in rural Nepal. *Social Science & Medicine* 33(10): 1141-1154.

Gittelsohn, J., Thapa, M. and Landman, L.T., 1997. Cultural factors, caloric intake and micronutrient sufficiency in rural Nepali households. *Social Science & Medicine* 44(11): 1739-1749.

Griffiths, P., Matthews, Z. and Hinde, A., 2002. Gender, family, and the nutritional status of children in three culturally contrasting states of India. *Social Science & Medicine* 55(5): 775-790.

Gross, R., Schoeneberger, H., Pfeifer, H. and Preuss, H.J., 2000. The four dimensions of food and nutrition security: definitions and concepts. *SCN News* 20: 20-25.

Haddad, L., 1999. Women's status: levels, determinants, consequences for malnutrition, interventions and policy. *Asian Development Review* 17(1-2): 96-131.

Haddad, L., Pena, C., Nishida, C., Quisumbing, A. and Slack, A., 1996. *Food security and nutrition implication of intrahousehold bias: a review of the literature*. FCND Discussion Paper No. 19. Washington DC: IFPRI.

Hadley, C., Lindstrom, D., Tessema, F. and Belachew, T., 2008. Gender bias in the food insecurity experience of Ethiopian adolescents. *Social Science & Medicine* 66(2): 427-438.

Harris-Fry, H., Shrestha, N., Costello, A. and Saville, N.M., 2017. Determinants of intra-household food allocation between adults in South Asia – a systematic review. *International Journal for Equity in Health* 16(1): 107-128.

Knoema, 2014. *Development statistics of Nepal, 2014*. Humla (Lali). Available at: http://tinyurl.com/ybssdfbd.

Levine, N., 1987. Caste, state, and ethnic boundaries in Nepal. *Journal of Asian Studies* 46(1): 71-88.

Luintel, Y.R., 2004. Agency, autonomy and the shared sexuality: gender relations in polyandry in Nepal Himalaya. *Contribution to Nepalese Studies* 31(1): 43-83.

Madjdian, D.S. and Bras, H., 2016. Family, gender and women's nutritional status: a comparison between two Himalayan communities in Nepal. *Economic History of Developing Regions* 31(1): 198-223.

Maharjan, K.L. and Khatri-Chhetri, A., 2006. Household food security in rural areas of Nepal: relationship between socio-economic characteristics and food security status. *Paper presented at the International Association of Agricultural Economics Conference*. August 2006. Queensland, Australia. Available at: http://tinyurl.com/y8rcnaep.

McKay, K.H., 2002. Health needs in two ethnic communities of Humla District, Nepal. *Contributions to Nepalese Studies* 29(2): 241-273.

McKay, K.H., Zahnd, A., Sanders, C. and Nepali, G., 2007. Responses to innovation in an insecure environment in rural Nepal. *Mountain Research Development* 27(4): 302-307.

Messer, E., 1997. Intra-household allocation of food and health care: current findings and understandings. *Social Science & Medicine* 44(11):1675-1684.

Miller, B.D., 1997. Social class, gender, and intrahousehold food allocations to children in South Asia. *Social Science & Medicine* 44(11): 1685-1695.

MOAD, CBS and FAO, 2016. *Food and nutrition security in Nepal: a status report* (TCP/RAS/3409). Kathmandu, Nepal: Ministry of Agricultural Development, Central Bureau of Statistics and Food and Agriculture Organization.

Nagoda, S. and Eriksen, S., 2014. The role of local power relations in household vulnerability to climate change in Humla, Nepal. In: Inderberg, T.H., Eriksen, S., O'Brien, K. and Sygna, L. (eds) *Climate change adaptation and development: transforming paradigms and practices*. New York: Routledge, pp. 200-218.

Narayanan, S., 2015. Food security in India: the imperative and its challenge. *Asia & The Pacific Policy Studies* 2(1): 197-209.

Niehof, A., 2011. Conceptualizing the household as an object of study. *International Journal of Consumer Studies* 35(5): 488-497.

Niehof, A. and Wahlen, S., 2016. Moralities of sharing and caring: gender and food in the moral household economy. In Bock, B. and Duncan, B. (eds) *Gendered food practices from seed to waste.* Yearbook of Women's History 36. Hilversum: Verloren, pp.141-155.

Noack, A.L. and Pouw, N.R.M., 2014. A blind spot in food and nutrition security: Where culture and social change shape the local food plate. *Agriculture and Human Values* 32(2): 169-182.

Rahman, A., 2002. *On measuring intrahousehold inequality in food distribution. Is the conventional calorie intake measure enough to understand individual wellbeing within the household*? London: University College, Department of economics.

Ramachandran, N., 2004. *Seasonal hunger: implications for food and nutritional security*. Towards Hunger Free India: From Vision to Action. Madras: East West Books.

Ramachandran, N., 2007. Women and food security in South Asia: current issues and emerging concerns. In: Guha-Khasnobis, B., Acharya, S.S. and Davis, B. (eds) *Food insecurity, vulnerability and human rights failure*. London: Palgrave Macmillan, pp. 219-240.

Ross, J.L., 1981. *Hindu and Tibetan reproduction and fertility in north-western Nepal: a study of population ecology and economics*. PhD thesis. Cleveland, Ohio: Department of Anthropology, Case Western Reserve University.

Rudie, I., 1995. The significance of eating: co-operation, support and reputation in Kelantan Malay households. In Karim, W. (ed.) *Male and female in developing Southeast Asia*. Washington DC: Berg Publishers, pp. 227-247.

Samir, K.C., Springer, M., Thapa, A. and Khanal, M.N., 2016. *Projecting Nepal's demographic future – how to deal with spatial and demographic heterogeneity* (IIASA Working Paper WP-16-021). Laxenburg, Austria: IASSA.

Sayer, A., 2005. Class, moral worth and recognition. *Sociology* 39(5): 947-963.

Sharma, P., Guha-Khasnobis, B. and Khanal, D.R., 2014. *Beyond geography, unlocking human potential.* Nepal Human Development Report 2014. Kathmandu, Nepal: Government of Nepal, National Planning Commission and United Nations Development Programme.

Skinner, G.W., 1997. Family systems and demographic processes. In: Kertzer, D.I. and Fricke, T. (eds), *Anthropological demography: toward a new synthesis.* Chicago and London: The University of Chicago Press, pp. 53-95.

Therborn, G., 2004. *Between sex and power: family in the world 1900-2000.* London and New York: Routledge.

Tiwari, N., 2007. *Women's agency in relation to population and environment in rural Nepal.* PhD thesis. Wageningen: Wageningen University and Research.

UN, 2014. Nepal information platform. District profile: Humla. Available at: http://un.org.np/district_profile/humla.

UNICEF, 1991. *Conceptual framework of malnutrition.* Available at: http://tinyurl.com/yam27dej.

UNICEF, 2012. *Multiple indicators cluster survey* (MICS). Retrieved from UNICEF website http://www.mics.unicef.org.

Ververs, M.T., Antierens, A., Sackl, A., Staderini, N. and Captier, V., 2013. Which anthropometric indicators identify a pregnant woman as acutely malnourished and predict adverse birth outcomes in the humanitarian context? *PLoS Currents Disasters* 1: 1-14.

WHO, 2004. Appropriate body-mass index for Asian populations and implications for policy and intervention strategies. *The Lancet* 363(9403): 157-163.

9. Parents' characteristics, food habits and physical activity of overweight schoolchildren in Bogor City, Indonesia

K.R. Ekawidyani[1*]*, I. Karimah*[2]*, B. Setiawan*[1] *and A. Khomsan*[1]

[1]*Department of Community Nutrition, Faculty of Human Ecology, Bogor Agricultural University, Indonesia;* [2]*Health Polytechnic Tasikmalaya, Ministry of Health of Indonesia, Indonesia; *karinare@apps.ipb.ac.id*

Abstract

Overweight and obesity is a serious problem among urban schoolchildren that continues into adulthood and leads to metabolic and degenerative diseases. In the context of urban development and social change, children's food habits and low physical activity put them at risk of childhood obesity. More specifically, factors such as relatively high educational level and income of parents, provision of pocket money, and easily accessible fast food restaurants, play a role in this. This study aimed to assess the impact of such factors on the nutritional status and physical fitness of middle-class schoolchildren in the city of Bogor, West Java, Indonesia. A cross sectional study involving 224 grade-5 elementary schoolchildren was done. Data collection included a mix of methods. Results shows that overweight risk is 4.5 higher for non-fit children compared to fit children. Overweight children had significantly lower physical fitness scores. Children with pocket money more than IDR 10,000/day (€ 0.66) had 2.7 times more risk to become overweight than children with less pocket money. Low education of the mother decreased overweight risk 0.4 times. Mothers with a high education mostly had an occupation outside the home, hence had less time for food care. High protein adequacy levels of more than 120% decreased overweight risk 0.4 times. These findings suggest that social change and development in the urban context have a negative impact on the nutritional status and fitness of schoolchildren.

Keywords: schoolchildren, overweight, urban environment, Indonesia

9.1 Introduction

Prevalence of overweight and obesity among children increases from year to year in both developed countries and developing countries. According to WHO (2000), one out of 10 children in the world is obese. In Indonesia, according to recent data the prevalence of overweight and obesity among children of 5 to 12 years old is high (18.8%). For adolescents of 13-15 years old it is 10.8%, and for the age group of 16 to 18 it is 7.3%. The prevalence for boys aged 5-12 years is higher (20.2%) than for girls of the same age (17.3%). Incidence of overweight including obesity is higher in urban areas (20.7%) than rural areas (17.2%) (MoH, 2013). Overweight is a condition where the body weight exceeds normal weight, by 20% for men and 25% for women. In the case of obesity, body weight is a result of fat accumulation.

A. Niehof *et al.* (eds) **Diversity and change in food wellbeing – Cases from Southeast Asia and Nepal**
DOI 10.3920/978-90-8686-864-3_9,

Overweight is a serious problem among schoolchildren because it continues when they grow up and puts them at risk of developing metabolic and degenerative diseases, such as cardiovascular diseases, diabetes mellitus, cancer, and osteoarthritis. Studies have shown that children and teenagers who are overweight have an increased risk of suffering from cardiovascular diseases in adulthood (Oren *et al.*, 2003; Wright *et al.*, 2001).

Incidence of obesity in primary schoolchildren has been associated with low physical activity and consumption of fast foods (Suryaalamsyah, 2009; Pramudita, 2011). There has been a shift in children's play activities. Instead of playing physical outdoor games, nowadays children watch television, play video games or online games on the computer or mobile phone. This limits their physical mobility and negatively affects their health (Kimbro *et al.*, 2011; Vandewater *et al.*, 2004).

Similarly, advancements in food technology have an impact on food choices among children. Children now are more likely to consume fast foods and soft drinks. Fast foods contain energy, fat, and high cholesterol, but have low fibre content. If habitually consumed, this may affect health adversely (Bowman *et al.*, 2004). As fast food consumption is strongly associated with weight gain and insulin resistance, it may contribute to the development of obesity and type 2 diabetes (Pereira *et al.*, 2005).

Social and cultural factors influence the consumption of fast food. A study in China found three factors associated with western fast food consumption among Chinese children: good accessibility of fast foods as in urban areas, college level education of mothers, and the caregiver's positive attitude to western food (Song *et al.*, 2015). Accessibility of fast foods relates to distance and affordability. Austin *et al.* (2005) found that fast-food restaurants are clustered within a short walking distance from schools. This exposes children to poor-quality food environments. Children cannot only reach the fast food restaurants easily, they also can afford buying the foods with their pocket money. In urban China, students with more pocket money consumed more sugary beverages, snacks, fast food or street food, and were more likely to be overweight or obese (Li *et al.*, 2017).

The changes in children's food consumption habits and physical activity reflect social change in urban areas, as indicated by relatively high educational level and high income of parents, provision of pocket money, and accessible fast food restaurants. Increasing work demands and time pressure in the urban context leave people little time for people to prepare food in a traditional manner; they rather buy pre-packaged meals and fast foods (Hill *et al.*, 2003). They also need to go everywhere faster, using any means of transportation instead of walking. All this puts children at risk of childhood obesity. This study aimed to analyse the effects of these factors on the nutritional status and physical fitness of middle-class schoolchildren in the city of Bogor.

9.2 Conceptual framework

Overweight is defined as children who have z-score of body mass index (BMI) for age more than 1 standard deviation (SD) including as well obese children (WHO, 2006; MoH, 2011). Overweight is the result of imbalance between nutrient intakes and physical activity, i.e. high macronutrient intakes combined with low physical activity.

Several factors may influence nutrient intake: household socioeconomic characteristics, characteristics of children, and food habits (Booth *et al.*, 2001; Patrick and Nicklas, 2005). Household characteristics (i.e. household size, parent's education, job and income) influence the amount of pocket money given to children. Children's characteristics and pocket money shape food habits. Suci (2009) found that parents give pocket money to their children because they are worried their children will get hungry at school and they do not have time to prepare meals for breakfast or lunch.

Crucial food habits comprise eating, breakfast in particular, and snacking. Eating habits are shaped by household characteristics. Parent's education, job, and income influence availability of food at home. Parents with low education usually have low-income jobs, which could prevent them to buy nutritious food. It is even more challenging when households are large and with many people to feed. Households with these characteristics usually spend more of their income on food with poor nutritional qualities. Instead of eating a balanced diet, they usually eat high-calorie food, which has immediate satiating effect.

Breakfast has been labelled the most important meal. Current evidence points at poor breakfast habits among schoolchildren. A study by Hardinsyah and Aries (2012) involving 35,000 Indonesian schoolchildren, found that 20 to 40% of the children did not customarily take breakfast in the morning, and that 44.6% of children aged 6-12 years old had breakfast with low nutrients quality. Children usually skip breakfast because they (or their parents) do not have enough time for it. This tempts children to buy foods that are available around the school. These foods tend to have high calorie and fat contents. Skipping breakfast and fast food consumption usually continue in adulthood. Both habits are associated with increased weight gain from adolescence to adulthood (Niemeier *et al.*, 2006). Children who missed breakfast are significantly less likely to consume fruits and vegetables and more likely to be frequent consumers of unhealthy snacks (Uter *et al.*, 2007).

Various types of snacks are being sold around school. Children are willing to buy these snacks because they are hungry and the snacks look attractive. These snacks generally are rich in calories but poor in fibre. This affects children's nutrient intake and nutritional status, i.e. overnutrition in calories and macronutrients but undernutrition in micronutrients. Obesity among children and adolescents is associated with taking carbohydrate-rich foods and sweet drinks (Gillis and Bar-Or, 2003), and obese children snack more (Pramudita, 2011).

Physical activity also has an impact on nutritional status. There is a positive association between physical activity levels and nutrient intake. Children now have lower activity levels than in the past, especially those who live in urban areas. Children prefer to play indoors with their gadgets and video games rather than playing outdoors. They rarely play outdoor games like hide and seek or ball games that burn a lot of energy. Because of high calorie intake and little physical activity, children become overweight and obese. Physical activity of children is positively associated with physical fitness (Boreham and Riddoch, 2001; Bovet *et al.*, 2007; Kim *et al.*, 2005). Finally, obesity in childhood can be carried on to adulthood and has long-term impacts on adult morbidity and mortality (Reilly and Kelly, 2011). However, this can be prevented if the modifiable risk factors are known. As shown in the conceptual framework (Figure 9.1), socio-economic profile, food habits and physical activity, are the risk factors that this study seeks to investigate. To do so, we formulated the following research questions:

1. How do the parent's characteristics relate to the weight of the schoolchildren?
2. How do the schoolchildren's food habits, nutritional status, physical activities, and fitness level correlate with their weight?
3. How do the parent's characteristics, children's food habits, and physical activities affect the weight of the schoolchildren?

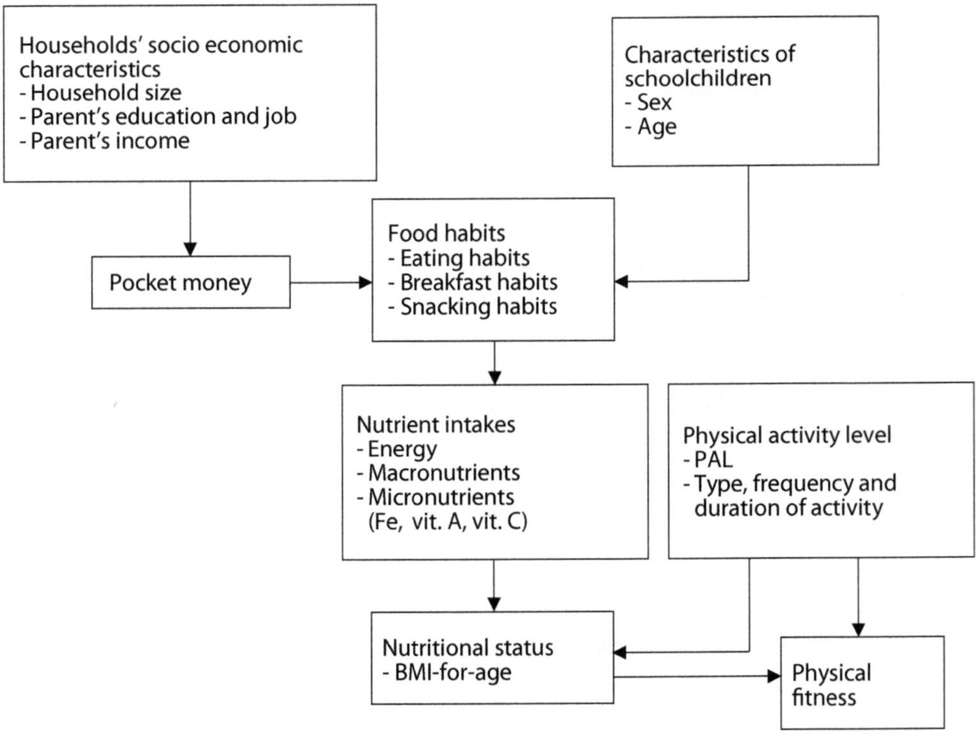

Figure 9.1 Conceptual framework.

4. What are the significant differences between normal and overweight middle-class schoolchildren?
5. How can the results be explained by 'social change and development'?

9.3 Research area and methodology

The study was conducted in Bogor City, West Java, Indonesia. Bogor City is an urban area located 60 km. south of Jakarta. Subjects involved in this study are fifth-grade elementary male and female schoolchildren, aged 10-12 years old. The subjects are overweight/obese and normal students, according to WHO standards. Students were categorised as overweight if the BMI/age +1SD<Z <+2 SD, obese if BMI/age >+2SD, and normal if BMI/age -2SD<Z <+1 SD (WHO, 2006; MoH, 2011).

Six elementary schools were purposively selected. The major consideration was their similarity in terms of the majority of the children coming from upper middle-class families. Research in Indonesia has shown that overweight and obesity are more prevalent in upper middle-class families and that prevalence increases with quintile of income (MoH, 2013). Using the Slovin formula, given a total of population of ±494 5th graders in the selected schools, the sample size was determined at 224. The 5th graders in the six schools were screened by measuring their body weight and height. Based on BMI for age, underweight children were excluded. The rest were grouped into normal and overweight, including obese, children. Simple random sampling was applied to each group.

Data collection was done by enumerators, supervised by the research team. All interviews and measurements (anthropometric nutritional, anaemia, and health status) were conducted in the schools, during March-August 2014. Primary data were collected through interviews using a questionnaire, anthropometry measurement, and observation. The schoolchildren reported their characteristics and food habits on a self-administered questionnaire after an explanation by the researcher. For the household characteristics, we used a self-administered questionnaire for the parents. Nutrient intake was measured for three days, two non-consecutive school days and one school holiday, using 24-hour food recall.

Physical Activity Level (PAL) was measured by interviewing schoolchildren about their activities on school days and holidays. Using a questionnaire, data on the frequency and duration of exercise habits and physical activity habits during leisure time and school days were collected. Physical fitness was assessed by several tests: 40 metres sprint, push up and sit up, straight jump, and running 600 metres. The children's speed ability was measured by their sprint performance, cardiovascular endurance through the 600-metres run. Endurance of shoulder and arm muscles was measured by the push-up test, the endurance of the abdominal muscles by the sit-up test. The straight jump tested the explosive power of the leg muscles.

We present descriptive statistics and inferential analysis, using the Mann-Whitney test and the Independent t-test. The first was used to analyse the differences between normal and overweight schoolchildren on variables that are not normally distributed, the t-test for normal distributions. Chi-square tests were done on categorical variables; the Spearman correlation test was applied to numeric variables. Finally, factors affecting overweight were analysed by a regression test.

Ethical clearance was obtained from the ethical committee of the Faculty of Medicine, Universitas Indonesia. The involvement of the schoolchildren was voluntary and permission was obtained from their parents or guardians. All subjects could withdraw from the study at any time. All data were treated confidentially and used for the purpose of this study only.

9.4 Results

9.4.1 Characteristics of the schoolchildren

The 224 schoolchildren were divided equally into two groups: the normal and the overweight group. The children were on average 11 years old. More than half of them are girls. Parents of overweight children gave their children more pocket money than the parents of the normal children. Group averages were IDR 8,036 and IDR 10,167, respectively. Almost half of the normal and overweight children are the first or eldest child in their family, and one-third is a second child.

9.4.2 Household characteristics

Table 9.1 presents the characteristics of the households and parents of the schoolchildren. The average household size in both groups was five. The age of the fathers and the mothers ranged from 39 to 43, the fathers being older than the mothers. Age of the parents did not significantly differ between the normal and overweight group. The t-test results on nutritional status as measured by BMI show that the fathers and mothers of the overweight children have a significantly higher BMI than those of the normal children. The fathers of the overweight children have overweight BMI, while the mothers have a normal BMI (≥18.5 – <24.9).

Most parents in both groups are *sarjana* (undergraduate). The percentage of the fathers who reached higher education (diploma, *sarjana*, and postgraduate) was slightly higher in the normal group (79.5%) than in the overweight group (75.9%). Conversely, more mothers reached higher education in the overweight group (73.2%) than in the normal group (66.3%). However, these differences are not significant.

The results show that 48.2% of the fathers and 16.1% of the mothers in the normal group work as private employees. The others work as civil servants, in the police or in the army, as labourers, have their own business, or work in the services sector (as tailors, drivers, or

Table 9.1. Characteristics of households and parents of school children.

Characteristics	Normal	Overweight	P^1
Household size (%)			0.819
Small (<4 persons)	7.1	8.9	
Medium (4-7 persons)	87.5	87.5	
Large (>7 persons)	5.4	3.6	
Parent's age (year)			
Father	43.9	43.9	0.859
Mother	40.0	40.0	0.844
Parent's nutritional status (BMI)			
Father	24.5	25.6	0.009*
Mother	23.4	24.4	0.019*
Father's education (%)			0.549
Elementary school	0.9	0.9	
Junior high school	1.8	0.9	
Senior high school	17.9	22.3	
Diploma	20.5	17.0	
Sarjana (undergraduate)	43.8	49.1	
Pascasarjana (postgraduate)	15.2	9.8	
Mother's education (%)			0.355
Elementary school	2.6	3.6	
Junior high school	2.6	1.8	
Senior high school	28.4	21.4	
Diploma	19.8	25.0	
Sarjana (undergraduate)	25.0	39.3	
Pascasarjana (postgraduate)	21.5	8.9	
Parent's monthly income (in thousand IDR)			
Father	7,138	6,812	0.550
Mother	2,145	2,002	0.772
Total	9,283	8,814	0.736

[1] * = significant at $P<0.05$.

mechanics). In this group, 55.4% of the mothers are housewives. Of the overweight children, the majority of the fathers are private employees (45.5%), the others working as civil servants or in the police or the army (22.3%), or are private businessmen (21.4%). Also in this group, most mothers are housewives. The parents' incomes in both groups ranged from IDR 8 to 9 million per month, which is much higher than Bogor City average (IDR 2.4 million). Although the average combined income of father and mother in the normal group was slightly higher than in the overweight group, the difference is not significant.

9.4.3 Food habits and nutritional status of schoolchildren

The frequency of the children's food consumption and eating together with their family were indicators of eating habits. More overweight children ate twice a day (16.1%) than children with normal nutritional status (10.7%). The majority of the children ate three times a day (76.8% in the normal group and 70.5% in the overweight group). According to the Chi-square test there is no association between eating frequency and nutritional status (P>0.05).

The proportion of children in the normal and the overweight group who stated 'always eating together with their family' was more or less the same (26.8 and 27.7%, respectively), but in the overweight group, there were more children who seldom ate together with their family (25.0%) than in the normal group (12.5%). The Chi-square test did not show a significant association between the practice of eating together and nutritional status (P>0.05). Eating together at dinner (65.3% and 62.0% in the normal and overweight group, respectively) was done more often than eating together at breakfast and lunch times. Dinner is usually the time when all family members are home and can eat together.

The number of schoolchildren in the normal group who always had breakfast (64.3%) was slightly higher than in the overweight group (56.3%). Overall, there was no significant association between the habit of having breakfast and nutritional status (P>0.05). Breakfast contributed approximately 26% of the total daily energy intake and 28% of the total protein intake, which can be considered sufficient. Plain water was the most frequently consumed beverage (median 5 times a day), then milk (0.7 – 1 times a day). Less frequently consumed beverages were sweetened tea, coffee, iced syrup, fruit juice, flavoured drinks, and soft drinks. Beverage consumption did not differ between the normal and the overweight group.

The fast foods consumed by the schoolchildren were pizza, burger, French fries, fried chicken, *kebab*, sandwich, etc. In general, fast foods were consumed 0.2 to 2.0 times a week. The frequency of fast food consumption between the normal and overweight groups did not show a difference. Junk food is a snack that is rich in calories but poor in nutrients. In this study, junk foods investigated were salty crispy fried snacks (*chiki-chikian*), wafers, chips, candies, and jelly. Junk food consumption ranged from one to three times per week. The mostly consumed junk food was candy (three times per week). There was no difference in frequency of junk food consumption between the two groups.

Table 9.2 shows that the adequacy levels of energy, protein, calcium, iron, vitamin A and vitamin C were higher in the normal group than in the overweight group, but not significantly so. The energy adequacy levels of both normal and overweight children were deficit (<90% recommended dietary allowances (RDA)). On the other hand, the protein adequacy levels were excessive (>110% RDA). The micronutrients adequacy levels of both groups were also deficit (<77% RDA) (Table 9.2).

Table 9.2 Nutrients adequacy level of schoolchildren by nutritional status.[1]

Nutrients	Normal			Overweight		
	Intake	RDA	% RDA	Intake	RDA	% RDA
Energy (kcal)	1,628	2,205	70.9	1,627	2,376	66.4
Protein (g)	40.6	30.1	134.0	38.9	33.9	118.2
Calcium (mg)	373.2	1,200.0	31.1	315.4	1,200.0	26.3
Iron (mg)	10.5	20.0	61.6	9.9	20.0	59.3
Vitamin A (RE)	384.0	600.0	64.0	323.5	600.0	53.9
Vitamin C (mg)	14.0	50.0	28.0	11.6	50.0	23.1

[1] RDA = recommended dietary allowances.

Most of the children (92.4%) had good eating habits, particularly having breakfast, of which the energy contribution was 26.2% and 25.6% for the normal group and the overweight group, respectively. Regarding household characteristics and the children's eating habits, only the association between the mother's working status and food habits of the children is significant (Table 9.3). The eating habit of the children whose mothers did not work was 19.4% better than that of the children of working mothers.

Table 9.3 Correlation between the household socio-economic factors and eating habits.

Variable		Eating habit				χ^2
		Good (n=200)		Bad (n=16)		
		n	%	n	%	
Fathers' education	Low (≤12 years)	48	21.4	2	0.9	0.277
	High (>12 years)	159	71.0	15	6.7	
Mothers' education	Low (≤12 years)	63	28.1	6	2.7	0.677
	High (>12 years)	144	64.3	11	4.9	
Pocket money/day	High (>IDR10,000)	41	18.3	6	2.7	0.132
	Moderate (≤IDR10,000)	166	74.1	11	4.9	
Income per capita	High (≥IDR1,500,000)	106	47.3	8	3.6	0.742
	Moderate(<IDR1,500,000)	101	45.1	9	4.0	
Mothers' working status	Not working	119	55.1	5	2.3	0.028
	Working	81	37.5	11	5.1	

Based on BMI-for-age (Table 9.4), the prevalence of overnutrition among the schoolchildren is at 38.5% (consisting of overweight at 20.8% and obesity at 17.7%). It was higher among boys (43.3%) than among girls (33.6%). The prevalence data were obtained from the screening data of the total population of the 5th grade in the selected schools.

Table 9.4 Distribution of children by nutritional status and sex.

BMI/Age	Nutritional status	Sex		Total (%)
		Boy (%)	Girl (%)	
Z-score <-3.0	Very thin	1.2	0.7	1.0
Z-score ≥-3.0 s/d <-2.0	Thin	3.6	4.4	4.0
Z-score ≥-2.0 s/d ≤1.0	Normal	51.9	61.2	56.5
Z-score >1.0 s/d ≤2.0	Overweight	20.1	21.5	20.8
Z-score >2.0	Obese	23.2	12.1	17.7
Total		100.0	100.0	100.0

9.4.4 Physical activities and fitness level of schoolchildren

Most children's activity levels during school days were considered light, for both groups of children. As Table 9.5 shows, the proportion of light activities was higher among overweight than among normal children (92.9 and 86.6%, respectively), the proportion of moderate and heavy activities higher (13.4%) in the normal than the overweight group (7.2%). The differences in activity levels between the normal and the overweight group during school days are significant ($P<0.05$). The median value of PAL among the overweight children was lower than that of the normal group (1.35 and 1.38, respectively). This shows that the children with normal nutritional status are slightly more active than the overweight children.

The level of the schoolchildren's activity during a holiday can be generally categorised as light. The proportion of heavy activities was higher among normal than overweight children, but the difference is not significant. The normal group has higher median PAL value (1.53) than the overweight group (1.48). The most frequently played sports among the school children were running, swimming, badminton, jogging, and football, which the children practised during holidays. Other sports played were basketball, futsal, aerobic, volleyball, and martial arts, such as karate and taekwondo. The average frequency of playing sports in a week was 1.5 times for both normal and overweight children. It was also quite similar in duration (64.1 and 63.5 minutes/week, respectively).

Table 9.5 Distribution of schoolchildren by levels of activity during school days and holiday.

PAL	Levels of activity	Normal (%)		Overweight (%)	
		School days	Holiday	School days	Holiday
1.40-1.69	Light	86.6	68.8	92.9	71.4
1.70-1.99	Moderate	11.6	16.1	4.5	18.8
2.00-2.40	Heavy	1.8	15.2	2.7	9.8
Med[1]	PAL	1.38[a]	1.53	1.35[b]	1.48

[1] [a,b] = Mann-Whitney test, significant at $P<0.05$.

We found that both overweight and normal children spend most of their leisure time watching TV or film. The duration of watching TV among the overweight children was slightly longer (126.7 minutes per week) than that of the normal children (122.0 minutes). Similarly, the overweight group showed a higher frequency (4.9 vs 4.7 times per week) and longer duration (82.7 vs 69.0 minutes per week) than the normal group of playing gadgets. The frequency of listening to music among normal children was higher than among the overweight ones. However, the average duration was longer (46.1 minutes per week) among the overweight children than among the normal children (32.9 minutes). The habit of going to the mall, in terms of weekly frequency and duration, was almost the same for the normal and overweight group: 1.4 times (192.5 minutes) and 1.3 times (168.7 minutes), respectively. Parents usually have their children participate in extra courses to supplement their learning at school. Children in this research were mostly enrolled in a twice-weekly additional course. The weekly frequencies and duration of additional courses among normal and overweight children were 2.4 times (106.9 minutes) and 2.8 times (110.5 minutes), respectively.

Regarding the results of the physical tests (Table 9.6), the normal children could run faster than the overweight ones, but the difference is not significant. For the other performances, the differences between the two groups are significant ($P<0.05$), with the children in the

Table 9.6 Results of the physical fitness tests.

Physical test	Normal	Overweight	*P*-value
40 m sprint (second)	9.81±2.49	10.41±2.37	0.065
Push up (times per 30 second)	21±6.6	18±6.1	0.000
Sit up (times per 30 second)	15±3.9	13±4.3	0.011
Straight jump (cm)	25±5.2	23±5.6	0.000
600 m run (minute)	3.59± 0.73	4.52±1.25	0.000

normal group performing better. The result shows that the level of cardiovascular endurance among the normal children is higher than among the overweight children. Likewise, the fitness level of the normal children is higher (25.0%) than that of the overweight children (6.2%).

9.4.5 Correlation of household factors and nutritional status

The Chi-square tests on socio-economic characteristics and nutritional status only yielded a correlation between the children's pocket money per day and their nutritional status (Table 9.7). Prevalence of obesity was found higher among the children whose pocket money was more than IDR 10,000/day. As observed, during breaks at schoolchildren would spend their pocket money buying junk foods such as salty crispy fried snacks, wafers, chips, candy, and jelly. Some schools are in walking distance of well-known fast food restaurants and there are students who would take their lunch there.

Table 9.7 Correlation between household socio-economic factors and nutritional status.

Variables		Nutritional status				χ^2
		Normal (n=112)		Overweight (n=112)		
		n	%	n	%	
Pocket money/day	High (>IDR10,000)	17	7.6	30	13.4	0.033
	Moderate (≤IDR10,000)	95	42.4	82	36.6	
Number of household members	Big (>4 persons)	64	28.6	63	28.1	0.893
	Small (≤4 persons)	48	21.4	49	21.9	
Fathers' education	Low (≤12 years)	23	10.3	27	12.1	0.521
	High (>12 years)	89	39.7	85	37.9	
Mothers' education	Low (≤12 years)	39	17.4	30	13.4	0.193
	High (>12 years)	73	32.6	82	36.6	
Income per capita	High (≥IDR 1,500,000)	53	23.7	61	27.2	0.285
	Enough(<IDR 1,500,000)	59	26.3	51	22.8	
Mothers' working status	Not working	62	28.7	62	28.7	0.874
	Working	47	21.8	45	20.8	

9.4.6 Factors causing overweight in schoolchildren

The results of the logistic regression analysis show that significant factors in the overweight status of school children are fitness, amount of daily pocket money, mothers' education, and protein adequacy level (Table 9.8). The children who daily had more than IDR 10,000 of pocket money were 2.7 times more at risk of being overweight than children getting less than IDR 10,000 (OR=2.756, CI 95%: 1.260-6.030).

Fitness status proved to be the strongest predictor of obesity. The risk of overweight is 4.5 times greater among the unfit children than the fit children (OR=4.499, CI 95%: 1.758-11.513). The data reflect that more children who were fit had a normal nutritional status, were more agile and moved more easily.

Children of mothers with a low level of education seemed to have lower chance of being overweight at 0.4 times (OR=0.400, CI 95%: 0.176-0.909). In addition, children with a protein adequacy level of more than 120% have a lower overweight risk at 45.8%. The adequacy levels of energy, calcium, iron, vitamin A and vitamin C did not affect overweight.

Table 9.8 Factors affecting overweight.

Variables	B	OR (95% CI)	*P*[1]
Fitness (1=unfit, 0=fit)	1.504	4.499 (1.758-11.513)	0.002*
Pocket money/day (1=>Rp.10,000; 0=≤Rp.10,000)	1.014	2.756 (1.260-6.030)	0.011*
Mothers' education (1=≤12 years, 0=>12 years)	-0.917	0.400 (0.176-0.909)	0.029*
Protein adequacy level (1=>120%, 0=≤120%)	-0.780	0.458 (0.239-0.879)	0.019*
Fathers' education (1=≤12 years, 0=>12 years)	0.413	1.512 (0.672-3.401)	0.318
Number of household members (1=≤4 people, 0=>4 people)	-0.130	0.878 (0.483-1.596)	0.670
Mothers' working status (1=not working, 0=working)	0.437	1.548 (0.808-2.967)	0.187
Habits of having breakfast (1=often, 0=seldom)	-0.315	0.730(0.237-2.245)	0.582
Meal-eating frequency per day (1=3 times, 0=others)	-0.121	0.886 (0.455-1.722)	0.720
Physical activities (1=heavy, 0=light)	-0.900	0.407 (0.048-3.426)	0.408
Energy adequacy level (1=>120%, 0=≤120%)	-0.108	0.898 (0.132-6.110)	0.912
Calcium adequacy level (1=>77%, 0=≤77%)	0.275	1.316 (0.171-10.151)	0.792
Iron adequacy level (1=>77%, 0=≤77%)	0.318	1.374 (0.700-2.698)	0.356
Vitamin A adequacy level (1=>77%, 0=≤77%)	-0.379	0.685 (0.337-1.389)	0.294
Vitamin C adequacy level (1=>77%, 0=≤77%)	0.365	1.441 (0.639-3.251)	0.379
Constant	-0.145		
R^2=0.186			

[1] * = significant at P<0.05.

9.5 Conclusion and discussion

This study found that overnutrition (overweight and obese) in the study group is higher than the national prevalence of 18.8% among 5-12 years old children (MoH, 2013). In this study, overweight among schoolchildren is strongly associated with low physical fitness, high amount of pocket money, high level of the mother's education and low protein adequacy. Among these variables, fitness status is the strongest predictor of overweight among schoolchildren. Overweight affects fitness status, as also found in the study conducted by Kim *et al.* (2005) who showed that physical fitness inversely correlates with overweight among schoolchildren. In this study, schoolchildren with BMI above 80th percentile (overweight) scored lower in fitness tests than those with a normal BMI.

On the other hand, this research could not prove that the level of overweight children's physical activity is lower than the normal children. This is probably because of the similarity in the children's school activities and schedule. Almost all children in the study have a light to moderate physical activities level. They spent most of their time on activities that require only small amounts of energy (like studying and watching TV). Moreover, most parents tend to channel their children's time and energy to activities to improve studies performance, like enrolling them in extra lessons or courses.

The low level of physical activity is also caused by lifestyle changes brought about by technological development. Technology has made life easier. For example, the washing machine and vacuum cleaner lighten the burden of household chores. Apart from this, the practice of hiring domestic help among middle-class families prevents the children from helping with the chores. Technological advancement created many gadgets that make children sit in front of screens. According to Laurson *et al.* (2008), children having little physical activity and more screen time than the recommended one to two hours per day (American Academy of Pediatrics, 2001) were three to four times more prone to be overweight than those having only one to two hours per day screen time. Both the normal and overweight schoolchildren in this study did not achieve the level of daily physical activity of 1.75 as recommended by WHO (1998).

We also found that the high amount of pocket money given to children on school days is a contributing factor to overweight. Kerkadi *et al.* (2005) observed that the prevalence of overweight children increases as the amount of pocket money goes up. Children who get more pocket money tend to spend it on buying street foods, which are calorie-rich but poor in nutrients. A main reason of the parents for giving pocket money to their children is to ensure that they not get hungry in school. Other reasons are reflections of social status. With the money, children can treat their friends and can show they get pocket money (Suci, 2009).

Clearly, the lifestyle changes that seem to engender overweight among children mirror social change and development processes. The parent's current habit to give pocket money to their children instead of sending them off with a meal box may be due to lack of time in combination with the abundance of fast foods in the vicinity. Similarly, the enabling environment for women's access to education has produced more highly educated women, especially in upper-middle class families. Gender equality in education is a global priority (UNESCO, 2011). It is thought that educating girls and women can break the cycle of poverty as they can work and earn money for their family. However, working mothers have been strongly linked to poor eating habits of children and less time for childcare. As this study shows, children of working mothers had poorer eating habits than those whose mothers did not work, which is a disturbing finding.

The children are also often sent to after-school additional lessons or courses. This limits their time for physical activity and lowers their physical fitness. It would be better if the after-school activity were in the form of a sports program. As Elkins *et al.* (2004) found, higher participation in athletic activities by adolescents lowers their BMI. Therefore, after-school sports activities would be a way of preventing overweight among adolescents.

To conclude, social change and development have an impact on the differences between normal and overweight schoolchildren. Overweight children tend to have higher educated mothers, who are likely to be working mothers. As both parents work and family income increases, parents can afford gadgets and after-school study activity for their children. These in turn decrease the children's physical activity level and physical fitness. In addition, parents give their children pocket money to compensate for the lack of time for preparing a healthy meal. All of these factors contributed to the child being overweight. Normal children tended to have lower educated mothers, little pocket money and high physical fitness.

The high prevalence of overnutrition among schoolchildren has to be a concern to both parents and schools. Initiatives to lower the prevalence of overnutrition are called for, such as increasing the children's physical activities at school. WHO recommends children of 5 to 17 years old do exercises at least three times a week for 30 to 60 minutes each time. Following up this recommendation would be a good start.

Acknowledgements

The authors would like to thank the Neys-Van Hoogstraten Foundation for funding the study. We also express gratitude to the principals, teachers, and students of SDN Polisi 1, SDN Polisi 4, SDN Polisi 5, SDN Bantarjati 9, SDIT at Taufiq, and SD Pertiwi for their support in this study. Many thanks go to Puspita Dewi, Rian Diana and Catur Dwi Anggarawati for assisting the project implementation.

References

American Academy of Pediatrics, 2001. Children, adolescents, and television. *Pediatrics* 107: 423-426.

Austin, S.B., Melly, S.J., Sanchez, B.N., Patel, A., Buka, S. and Gortmaker, S.L., 2005. Clustering of fast-food restaurants around schools: a novel application of spatial statistics to the study of food environments. *American Journal of Public Health* 95(9): 1575-1581.

Booth, S.L., Sallis, J.F., Ritenbaugh, C., Hill, J.O., Birch, L.L., Frank, L.D. and Hays, N.P., 2001. Environmental and societal factors affect food choice and physical activity: rationale, influences, and leverage points. *Nutrition Reviews* 59(3): S21-S36.

Boreham, C. and Riddoch, C., 2001. The physical activity, fitness and health of children. *Journal of Sports Sciences* 19: 915-929.

Bovet, P., Auguste, R. and Burdette, H., 2007. Strong inverse association between physical fitness and overweight in adolescents: a large school-based survey. *International Journal of Behavioral Nutrition and Physical Activity* 4(1): 24-31.

Bowman, S.A., Gortmaker, S.L., Ebbeling, C.B., Pereira, M.A. and Ludwig, D.S., 2004. Effect of fast food consumption on energy intake and diet quality among children in a national household survey. *Pediatric* 113(1): 112-118.

Elkins, W.L., Cohen, D.A., Koralewicz, L.M. and Taylor, S.N., 2004. After school activities, overweight, and obesity among inner city youth. *Journal of Adolescence* 27(2): 181-189.

Gillis, L.J. and Bar-Or, O., 2003. Food away from home, sugar-sweetened drink consumption and juvenile obesity. *Journal of American Nutrition* 22(6): 539-545.

Hardinsyah, and Aries, M., 2012. Jenis pangan sarapan dan perannya dalam asupan gizi harian anak usia 6-12 tahun di Indonesia [Type of food for breakfast and its role in daily nutrient intake of school age children 6-12 years in Indonesia]. *Jurnal Gizi dan Pangan* 7(2): 89-96

Hill, J.O., Wyatt, H.R., Reed, G.W. and Peters, J.C., 2003. Obesity and the environment: where do we go from here? *Science* 299: 853-855.

Kerkadi, A., Abo-Elnaga, N. and Ibrahim, W., 2005. Prevalence of overweight and associated risk factors among primary female schoolchildren in Al-Ain city United Arab Emirates. *Emirates Journal of Agriculture Science* 17(1): 43-56.

Kim, J., Must, A., Fitzmaurice, G.M., Gillman, M.W., Chomitz, V., Kramer, E. and Peterson, K.E., 2005. Relationship of physical fitness to prevalence and incidence of overweight among schoolchildren. *Obesity Research* 13(7): 1246-1254.

Kimbro, R.T., Brooks-Gunn, J. and McLanahan, S., 2011. Young children in urban areas: Links among neighborhood characteristics, weight status, outdoor play, and television-watching. *Social Science & Medicine* 72(5): 668-676.

Laurson, K.R., Eisenmann, J.C., Welk, G.J., Wickel, E.E., Gentile, D.A. and Walsh, D.A., 2008. Combined influence of physical activity and screen time recommendations on childhood overweight. *The Journal of Pediatrics* 153(2): 209-214.

Li, M., Xue, H., Jia, P., Zhao, Y., Wang, Z., Xu, F. and Wang, Y., 2017. Pocket money, eating behaviors, and weight status among Chinese children: The childhood obesity study in China mega-cities. *Preventive Medicine* 100: 208-215.

MoH, 2011. *Keputusan Menteri Kesehatan RI Nomor: 1995/MENKES/SK/XII/2010 tentang Standar Antropometri Penilaian Status Gizi Anak*) [Standard anthropometric measurement for children nutritional status]. Jakarta: Ministry of Health.

MoH, 2013. *Riset Kesehatan Dasar 2013* [Results of 2013 Basic Health Research]. Jakarta: Ministry of Health.

Niemeier, H.M., Raynor, H.A., Lioyd-Richardson, E.E., Rogers, M.L. and Wing, R.R., 2006. Fast food consumption and breakfast skipping: predictors of weight gain from adolescent to adult. *Journal of Adolescent Health* 39(6): 842-849.

Oren, A., Vos, L.E., Uiterwall, C.S., Gorissen, W.H., Grobbee, D.E. and Bots, M.L., 2003. Change in body mass index from adolescence to young adulthood and increased carotid intima-media thickness at 28 years of age: the atherosclerosis risk in young adults study. *International Journal of Obesity and Related Metabolic Disorders* 27(11): 1383-1390.

Patrick, H. and Nicklas, T.A., 2005. A review of family and social determinants of children's eating patterns and diet quality. *Journal of the American College of Nutrition* 24(2): 83-92.

Pereira, M.A., Kartashov, A.I., Ebbeling, C.B., Horn, L.V., Slattery, M.L., Jacobs Jr, D.R. and Ludwig, D.S., 2005. Fast-food habits, weight gain, and insulin resistance (the CARDIA study): 15-year prospective analysis. *The Lancet* 365(9453): 36-42.

Pramudita, R.A., 2011. *Faktor risiko obesitas pada anak sekolah dasar di kota Bogor* [Obesity risk factors of elementary school children in Bogor]. [Unpublished thesis]. Bogor: Institut Pertanian Bogor.

Reilly, J.J. and Kelly, J., 2011. Long-term impact of overweight and obesity in childhood and adolescence on morbidity and premature mortality in adulthood: systematic review. *International Journal of Obesity* 35: 891-898.

Song, C., Goto, K., Hu, M. and Wolff, C., 2015. The impact of social and cultural factors on the consumption of western fast food among Chinese children. *Community Development Journal* 50(4): 709-723.

Suci, E.S.T., 2009. Gambaran perilaku jajan murid sekolah dasar di Jakarta [The snack behaviour of elementary school students in Jakarta]. *Psikobuana* 1(1): 29-38.

Suryaalamsyah, I.I., 2009. *Konsumsi* fast food *dan faktor-faktor yang berhubungan dengan kegemukan anak sekolah di SD Bina Insani Bogor* [Fast food consumption and factors associated with obesity in schoolchildren of SD Bina Insani Bogor] [unpublished thesis]. Bogor: Institut Pertanian Bogor.

UNESCO, 2011. Education and gender equality. Paris: United Nations Educational, Scientific and Cultural Organization. Available at: http://tinyurl.com/y78k47fq.

Uter, J., Scragg, R., Mhurchu, C.N. and Schaaft, D., 2007. At-home breakfast consumption among New Zealand children: Associations with body mass index and related nutrition behaviours. *Journal of the American Dietetic Association* 107(4): 570-576.

Vandewater, E.A., Shim, M. and Caplovitz, A.G., 2004. Linking obesity and activity level with children's television and video game use. *Journal of Adolescence* 27: 71-85.

WHO, 1998. *Obesity preventing and managing the global epidemic*. Geneva: World Health Organization. Available at: http://apps.who.int/iris/handle/10665/63854.

WHO, 2000. *Obesity: preventing and managing the global epidemic. WHO Obesity Technical Report series 894*. Geneva: World Health Organization. Available at: http://www.who.int/nutrition/publications/obesity/WHO_TRS_894/en.

WHO, 2006. *WHO child growth standards: length/ height-for-age, weight-for-age, weight-for-length, weight-for-height and body mass index-for-age: methods and development.* Geneva: World Health Organization. Available at: http://www.who.int/childgrowth/standards/technical_report/en.

Wright, C.M., Parker, L., Lamont, D. and Craft, A.W., 2001. Implications of childhood obesity for adult health: findings from a thousand families cohort study. *British Medical Journal* 323: 1280-1288.

10. Climate change and nutritional status of preschool children living in Infanta, Quezon, and Los Baños, Laguna, Philippines

M.T.M. Talavera[1], M.R.L. Pangan[2], J.O. Colladilla[3], J.N. Sarol[4], I.A. Agdeppa[5] and F.M.E. Lorenzo[2]*

*[1]Institute of Human Nutrition and Food, College of Human Ecology, University of the Philippines Los Baños, The Philippines; [2]College of Public Health, University of the Philippines, Manila, The Philippines; [3]School of Environmental Science and Management, University of the Philippines Los Baños, The Philippines; [4]National Teacher Training Center for the Health Professions, University of the Philippines Manila, The Philippines; [5]Food and Nutrition Research Institute, Department of Science and Technology, Manila, The Philippines; *mmtalavera@up.edu.ph*

Abstract

Both climate change and undernutrition affect young children. Climate change can contribute to the progression of undernutrition and can exacerbate existing conditions. Millions of people are affected when these phenomena occur. This study explored the relationship between climate variables and the nutritional status of young children from farming households in Infanta (Quezon) and Los Baños (Laguna) in the Philippines. Rainfall and temperature data were used to characterise the long-term climate conditions in the study areas. In addition, we collected data on children's nutritional status and household characteristics through interviews with primary caregivers. The multiple logistic regression test showed that rainfall and temperature are not significantly related to stunting of children; location was the only variable that could explain stunting. Likewise, location was identified a factor that explains both underweight among children and rainfall. Children in Infanta were more at risk of being stunted and underweight than those in Los Baños. Infanta has a climate characterized by lower temperatures and higher rainfall than Los Baños. This study shows that climate change, particularly increased rainfall, can affect the nutritional status of children through higher incidence of illness. Thus, knowledge on climate variables is important to select and design appropriate strategies that take into account the effects of rainfall to help improve child nutrition.

Keywords: climate change, undernutrition, climate change adaptation

10.1 Introduction

Undernutrition and climate change are global problems that are intricately linked and affect millions of people. According to the Global Humanitarian Forum (GHF, 2009), climate change is estimated to affect 325 million people and cause 300,000 deaths every year. By

DOI 10.3920/978-90-8686-864-3_10,

2030 and 2050, climate change is expected to cause approximately 250,000 additional deaths per year due to malnutrition, malaria, diarrhoea, and heat stress. Approximately 155 million children under five are estimated to be stunted and 52 million wasted (UNICEF/WHO/World Bank Group, 2017). It is expected that climate change will lead to an increase in moderate stunting of 1-29% in 2050 and that it will have a greater impact on rates of severe stunting with an estimated increase of 23 to 62% in sub-Saharan Africa and South Asia, respectively (Lloyd *et al.*, 2011). The prevalence of undernutrition varies according to geographical location. Populations and especially children in developing countries with weak health infrastructures will be affected the most by climate change (Ebi and Paulson, 2007; Wijaya, 2014).

In the Philippines, climate change and malnutrition are serious problems. On average, the country has 20 typhoon occurrences per year, resulting in casualties and damage to property. It ranks eighth in the world with respect to the absolute number of people annually exposed to floods, and tenth on the Mortality Risk Index for landslides, earthquakes, and volcanic eruptions (ISDR, 2009). In terms of malnutrition, recent surveys showed that 19.9% of children under five were underweight, 30.3% were stunted and 7.9% wasted (FNRI, 2013). There is a decline in the prevalence of malnutrition, but it is slow, particularly regarding stunting. Factors known to cause malnutrition in children are low socio-economic status, poorly educated parents, and poor environmental conditions including climate change. Climate factors can exacerbate the nutrition situation through different pathways such as through food security, sanitation, water and food safety, and child health and care practices (Krishnamurthy *et al.*, 2015).

The relationship between undernutrition and climate change has not received much attention, even though determinants of undernutrition such as food, health, sanitation, water, and care practices, are known to be affected by climate change (ACF, 2012; Crahay *et al.*, 2010). Moreover, the nutrition sector is poorly linked to emerging policies and practices that address climate change. Nutrition also remains a secondary theme in the scientific discourse on climate change, agriculture and food security, risk management and social protection (Tirado *et al.*, 2015). So far, studies on establishing evidence on the occurrence, projection, and impacts of climate change, were conducted mostly at the global and national levels. Hence, there is a need for data that assess the situation at the household level. As yet, there is also scant research about the effects of climate change on Filipino children (Urbano *et al.*, 2011).

This study aimed to addresses these shortcomings by investigating the following questions: How do climate change variables relate to the nutritional status of preschool children in the municipalities of Infanta, Quezon and Los Baños, Laguna in the Philippines? How does this relationship differ between the two areas and why? What are the possible adaptation measures to augment climate change impacts on preschool children?

10.2 Conceptual framework

This study was guided by the conceptual framework shown in Figure 10.1.

There are several inter-related factors that are immediate, underlying, or basic causes of undernutrition (UNICEF, 1990). At the household level, these include food insecurity, inadequate childcare, poor health services, and an unhealthy environment. At the individual level, causal factors relate to food intake, health status and hygiene and sanitation practices. It is also recognized that deficient human and environmental resources contribute to malnutrition. Climate change can exacerbate hunger and undernutrition (ACF, 2012). Food and nutrition insecurity make people vulnerable, and thus less able to cope and adapt to climate change. Climate is defined as the average of weather over time and space (SEAMEO, 2010). Weather and climate have different time frames. Weather reflects the short-term conditions of the atmosphere, whereas climate is the average daily weather for an extended period at a certain location (NASA, 2005). Climate change is defined by the Intergovernmental Panel on Climate Change as a change in the state of the climate that can be identified by changes in the mean and/or the variability of its properties and that persist for an extended period, typically for decades or longer (IPCC, 2012).

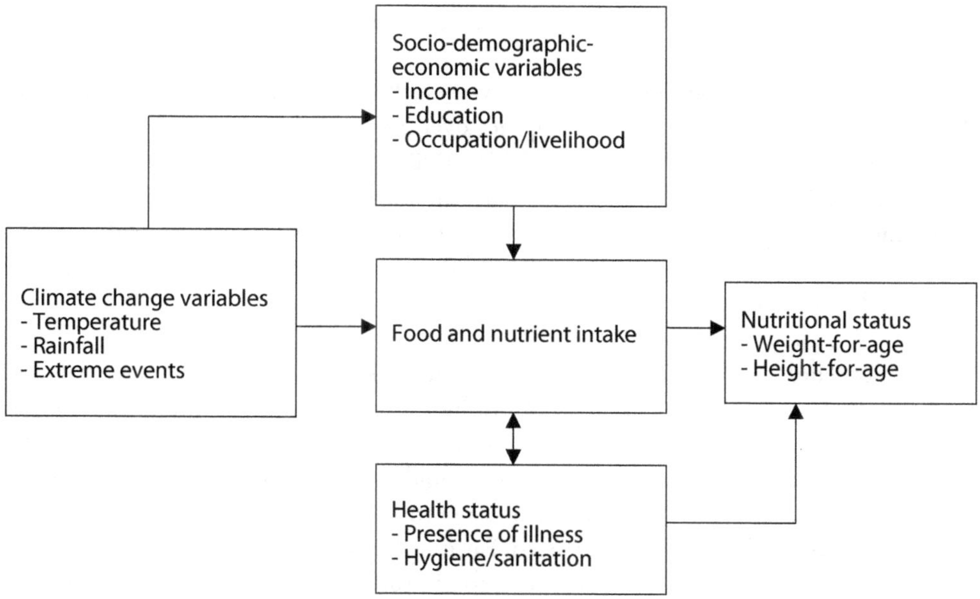

Figure 10.1. Conceptual framework.

As in many parts of the world, also in the Philippines climate change can be noted. In terms of temperature, the mean annual temperature in the Philippines is 26.6 °C (PAGASA, 2011). The coolest month is January, with a mean temperature of 25.5 °C and the warmest month is May, with a mean temperature of 28.3 °C. In the last 50 years, a mean temperature increase of 0.01 °C per year from 1951 to 2010 was observed. Moreover, the maximum (daytime) and minimum (night time) temperatures also increased by 0.36 °C and 0.1 °C, respectively. Abrupt temperature increases began in the second half of the 1980s and significantly high temperatures were recorded since 1995 (Thomas *et al.*, 2013). Using the Providing Regional Climates for Impact Studies (PRECIS) model, a regional climate modelling system, predicted a likely temperature increase of 0.9-1.1 °C by 2020 and of 2.0-2.2 °C by 2050 (PAGASA, 2011). The projection also indicates that Luzon (where the study areas are located) will not be as much affected as Mindanao. Mean annual rainfall in the Philippines varies from 965 millimetres (in sheltered valleys) to 4,064 millimetres (in mountainous areas). To date, the highest recorded annual rainfall was 1,085.8 mm during typhoon Feria in 2001 (De Guzman, 2014). The variation in rainfall distribution across the country is explained by the direction of the moisture-bearing winds and the location of the mountain systems (De Guzman, 2014). PAGASA (2011) forecasted a change in annual precipitation from -0.5 to 17.4% in 2020 and from -2.4 to 16.4% in 2050. The average annual rainfall increase over most parts of Luzon and Visayas is expected to be at 2-17% by 2020 and 1-16% by 2050. Luzon and the Visayas are expected to experience increased rainfall, Mindanao a reduction in rainfall by about 0.5%.

On average, about 20 typhoons occur every year in the Philippines. The average number of tropical cyclones entering the Philippines over 30 years showed variability: 19.8 in 1951-1980, 20.2 in 1961-1990, 19.5 in 1971-2000, and 18.8 in 1981-2000. In the last two decades (1991 to 2010), typhoons have shown increased gustiness (intensity) that has caused more damage to the population, agriculture, and infrastructure. Rainfall brought about by typhoons leads to flooding and landslides. This situation is compounded by the effects of climate change.

Farming households produce and consume food, but are known to experience seasonal food insecurity. Though having the capacity to produce food for others as well as themselves, farming households have been observed as nutritionally at-risk in terms of having underweight children. Food security at the household level does not automatically translate into the nutritional security of each member; even in a food-secure household, not all members always get the nutrients they need. Farming households have to be ready to adapt to climate change and its various impacts on food production, health, and nutrition. The young and the elderly are especially vulnerable to sudden changes in temperature, rainfall, and extreme weather events. Changes in the livelihood and food production capacity of households compounded by the onset of changes in climate variables can lead to food insecurity, poor feeding, and consequently, low nutritional outcomes among household members, with young children most readily affected.

10.3 Research area and methodology

10.3.1 Study areas

The two provinces of Laguna and Quezon were selected, based on climate type. In each province, one municipality was selected based on proximity to a PAGASA station where climate data are collected. In Laguna, the station is in the municipality of Los Baños, in the Los Baños Campus of the University of the Philippines. In Quezon, it is located in Infanta municipality. Proximity to the stations helped with data collection. The Philippines have four types of climate. Laguna has a Type I climate with two pronounced seasons, dry from November to April and wet during the rest of the year. The maximum rain period is from June to September. Quezon has a Type II climate: no dry season, with a very pronounced maximum rain period from December to February and minimum monthly rainfall from March to May.

10.3.2 Study design, participants and sampling

This study used a cross-sectional design. A basic limitation of this design is that it is difficult to prove causal sequences, since exposure to the causal factors occurred before the collection of outcome data. By collecting data in two different climate zones (see above), we tried to strengthen the explanatory power of the study.

Participants of the study were farming households, with the primary caregivers as the respondents. The primary caregivers were either mothers or somebody who prepares the food and feeds the child when the mother was not available (grandmother, aunt, or a sibling). The households were randomly selected from a list kept by the municipal nutrition officer.

Data were collected using a combination of techniques. Climate change variables, that is, temperature, rainfall, and extreme weather events were collected from the PAGASA stations. A structured questionnaire was used collect data on socio-demographic, economic status, and illness experience. The nutritional status of children was assessed by measuring their weight and height using standard techniques, which was done monthly during six months. Prior to data collection, the questionnaire was pretested and revised. The enumerators were trained to ensure that they would use the techniques correctly. During the data collection, monitoring visits were conducted to check on the correct measurement of the children's weight and height.

Data were analysed using the analytical computer package SPSS version 16.0 and Microsoft Office Excel 2003. Descriptive statistics were first computed, followed by tests of associations and then multiple logistic regression. Analytical statistics were employed to determine the relationship among the variables studied. Descriptive statistics were used to document the temperature and rainfall in the two areas. The nutritional status of children was determined using weight-for-age (which reflects both past and present undernutrition) and height-for-age (index of stunting or past growth failure). Weight-for-age z-scores and height-for-age z-scores were determined using the WHO-CGS reference charts. Children with weight-for-age z-scores and height-for-age z-scores below -2SD were classified as underweight and stunted, respectively. The children's health status was determined by asking the mothers/caregivers if the child got sick last month. The percentage of children with/without illnesses was computed monthly.

Pearson's correlation (r) was used to test the association between: (1) rainfall/temperature and health status; (2) rainfall/temperature vs stunting and underweight. Fisher's exact probability test was used for variables such as family size, mothers' and fathers' age, educational attainment, employment status, primary occupation, and monthly income. The correlation coefficient (r) ranges from +1.0 to -1.0, r>0 indicating a positive linear relationship, r<0 a negative one, and r=0 no relationship. The Pearson correlation coefficients were interpreted as follows: 0 no linear relationship; 0.01-0.20 very weak linear relationship; 0.21-0.40 weak linear relationship; 0.41-0.60 moderate linear relationship; 0.61-0.80 strong linear relationship; 0.81-0.99 very strong linear relationship; and 1 perfect relationship. Multiple Logistic Regression with multiple predictors was conducted following the model below.

Given:
Multiple response variable – Y (nutritional status – normal, underweight)
K explanatory variables X1, ...Xj, Dj+1, ..., Dk (continuous and categorical)
where:

$$\pi(Y=1) = \frac{e^{\alpha+\beta_1 x_1 + \dots + \beta_j x_j + \beta_{j+1} D_j + 1 + \dots + \beta_k D_k}}{1 + e^{\alpha+\beta_1 x_1 + \dots + \beta_j x_j + \beta_{j+1} D_j + 1 + \dots + \beta_k D_k}}$$

Y = nutritional status (0=normal, 1=underweight)
= probability that nutritional status = 1
α = regression constant
βi = regression coefficient
xj = j^{th} independent variable
Dk = k^{th} dummy variable

10.4 Results

10.4.1 Climate situation in the study sites

Mean annual and monthly temperature

The PAGASA dataset for temperature revealed that the 36-year mean annual temperature was almost the same at 27.9 °C in Los Baños and 27.8 °C in Infanta. In Los Baños, the lowest mean annual temperature was 26.6 °C recorded in 1989, and the highest was 28.0 °C in 1998. The lowest mean annual temperature recorded in Infanta was 26.6 °C from 1977 to 1981, and the highest was 27.8 °C reported in 1998 and 2005.

As shown in Figure 10.2, the mean annual temperatures in the two areas have inter-annual (changes in temperature from year to year) and inter-decadal (changes in temperature between decades) variability. High temperatures were noted in 1987-1988 and 1998 in both areas. The increasing trend was more apparent in Infanta. Between 2000 and 2010, the mean temperature in Infanta was notably higher than in Los Baños. The increasing trend in temperature was more evident in Infanta than in Los Baños based on the linear trend line. The 5-year average trend in Los Baños and Infanta is increasing. Los Baños consistently had higher temperatures than Infanta; however, from 2001 to 2009, Infanta's temperatures were higher than in Los Baños.

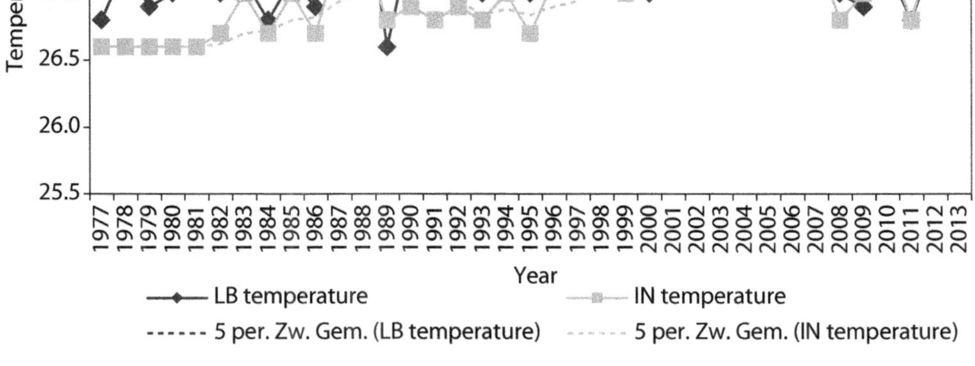

Figure 10.2. Mean annual temperature, 36-year period, Los Baños and Infanta.

The mean monthly temperatures in the 36-year period of the two study areas are shown in Figure 10.3. The highest temperature in Los Baños was 29.3 °C in April, the lowest 24.7 °C in December. In Infanta, the highest temperature noted was 28.0 °C in October, the lowest 24.7 °C in January. The trend shows that Los Baños is warmer than Infanta around the months of November to April, but from May to October, however, this was reversed with Infanta being warmer.

Figure 10.4 and 10.5 compare the monthly temperatures over a 6-month period and those over a 36-year period. In Los Baños, the monthly temperature over 6 months was consistently lower than that of the mean monthly temperature over 36 years, except in December (26.0 °C) and April (29.7 °C). January was observed to have the lowest mean temperature (23.8 °C) over the 6-month period and December (24.7 °C) over the 36-year period. By February, it was observed that the temperature started to increase until April.

In Infanta, the 6-month mean monthly temperature closely approximated that of the 36-year mean monthly temperature. The lowest mean temperature was recorded in January in both periods (24.1 °C for a 6-month period and 24.7 °C for 36-year monthly period). In both areas, the mean temperature during the 6-month study period closely resembled that of the 36-year mean monthly temperature. The 6-month temperature was reflective of the 36-year change in temperature. During the 6-month study period, April had the highest reported monthly temperature at 29.7 °C in Los Baños and 27.6 °C in Infanta (Figure 10.6). In both areas, January was observed to have the lowest temperature recorded (23.8 °C in Los Baños and 24.1 °C in Infanta). Infanta tended to have a lower mean monthly temperature than

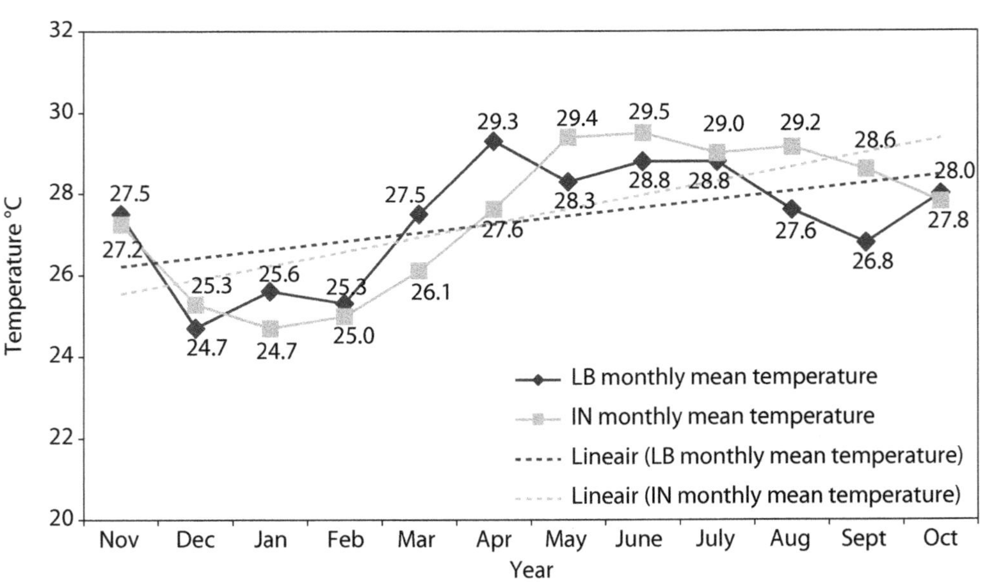

Figure 10.3. Mean monthly temperature, 36-year period, Los Baños and Infanta.

Los Baños based on the linear trend line. Starting in February, in both areas the temperature increased until April 2014.

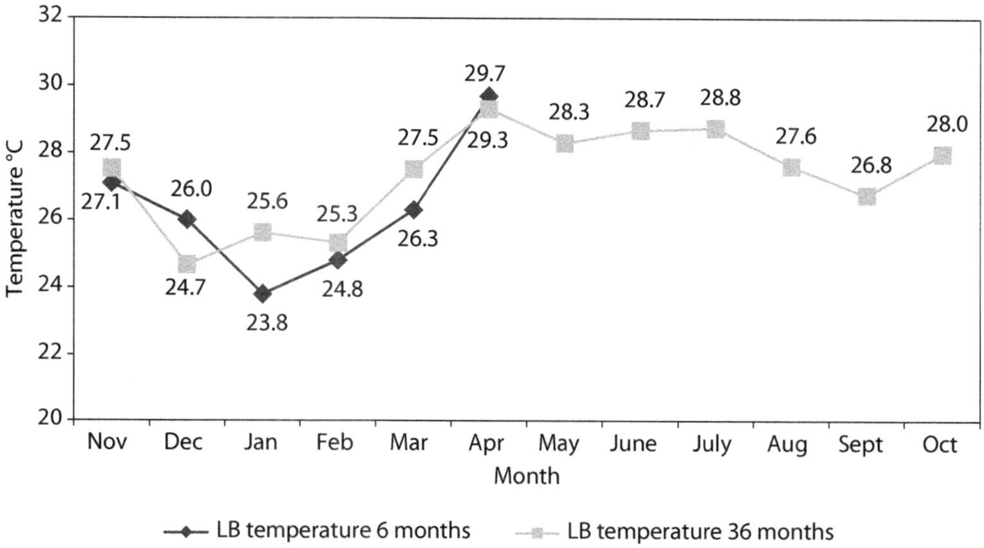

Figure 10.4. Mean monthly temperature, 36-year and 6-month periods, Los Baños.

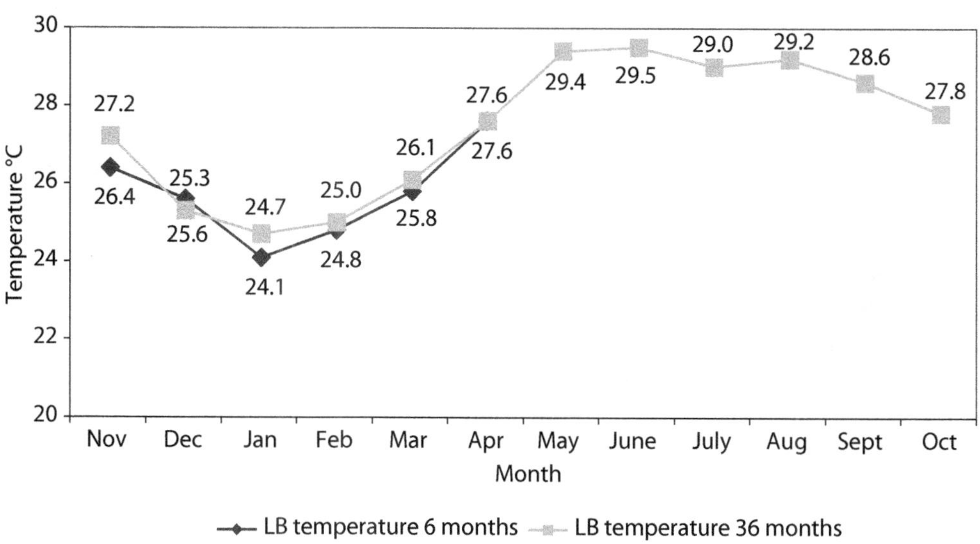

Figure 10.5. Mean monthly temperature, 36-year and 6-month periods, Infanta.

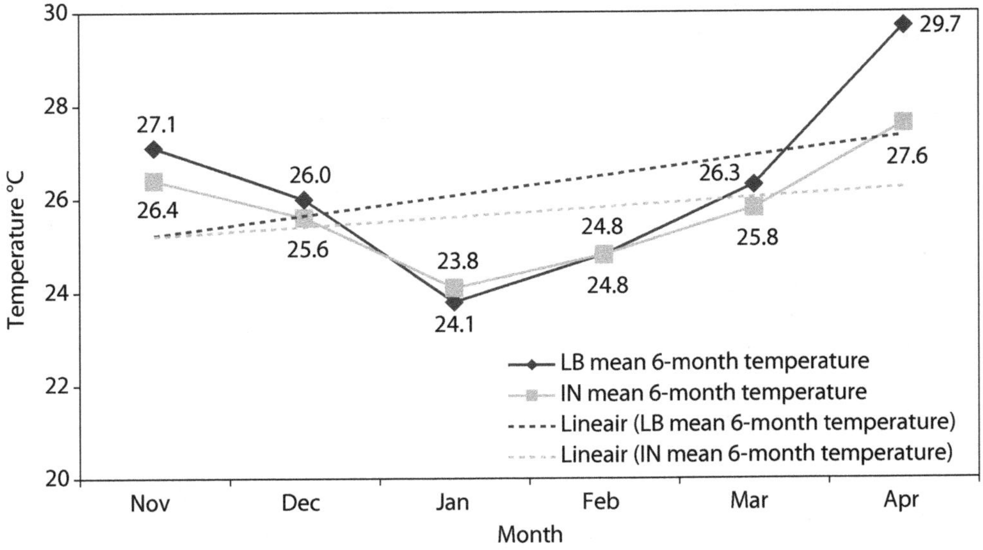

Figure 10.6. Mean monthly temperature, 6-month period, Los Baños and Infanta.

Mean annual and monthly rainfall

The mean annual rainfall during the 36-year period was 173.5 mm in Los Baños and 349.9 mm in 1987. In Infanta, the mean annual rainfall was lowest in 2004 (197.9 mm) and highest (507.1 mm) in 2008. As seen in Figure 10.7, a similar pattern can be noted on a yearly basis. The amount of rainfall was always higher in Infanta than in Los Baños. The 5-year average trend indicates the increasing rainfall in both areas. Similar to temperature, inter-annual and inter-decadal variations were noted. The 5-year average temperature is also presented to show the normal trend in temperature.

As seen in Figure 10.8, the highest amount of rainfall in Infanta was 655.8 mm recorded in December, whereas the lowest was in August at 193.6 mm. In Los Baños, the highest amount of rainfall was 293.5 mm in July and the lowest was 32.7 mm in February. The amount of rainfall on a per month basis was higher in Infanta than in Los Baños, except in July and August. In both areas, high amounts of rainfall were experienced in November and December, followed by a decline from January to April, and then an increasing trend by May. The linear trend in Los Baños is increasing from November to October, whereas a decreasing trend can be observed in the case of Infanta.

During the 6-month study period, the mean monthly rainfall in Los Baños was generally lower than the values obtained for Infanta (Figure 10.9). Unlike for temperature, in which the differences between the two study areas were small, the differences in amount of

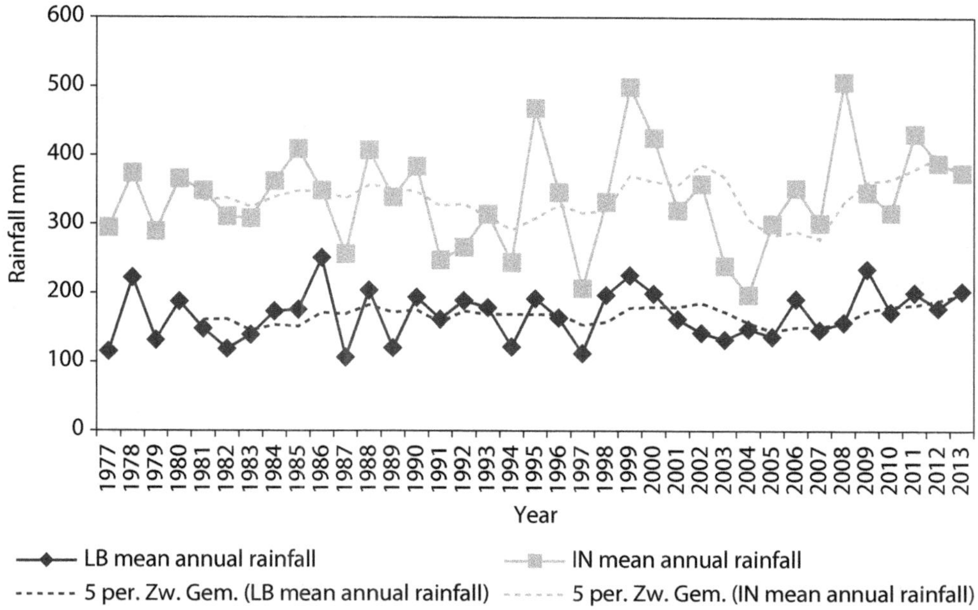

Figure 10.7. Mean annual rainfall, 36-year period, Los Baños and Infanta.

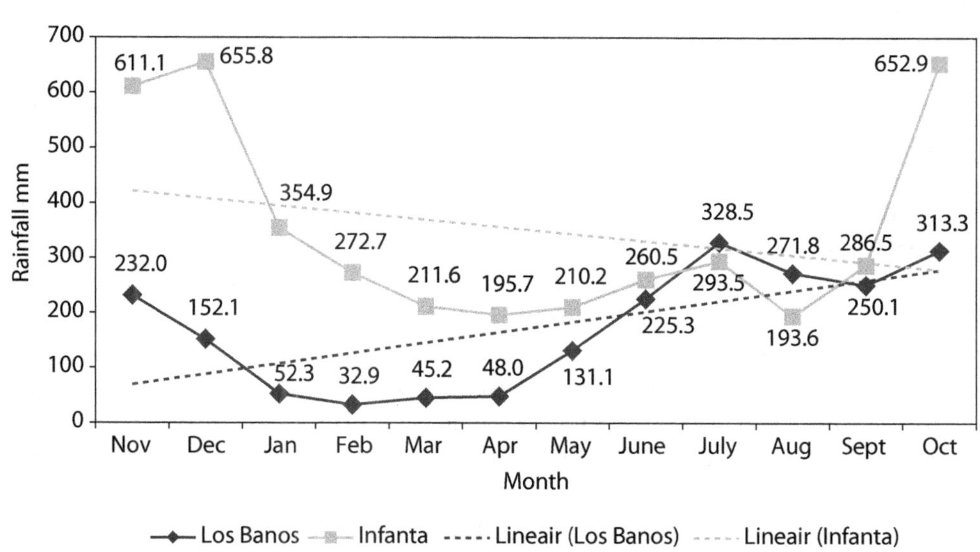

Figure 10.8. Mean monthly rainfall, 36-year period, Los Baños and Infanta.

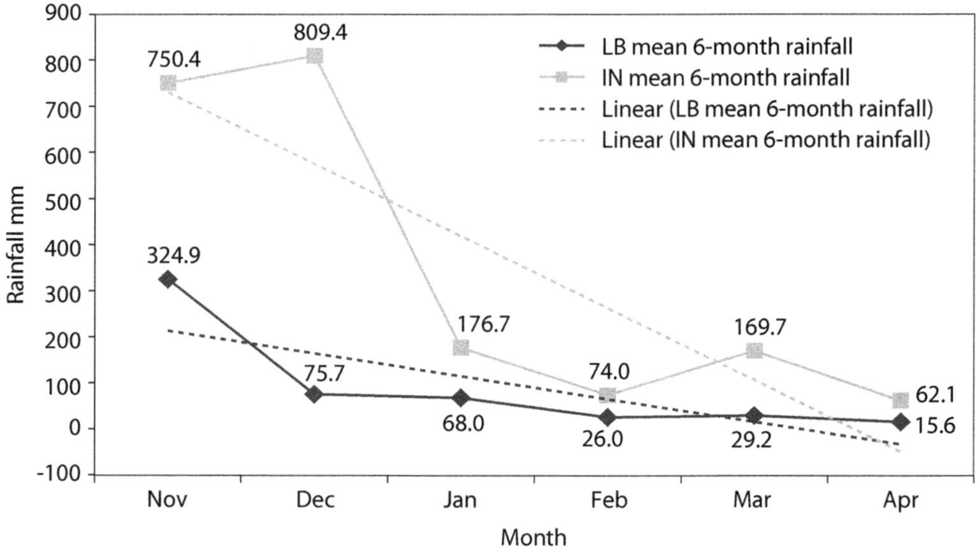

Figure 10.9. Mean monthly rainfall, 6-month period, Los Baños and Infanta.

rainfall were markedly high. In December 2013, the mean rainfall was 809.4 mm in Infanta compared to only 75.7 mm in Los Baños. The lowest means of rainfall for Los Baños (15.6 mm) and Infanta (62.1 mm) were recorded in the same month, April 2014.

Figure 10.10 and 10.11 show a comparison of the mean monthly rainfall of the two periods in the two study areas.

The 6-month mean monthly rainfall in Los Baños followed the trend of the mean monthly rainfall from November to April over a 36-year period, which is characterised by decreasing amount of rainfall during those months. The amount of rain in the 6-month period (324.9 mm) was higher than that in the 36-year period in November (232.0 mm). A sudden drop in the amount of rainfall in December was noted during this period (75.7 mm) compared with the 36-year period (152.1 mm). The lowest amount of rainfall was recorded at 15.6 mm in April (6-month period) compared to 32.9 mm in February (36-year period). A similar trend can be noted in the amount of rainfall in Infanta from November to April during both periods. There was an increase in rainfall in November from 611.1 mm to 655.8 in December, followed by a sharp decline in January from 655.8 mm to 354.9 mm. However, the amount of rainfall over the 6-month period was lower from January to April compared to that over the 36-year period. In both areas, the mean amount of rainfall during the 6-month study period closely approximated that of the 36-year mean monthly rainfall. The amount of rainfall over the 6-month period reflected the 36-year change in the amount of rainfall experienced, characterised by decreasing amounts of rainfall during those months.

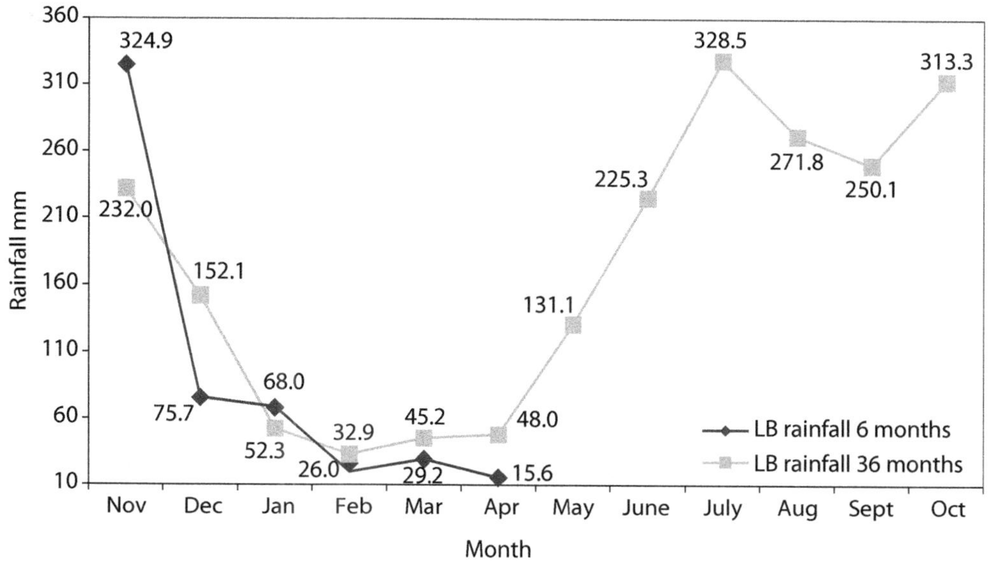

Figure 10.10. Mean monthly rainfall, 36-year and 6-month periods, Los Baños.

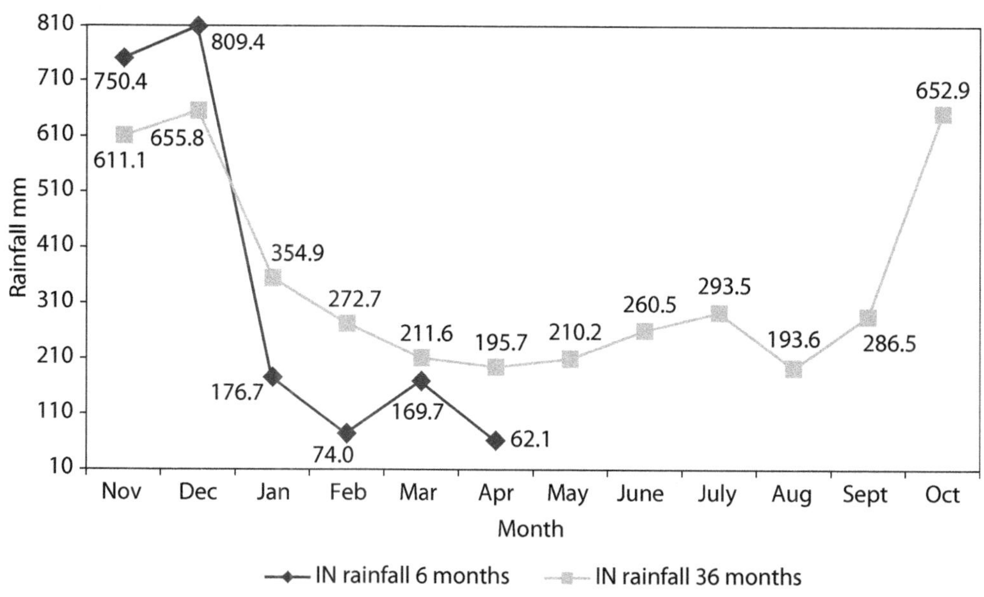

Figure 10.11. Mean monthly rainfall, 36-year and 6-month periods, Infanta.

The amount of rain in the 6-month period (324.9 mm) was higher than that in the 36-year period in November (232.0 mm). A sudden drop in the amount of rainfall in December was noted during this period (75.7 mm) compared with the 36-year period (152.1 mm). The lowest amount of rainfall was recorded at 15.6 mm in April (6-month period) compared to 32.9 mm in February (36-year period).

10.4.2 Undernutrition situation in the study areas

Table 10.1 presents some background indicators of the children in the study. There were more Los Baños households (69.3%) in the income bracket of Php1000 to Php5,000 (US$20-100) than in Infanta (58.7%). Household size was larger in Infanta than in Los Baños. Household appliances are indicators of economic status. There were more of these in Infanta than Los Baños households, with appliances such as television (81.1 vs 65.3%), VCD/DVD player (42.3 vs 29.7%), and refrigerator (18.6 vs 5.0%).

Table 10.1. Percentage distribution of socio-demographic characteristics of households.

Socio-demographic characteristics	Los Baños (%)	Infanta (%)	Total (%)
Household size			
<5	56.7	56.7	54.0
6-10	45.5	1.0	23.7
>10	2.9	42.3	22.2
Total	100.0	100.0	100.0
HH average monthly income			
0	1.9	1.0	1.5
1-999	27.7	39.2	33.3
1000-5,000	69.3	58.7	64.1
5,001-10,000	0.9	1.0	1.0
Total	100.0	100.0	100.0
Type of household			
Nuclear	71.3	71.1	71.2
Extended	28.7	28.9	28.8
Total	100.0	100.0	100.0
With household appliances (multiple responses)			
TV	65.3	81.4	73.2
Refrigerator	5.0	18.6	11.6
Cable TV	3.0	11.3	7.1
VCD/DVD player	29.7	42.3	35.9
Cell phone	80.2	86.6	62.1

Table 10.2 presents data on educational attainment, employment status and income of the parents of the children in Infanta and Los Baños. In Infanta, there were more mothers and fathers with high school education (55%) than in Los Baños (33%), but also more unemployed mothers (81.4%) than in Los Baños (75.2%). Among those who were employed in Los Baños, 11% worked in farms, 6% in the fishing sector, and 7% in vending. The rest were employed as cook, factory worker, secretary, or *barangay* employee. In Infanta, among fathers, all except a few had work related to farming. Infanta fathers also had higher incomes than Los Baños fathers did.

Table 10.3 shows the profile of the children under study. The mean ages of the Los Baños and Infanta children were 21.7 months and 20.3 months, respectively. Infanta had more children in the age group between 12 and 23 months (69.1%) than Los Baños (55.5%) and a higher percentage of female children (53.6 vs 48.5%, respectively). Children in the age group 12 to 35 months are vulnerable to malnutrition, varying according to personal characteristics, exposure to risks, levels of resilience to climate effects, and care (Confalonieri

Table 10.2. Percentage distribution of children's parents educational attainment and income.

Educational attainment and income	Mother		Father	
	Los Baños	Infanta	Los Baños	Infanta
Educational attainment				
Elementary	31.6	18.6	39.7	24.7
High School	59.4	57.7	47.5	59.8
College	9.0	23.7	10.8	15.5
Vocational	0	0	20.0	0
Total	100.0	100.0	100.0	100.0
Employment status				
Employed	24.8	18.6	98.0	100.0
Unemployed	75.2	81.4	2.0	0
Total	100.0	100.	100.0	100.0
Monthly income (Php)				
0	74.0	82.5	0.9	1.0
1-999	4.0	0	11.8	0
1000-5,000	13.0	16.5	60.4	3.1
5,001-10,000	7.0	0	25.7	82.5
10,001-15,000	1.0	0	0	0
15,001-20,000	1.0	1.0	0.9	13.4
Total	100.0	100.0	100.0	100.0

et al., 2007; Ebi and Paulson, 2007; Llamas-Clark, 2011; PAGASA, 2011). They are also at risk of faltering growth, which occurs between the ages of 6 and 24 months when the child is no longer protected by exclusive breastfeeding and is more exposed to disease and infection through contaminated food or water (UNICEF, 1990). Even a child who is adequately nourished at a later age, is unlikely to recover the growth that was lost in the first two years as a result of malnutrition.

Figure 10.12 presents the data on underweight and stunted children in the two locations. The figure shows that the prevalence of children with stunted growth in Los Baños (48.7%) was almost twice that in Infanta (27.0%). Likewise, the prevalence of underweight children in Los Baños (16.3%) was higher than in Infanta (9.6%). The highest prevalence of stunting (55.8%) was recorded in Los Baños in the month of February. In Infanta, the highest prevalence rates of 28.8 and 28.1% were recorded in November and January, respectively. Monthly prevalence of stunting in Los Baños was higher than that in Infanta. The lowest rate of prevalence of underweight children in Infanta (6.7%) was recorded in March, the highest (13.9%) in November. In Los Baños, the lowest rate was in January (12.4%), the highest rate in April (22.0%).

Table 10.4 shows that the differences noted in the prevalence of stunting and being underweight among children in the two areas during the 6-month period were statistically significant. Prevalence of stunting in Los Baños and Infanta significantly differed during the study period. However, the prevalence of underweight children was significantly different during the months of December, February, March, and April.

Table 10.3. Characteristics of children, Los Baños and Infanta.

Age and sex of the children	Los Baños, Laguna		Infanta, Quezon		Total	
	No.	%	No.	%	No.	%
Age in months						
12-23	55	55.5	67	69.1	122	62.2
24-35	44	44.5	30	30.9	74	37.8
Total	99	100.0	97	100.0	196	100.0
Sex						
Girls	48	48.5	52	53.6	100	51.0
Boys	51	51.5	45	46.4	96	49.0
Total	99	100.0	97	100.0	196	100.0

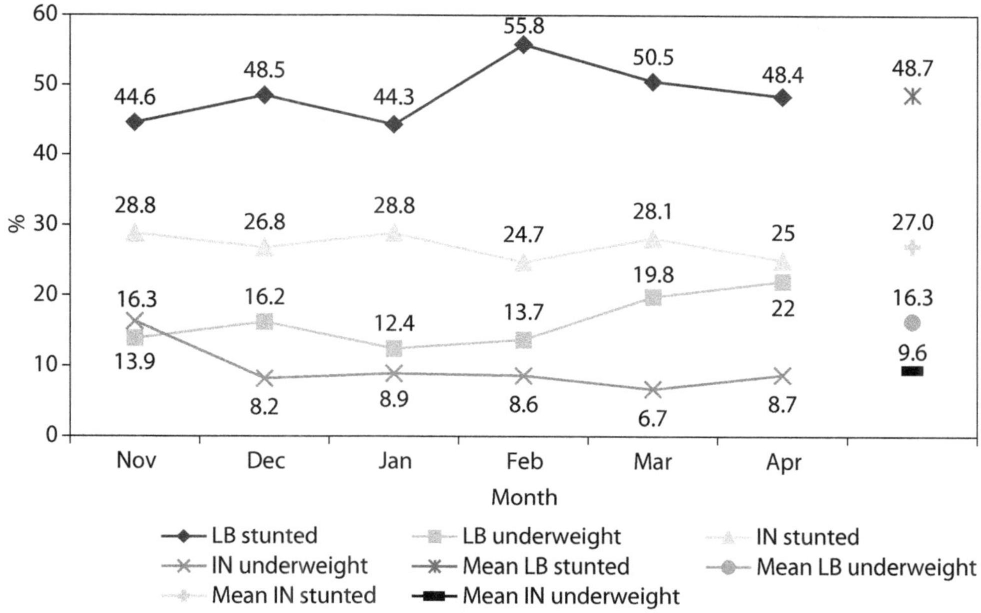

Figure 10.12. Percentages underweight and stunted children over 6 months, Los Baños and Infanta.

Table 10.4. Differences in the prevalence of stunting and being underweight.[1]

Months	Stunting			Underweight		
	Los Baños	Infanta	*P*-value	Los Baños	Infanta	*P*-value
November	0.4639	0.2929	0.0135*	0.2400	0.1667	0.1936
December	0.5000	0.2688	0.0011*	0.1753	0.0842	0.0615**
January	0.4674	0.2360	0.0011*	0.1771	0.1099	0.1902
February	0.5699	0.2556	0.0000*	0.2021	0.0879	0.0278*
March	0.5227	0.2644	0.0005*	0.1978	0.0690	0.0117*
April	0.5057	0.2556	0.0006*	0.2198	0.0879	0.0135*

[1] * = correlation significant at the 0.05 level of significance (2-tailed); ** = correlation is significant at the 0.10 level of significance (2-tailed).

Table 10.5 presents the results of Pearson's correlation test to determine the association between rainfall and temperature and nutritional status. In Los Baños, rainfall is associated with underweight but not in Infanta.

Table 10.5. Relationship between climate variables and nutritional status.[1]

Nutritional status	Los Baños				Infanta			
	Rainfall		Temperature		Rainfall		Temperature	
	R	*P*-value	R	*P*-value	R	*P*-value	R	*P*-value
Stunting	-0.025	0.557	0.003	0.935	0.034	0.423	0.008	0.848
Underweight	-0.118	0.005**	-0.017	0.68	-0.02	0.64	0.027	0.531

[1] ** = correlation is significant at the 0.01 level of significance (2-tailed).

Data on illness as reported by the mothers over the 6-month period are shown in Figure 10.13. In Los Baños, the percentage of children who became ill was highest in January (71.7%) and lowest in March (36.8%). In contrast, the peak month for Infanta was in November (62.5%) and the lowest was in April (29.7%). More children fell ill in Los Baños than in Infanta on a monthly basis except for the months of November (59.4% in Los Baños, 62.5% in Infanta) and March (36.8% in Los Baños, 45.2% in Infanta.

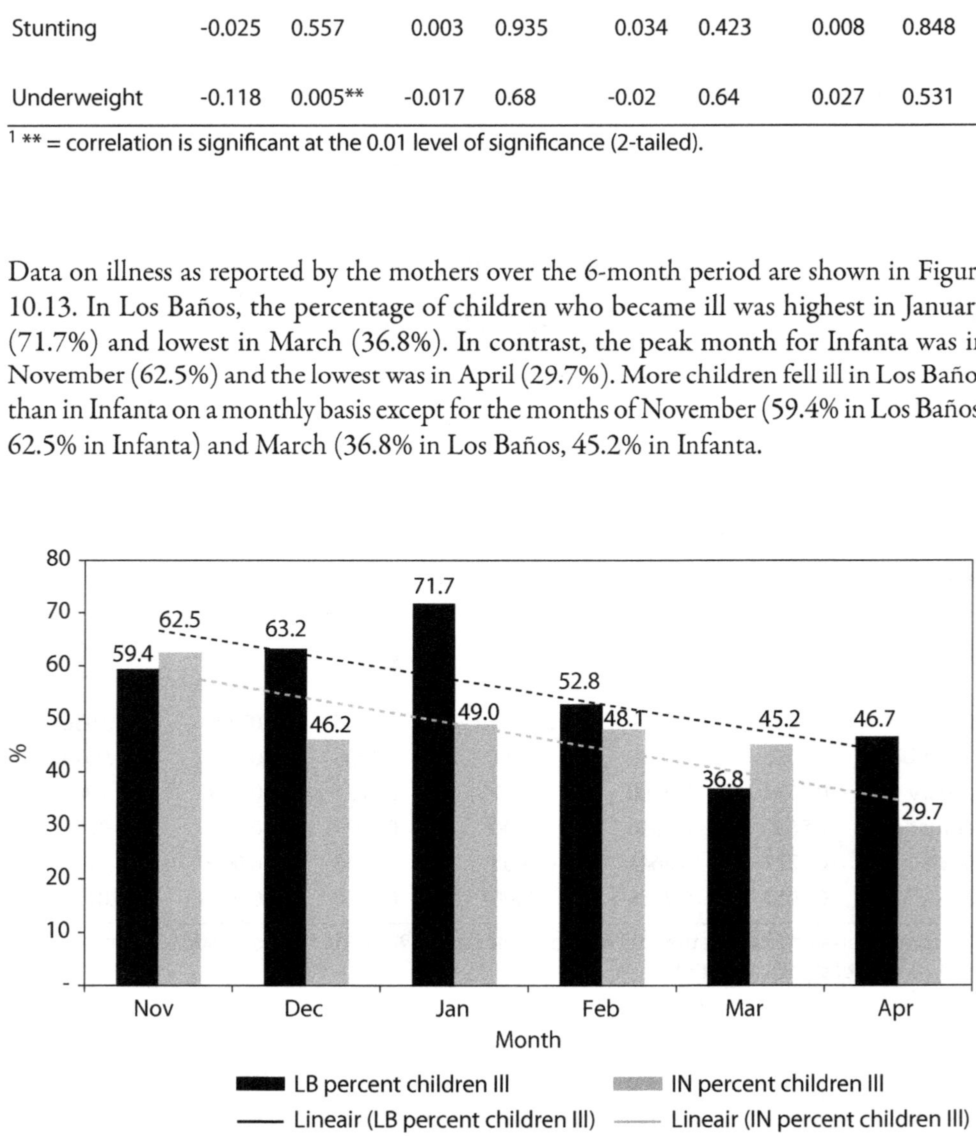

Figure 10.13. Percentage of children with illness, Los Baños and Infanta

Results of the Pearson's correlation test indicated that rainfall and temperature were correlated with the health status of children in both study areas. This could indicate higher prevalence of common illnesses such as cough, colds, and fever due to exposure to rainy and cold weather (Table 10.6).

Table 10.7 presents the children's intake of energy, iron, and vitamin A in Los Baños and Infanta.

Children in both areas had inadequate intakes of iron and vitamin A and are at risk of developing iron and vitamin A deficiencies. In particular, more Infanta children had inadequate iron intakes than Los Baños children (79.4%, 58.6%). In the case of vitamin A, a high 87% of the children in both areas had inadequate intakes of this nutrient.

Figure 10.14 presents the monthly intakes of energy, iron and vitamin A. In Los Baños, the children's energy intake decreased from November to January and picked up in February to April. Infanta children had lower intakes of energy from November to April. Children's intake of iron was low in both areas with a slight increase among Infanta children in April.

Table 10.6. Relationship between climate change variables and health status.[1]

	Los Baños		Infanta	
	Rainfall	Temperature	Rainfall	Temperature
	P-value	*P*-value	*P*-value	*P*-value
Health status	<0.000*	<0.000*	0.044*	0.044*

[1] * = correlation significant at the 0.05 level of significance (2-tailed).

Table 10.7. Children's mean intake and adequacy of energy, iron and vitamin A.

Nutrients	RENI	Los Baños			Infanta		
		Mean intake	Adequate (%)	Inadequate (%)	Mean intake	Adequate (%)	Inadequate (%)
Energy, kcal	1,070	1,611.90	52.5	47.5	1,114.7	50.5	49.5
Iron, mg	8	7	35.4	58.6	13.1	20.6	79.4
Vitamin A, ug RE	400	252.1	13.1	86.9	305.6	13.4	86.6

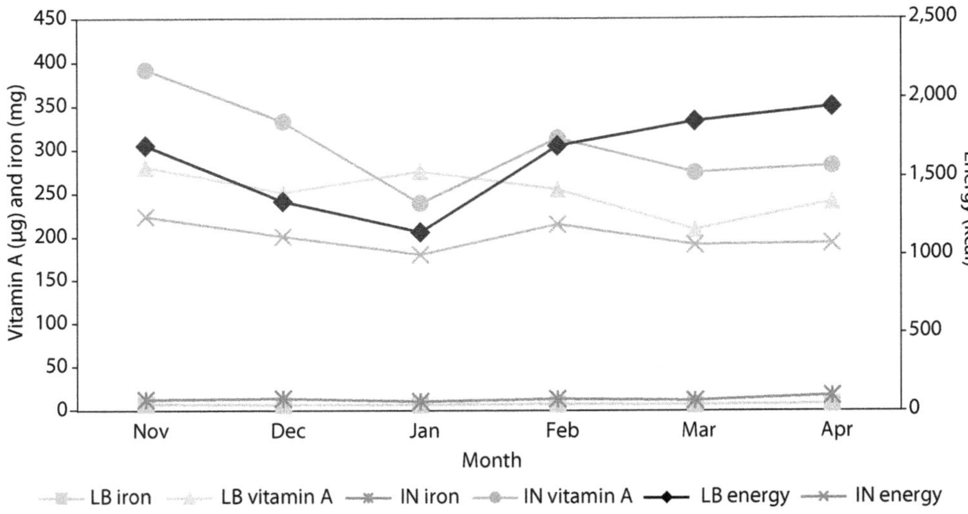

Figure 10.14. Children's nutrient intake, 6-month period, Los Baños and Infanta.

Pearson's correlation test was employed to determine the association between climate change variables (i.e. rainfall and temperature) and nutrient intake (energy, iron, vitamin A). Results showed that there is a positive and significant association between rainfall and energy and vitamin A intake of Infanta children (P=0.099) but no other significant associations (Table 10.8).

Table 10.8. Relationship between climate change variables and nutrient intake.[1]

Nutrient intake	Climate variables							
	Los Baños				Infanta			
	Rainfall		Temperature		Rainfall		Temperature	
	R	*P*-value	r	*P*-value	r	*P*-value	r	*P*-value
Energy (kcal)	0.014	0.746	0.048	0.257	0.076	0.079*	-0.003	0.95
Iron (mg)	-0.022	0.602	0.046	0.276	0.015	0.731	0.059	0.17
Vitamin A (ug)	0.002	0.971	-0.042	0.316	0.071	0.099*	-0.011	0.796

[1] * = correlation is significant at the 0.10 level of significance (2-tailed).

The computed food consumption score shows that majority of the children scored 'acceptable' in food consumption (Table 10.9). However, in January 50% of the children in Los Baños had poor scores.

Table 10.9. Percent distribution of children's food consumption score.

Month	Los Baños		Infanta	
	Acceptable	Poor	Acceptable	Poor
November	91.3	0.0	93.4	0.0
December	91.5	0.0	91.5	0.0
January	35.6	50.0	95.3	0.0
February	86.5	0.0	89.6	0.0
March	90.4	0.0	85.8	0.0
April	90.4	0.0	85.8	0.0

Table 10.10 summarises the results on the relationships between nutritional status and socio-economic-demographic variables, nutrient intake, feeding practices, access to health care, and sanitation and hygiene. Fisher's exact test was used to determine the association between nutritional status and family size, mothers' and fathers' age, employment status, primary occupation, and monthly income; feeding practices; access to health care; and hygiene and sanitation. Pearson's correlation test was employed to determine the association between nutritional status and nutrient intake and health status.

Factors found to be associated with stunting of children in Los Baños include mother's age, father's age, intake of energy and vitamin A, feeding practices such as the age the child was given its first food, first liquid given to the child, and utensils used to feed child, and access to health care. In Infanta, the factors associated with stunting were father's monthly income, intake of energy and iron, access to health care, sanitation and hygiene practices such as material used to wash child's hands. Not a single feeding practice was significantly correlated with stunting among the Infanta children. The two factors associated with stunting in both Los Baños and Infanta children were intake of energy and iron and kind of health facilities available. Health status was correlated with stunting and underweight among children in both areas ($P<0.000$).

Factors found to be associated with underweight children in Los Baños were mother's employment status, intake of energy, iron and vitamin A, and sanitation and hygiene practices such as type of toilet available and disposal of faeces. In Infanta, the factors associated with

Table 10.10. Factors associated with children's nutritional status.[1]

	Los Baños		Infanta	
	Stunting *P*-value	Underweight *P*-value	Stunting *P*-value	Underweight *P*-value
Mothers				
Family size	0.75	0.411	0.345	0.071**
Age	0.032*	0.267	0.575	0.316
Employment status	0.765	0.084**	0.605	0.533
Primary occupation	0.893	0.739	0.4	0.079**
Fathers				
Age	0.033*	0.137	0.557	0.239
Monthly income	0.27	0.384	0.019*	0.247
Health status	<0.000*	<0.000*	<0.000*	<0.000*
Nutrient intake				
Energy (kcal)	0.019**	0.274	0.015*	<0.000**
Iron (mg)	0.909	0.043*	0.008**	0.012*
Vitamin A (µg)	0.188	0.044*	0.795	0.009**
Feeding practices				
Age child given first food	0.066**	0.177	0.733	0.504
First liquid given to the child	0.047*	0.893	0.674	0.928
Utensil used to give first liquid to child	0.094**	0.11	0.708	0.156
Access to health care				
Kind of health facilities available	0.061**	0.565	0.051**	0.219
Health facility used	0.117	0.761	0.088**	0.272
Means of transportation to health facility	0.045*	0.412	0.433	0.207
Sanitation and hygiene				
Type of toilet	0.635	0.069**	0.789	0.768
Disposal of faeces	0.479	0.012**	0.551	1
Material used to wash children's hands	0.886	0.113	0.093**	0.175

[1] * = correlation significant at the 0.05 level of significance (2-tailed); ** = correlation is significant at the 0.01 level of significance (2-tailed).

underweight children were family size, mother's primary occupation and intake of energy, iron, and vitamin A. The factor common to Los Baños and Infanta children's underweight status were intakes of iron and vitamin A only. Not a single feeding practice was found to be significantly associated with underweight in both areas. In addition, there were no hygiene and sanitation variables found to be correlated with underweight for Infanta children

In Los Baños, rainfall was positively and significantly correlated with underweight children (P=0.005) but not with stunting. Underweight reflects both the long- and short-term nutritional status of children. It also can be a short-term response to climate seasonal flux or shocks. In Infanta, no climate variables were found to be associated with the nutritional status of children. This study's finding that geographical location and associated climate type is one factor that could explain underweight and stunting is notable, since it is one of several dimensions of risks related to climate change.

Multiple logistic regression analysis was done to determine the effects of climate variables (temperature and rainfall) in predicting the nutritional status of children. The dependent variable was the nutritional status of children (whether stunted or underweight), whereas the independent variables included area, rainfall, temperature, nutrient intake (energy, vitamin A, iron), food consumption score, and health status. In Model 1, rainfall and temperature are not significantly related to stunting in children. In fact, area was the only variable that could explain stunting. The results also show that the log odds of being stunted is higher for children in Infanta than children in Los Baños. Infanta has a climate characterised by a wet season throughout the year and lower temperature and higher rainfall than Los Baños. Similarly, the factors that could explain being underweight among children were area and rainfall (Model 2). The log odds of being underweight were higher for children in Infanta relative to children in Los Baños.

While the mechanism of the associations cannot be explained yet, the results of this study imply that climate change variables can affect the nutritional status of children. Thus, knowing the areas where rainfall is heavier or temperatures are high can help the local government better prepare in terms of food production (e.g. crops resilient to flooding or that need little water), providing health and nutrition services especially for climate sensitive diseases, and improving infrastructure and transportation to enhance accessibility to health

Model 1. Dependent variable: stunting.

	B	S.E.	Wald	df	Sig.	Exp(B)
Area	1.02500	0.15200	45.76400	1.00000	0.00000	2.78800
Rainfall (mm)	-0.00016	0.00032	0.26800	1.00000	0.60500	1.00000
Temperature, °C	-0.01300	0.03300	0.15900	1.00000	0.69000	0.98700
Energy (kcal)	0.00017	0.00014	1.65900	1.00000	0.19800	1.00000
Iron (mg)	0.01100	0.01200	0.81300	1.00000	0.36700	1.01100
Vitamin A (µg)	-0.00003	0.00013	0.06800	1.00000	0.79500	1.00000
FCS	-0.00015	0.00200	0.00700	1.00000	0.93400	1.00000
Health status	0.16300	0.13200	1.52200	1.00000	0.21700	1.17700
Constant	-0.04000	0.89100	0.00200	1.00000	0.96400	0.96100

Model 2: Dependent variable: underweight.

	B	S.E.	Wald	df	Sig.	Exp(B)
Area	0.92400	0.21200	18.91300	1.00000	0.00000	2.51800
Rainfall amount (mm)	-0.00092	0.00042	4.85200	1.00000	0.02800	0.99900
Temperature, °C	0.01600	0.04200	0.14800	1.00000	0.70000	1.01600
Energy (kcal)	0.00003	0.00004	0.55700	1.00000	0.45600	1.00000
Iron (mg)	0.00200	0.00400	0.22000	1.00000	0.63900	1.00200
Vitamin A (μg)	0.00008	0.00019	0.18000	1.00000	0.67100	1.00000
FCS	-0.00400	0.00200	2.19200	1.00000	0.13900	0.99600
Health status	0.08900	0.17400	0.26000	1.00000	0.61000	1.09300
Constant	1.26900	1.15400	1.20900	1.00000	0.27200	3.55600

and nutrition services. In a study conducted in eastern Ethiopia among 6-36 months old children by Egata *et al.* (2013), the prevalence of undernutrition was found to be higher, though not statistically significant, during the dry season or period with high temperatures. However, poverty and poor access to health services were identified as predictors of wasting.

Studies in other countries have shown similar results about children's sensitivity to rainfall. Hagos *et al.* (2014) showed that among Ethiopian children rainfall and temperature were partly predictive of the variation in children's stunting and underweight conditions. Among Rwandan children rainfall patterns affected the nutritional status (under-nutrition) of children under five years of age, implying that estimating the effect of rainfall on malnutrition may help to better identify regions of vulnerability to undernutrition (Mukabutera *et al.*, 2016). In the study of Shively (2017), nutrition sensitivity to rainfall was found to be greater in Nepal, where rainfall is lower on average and wider ranging, than in Uganda. Good health and transport infrastructures could help buffer children against the deleterious nutritional effects of precipitation shortfalls.

Between temperature and rainfall, the latter seems to be the one that has a more negative effect to child's nutritional status. Thus, to help improve child nutrition, appropriate interventions and innovative strategies should be developed that mitigate the effects of rainfall. However, one should not overlook other factors than climate change such as inadequate food intake, illness, and childcare (UNICEF, 2015).

10.5 Discussion and conclusion

The tests of association and regression show that the variables rainfall and location of study site are related to undernutrition, indicating that children do react to changes in climate and their environment. In Infanta, with a Type II climate, the odds of stunting in children are higher than in Los Baños. This finding is similar to the study of Hagos *et al.* (2014), which reported that rainfall and temperature partially predicted the variation in stunting and underweight among Ethiopian children. Regarding location, Hess *et al.* (2012) identified this variable as a major exposure determinant. The public health threats related to climate change are 'location-specific and path-dependent' (Yohe and Tol, 2002). Thus, identifying locations with increased human health risk would facilitate vulnerability mapping and enhance public health preparedness in such places, which – in turn – would stimulate adaptation to climate change. In view of the impact of a changing climate on geographical location, stunting and underweight in children would continue to be a problem in the study areas, and in the country as a whole.

Relative to adaptation measures, a review of the relevant documents showed that laws have been passed related to climate change adaptation, and agencies have been given authority to act on climate-related matters. These policies need to be translated into programs that target farming households to enable them to adapt to climate change and be self-reliant in meeting their food and nutrition needs. It is important to review the policies and institutions with work on climate change as they influence the range of livelihood options and the access to assets and vulnerability to shocks, including climate change itself (cf. Carloni and Crowley, 2005).

The Philippines is a signatory to the Paris Agreement on Climate Change 2016. Since then, government agencies have been working toward the adaptation and mitigation of climate change. PAGASA, created through Presidential Decree 1149, is tasked to perform weather forecasting, long-term climate monitoring, and to collect meteorological, seismic, oceanographic, and astronomical data to make these available for the benefit of agriculture, commerce, and industry. PAGASA conducts continuing studies of Philippine climatology, agrometeorology, solar radiation and other issues that are vital to national progress. The Climate Change Commission (CCC), the lead policy-making and coordinating body of the government on climate change initiatives, formulates the National Strategic Framework on Climate Change and the National Climate Change Action Plan (NCCAP). However, the CCC has limited local presence, thus limited capacity to engage with local government units (World Bank, 2013). The NCCAP serves as the Government's road map for climate action and is the lead policy document guiding the climate agenda of the government (World Bank, 2013). The Department of Environment and Natural Resources (DENR) is also involved in climate-change related adaptation measures. It has created a Climate Change Office that has a coordinating role for the DENR offices and for other national government agencies, NGOs, and local government units on matters related to climate change. The Department of Health is mandated to develop national action plans for climate change health sector adaptation and mitigation, to implement adaptation measures specific to local health determinants and outcomes, and to facilitate community-based resource management. The National Disaster Risk Reduction Management Council's current efforts focus on preventing casualties, particularly regarding typhoons and earthquakes.

At the local government level, climate variables data are not yet widely used. However, though farmers could use PAGASA data to guide in food production, they lack the necessary knowledge and access to those data. Technologies could be made available to them to inform them on local climate changes. The Food and Agriculture Organization and National Nutrition Council have included climate variables in their Early Warning Systems (National Nutrition Council, 2016). Climate change could be also mainstreamed in the local nutrition planning. Each municipality is mandated to formulate and implement a local nutrition action plan, which aims to address the malnutrition problem in the locality. It is prepared by the local nutrition committee, in which different sectors are represented. A recent study however, revealed that although the national plan for nutrition is a well-prepared document, it has not been integrated into development planning at the national and local levels, which results in nutrition being missed out in local programming and budgeting (Briones *et al.*, 2017). The study also revealed the weak local government unit (LGU) mobilisation for nutrition programs wherein many local nutrition committees are not functional. The study of Otieno *et al.* (2013) suggested that incorporating the nutrition situation in local climate change programmes could contribute to reducing malnutrition and deprivation due to climate change.

Some nutrition-sensitive interventions have been put in place by NGOs, such as the school gardening project by the Southeast Asian Regional Center for Graduate Study and Research in Agriculture (SEARCA) that includes the integration of climate change topics and nutrition in the school curriculum (IIRR, 2014; SEAMEO, 2010). In addition, the Department of Education sought ways to integrate climate change adaptation and disaster risk reduction into the public school system.

There are moves to integrate aspects of food and nutrition security in local climate action plans (LCCAP), though progress is slow. For climate change to be integrated in local development planning, the local nutrition workers have to be trained to mainstream climate change in the planning process. This has not yet been done and would present a challenge, because as the study of Briones *et al.* (2017) showed, the capacity levels of local nutrition workers vary and they lack adequate training for handling malnutrition cases in their communities. Another obstacle to climate change mainstreaming is lack of awareness on the climate change impacts on sustainable development (Lasco *et al.,* 2008). It has been suggested to involve children in planning for climate change adaptation by capitalising on their varied experiences. Lawler and Patel (2012) pointed out that children are exposed to conditions such as livelihood insecurity in Mongolia, sea level rising in the Pacific, flooding in the Philippines, and crop failures in Indonesia. Therefore, involving children in policy or program formulation could lead to better designed programmes.

At the household level, not many households, whether or not they have malnourished children, are knowledgeable about climate change and its adverse effects. A study commissioned by the World Bank (2013) showed that many adults still do not fully understand climate change and its impacts. About 38% have 'only little' and 14% have 'almost no understanding' of climate change. This may be different for farmers, since they experience the impact of climate change more directly. The study of Ngilangil *et al.* (2013) in the northern Philippines revealed that farmers were highly aware of climate change effects. They know that crop production is at great risk during extreme weather events, but they not do not link this to human activities that increase greenhouse gas emissions.

Based on the results of this study, adaptation measures should include increasing awareness, use of climate change-resilient crops, and coping strategies to reduce the negative effects of climate change especially among farming households. Different coping strategies have been documented in the literature. These include accepting support from social networks, selling agricultural and livestock products, collecting wild foods, using savings, relying on less expensive foods, limiting food consumption, borrowing from friends, buying food on credit, relying on food aid, withdrawing children from school to save money, and sending children to work (Fletcher *et al.*, 2013; Israel and Briones, 2014; Talavera *et al.*, 2008). By living in a certain area with geographic and climate features that are susceptible to climate change effects, the farming households in that area would remain to be vulnerable if they are not able to adapt.

A limitation of this study is the use of the cross sectional study design that cannot directly attribute climate change variables as the causes of malnutrition. Additionally, the data on illness and frequency and food intake were collected based on the reports of mothers and could have a recall bias.

This study explored the association between climate change variables and nutritional status of preschool children in two locations. However, climate change variables (i.e. rainfall) can only partly explain undernutrition of children in the study areas. Irrespective of climate change, stunting and underweight will continue to be problems if other factors such as income, education, and climate-resistant food production methods, are not addressed. With climate change, the malnutrition problem is likely to worsen, particularly with the occurrence of natural disasters such as typhoons and flooding. In addition to climate change, children would still be exposed to the various causes of malnutrition such as poverty, inadequate food intake, and poor health services and water and sanitation facilities. Local climate and geographic characteristics could pose additional challenges to the local government units (LGUs). These should have the capacity to predict rainfall and temperature in the short and long term to better plan for food production, livelihood, and health and nutrition services in their location.

Based on the results of this study, we recommend the following. First, to integrate topics on the effects of climate change on the nutritional status of children in risk communication and behaviour change programs. Second, to continue the implementation of existing nutrition interventions, such as promotion of optimal feeding practices, correct hygiene and sanitation practices, and micronutrient supplementation, since undernutrition and stunting will continue to prevail amid different climate types and more so in the occurrence of extreme events. Third, to improve the adaptability of households to climate change factors through livelihood projects. And, finally, to design and implement capacity building for LGUs and local nutrition workers to enhance their understanding of the association between climate change and nutritional status.

Acknowledgements

This study would not have been completed without the support of the households and children; *barangay* nutrition scholars; local nutrition action officers; the municipal governments of Infanta and Los Baños; and the Neys-Van Hoogstraten Foundation, for which the authors are grateful.

References

ACF, 2012. *Enhancing climate resilience and food & nutrition security ACF approach to face climate change, hunger and undernutrition in at-risk communities*. Paris: ACF International.

Briones, R., Antonio, E., Habito, C., Porio, E. and Songco, D., 2017. Strategic review: food security and nutrition in the Philippines. An independent review commissioned by the World Food Programme. Available at: https://docs.wfp.org/api/documents/WFP-0000015508/download.

Carloni, A.S. and Crowley, E., 2005. *Rapid guide for missions: analysing local institutions and livelihoods*. Rome: Food and Agriculture Organization of the United Nations.

Confalonieri, U., Menne, B., Akhtar, R., Ebi, K.L., Hauengue, M., Kovats, R.S., Revich, B. and Woodward, A., 2007. *Human health. Climate change 2007: impacts, adaptation and vulnerability*. Contribution of Working Group II to the Fourth Assessment Report of the Intergovernmental Panel on Climate Change, M.L. Parry, O.F. Canziani, J.P. Palutikof, P.J. van der Linden and C.E. Hanson (eds). Cambridge UK: Cambridge University Press, pp. 391-431.

Crahay, P., Mitchell, A., Gomez, A., Israël, A-D., Salpeteur, C., Mattine, H., Dere, H., Lapegue, J., Grosjean, L., Aissa, M.A., Brown, R., Swan, S.H., Pietzsch, S. and Dufour, C., 2010. *The threats of climate change on undernutrition – a neglected issue that requires further analysis and urgent actions*. SCN, 2010. Newsletter #38 – Climate Change: Food and Nutrition Security Implications.

De Guzman, R., 2014. *Climate change impacts in the Philippines and current policies and initiatives on climate change adaptation*. Latest Scientific Knowledge on Climate Change and Comprehensive Countermeasures, IPCC Yokohama Session, Yokohama, Japan.

Ebi, K.L. and Paulson, J.A., 2007. Climate change and children. *Pediatric Clinics of North America*. 54(2): 213-226.

Egata, G., Berhane, Y. and Worku, A., 2013. Seasonal variation in the prevalence of acute undernutrition among children under five years of age in east rural Ethiopia: a longitudinal study. *BMC Public Health* 13: 864.

Fletcher, S.M., Thiessen, J., Gero, A., Rumsey, M., Kuruppu, N. and Willetts, J., 2013. Traditional coping strategies and disaster response: examples from the South Pacific Region. *Journal of Environmental and Public Health*. Article ID 264503, http://dx.doi.org/10.1155/2013/264503.

FNRI, 2013. *2013 National nutrition survey*. Manila: Food and Nutrition Research Institute.

GHF, 2009. *Human impact report: climate change – the anatomy of a silent crisis*. Geneva: Global Humanitarian Forum.

Hagos, S., Lunde, T., Mariam, D.H., Woldehanna, T. and Lindtjør, B., 2014. Climate change, crop production and child under nutrition in Ethiopia; a longitudinal panel study. *BMC Public Health* 14: 884.

Hess, J.J., McDowell, J.Z. and Luber, G., 2012. Integrating climate change adaptation into public health practice: Using adaptive management to increase adaptive capacity and build resilience. *Environmental Health Perspectives* 120(2): 171-179.

IIRR, 2014. International Institute of Rural Reconstruction. Available at: http://iirr.org/our-impact/education/philippines.

IPCC, 2012. Glossary of terms. In: Field, C.B., Barros, V., Stocker, T.F., Qin, D., Dokken, D.J., Ebi, L., Mastrandrea, M.D., Mach, K.J., Planer, G.K., Allen, S.K., Tignor, M. and Mikley, P.M. (eds) *Managing the risks of extreme events and disasters to advance climate adaptation*. A Special Report of Working Groups I and II of the Intergovernmental Panel on Climate Change (IPCC). Cambridge UK and New York: Cambridge University Press, pp. 555-564.

ISDR, 2009. *Global assessment report on disaster risk reduction*. Geneva: United Nations.

Israel, D.C. and Briones, R.M., 2014. *Disasters, poverty and coping strategies: the framework and empirical evidence from micro/household data – Philippine case*. Discussion Paper Series No. 2014-06. Manila: Philippine Institute for Development Studies.

Krishnamurthy, P.K., Lewis, K., Kent, C. and Aggarwal, P. 2015. Climate impacts on food security and livelihoods in Asia: a review of existing knowledge. Report prepared by United Nations World Food Programme Regional Bureau for Asia, the UK Met Office Hadley Centre, and the International Water Management Institute (IWMI/CCAFS) with financial support from the Government of Sweden. Available at: http://tinyurl.com/y8za9k4z.

Lasco, R.D., Pulhin, F.P., Jaranilla-Sanchez, P.A., Garcia, K.B. and Gerpacio, R.V., 2008. Mainstreaming climate change in the Philippines. Working Paper 62. Los Baños, Philippines: World Agroforestry Centre.

Lawler, J. and Patel, M., 2012. Exploring children's vulnerability to climate change and their role in advancing climate change adaptation in East Asia and the Pacific. *Environmental Development* 3: 123-136.

Llamas-Clark, E.F., 2011. Extreme weather events and preserving child nutrition: Revisiting concepts for policy and programs. *Philippine Journal of Nutrition* 58: 95-102.

Lloyd, S.J., Kovats, R.S. and Chalabi, Z, 2011. Climate change, crop yields, and undernutrition: development of a model to quantify the impact of climate scenarios on child undernutrition. *Environmental Health Perspectives* 119(12): 1817-1823.

Mukabutera, A., Forrest, J.I., Nyirazinyoye, L., Marcelin, H. and Basinga, P., 2016. Associations of rainfall with childhood under-nutrition in Rwanda: an ecological study using the data from the Rwanda Meteorology Agency and the 2010 Demographic and Health Survey. *Asian Journal of Agriculture and Food Sciences* 4(1): 35-45.

NASA, 2005. What's the difference between weather and climate? Available at: http://www.nasa.gov/mission_pages/noaa-n/climate/climate_weather.html.

National Nutrition Council, 2016. Briefer on early warning system on food and nutrition security (FNS-EWS). Manila: National Nutrition Council.

Ngilangil, L.E, Olivar, S.O. and Ballesil, M.L., 2013. Farmers' awareness and knowledge on climate change adaptation strategies in Northern Luzon, Philippines. *International Scientific Research Journal* 5(3).

Otieno, A.C., Mugalavai, V. and Kamau, E.H., 2013. A study on impact of climate change on the food and nutrition security of children in small-scale farmer households of Migori Area. *Global Journal of Biology, Agriculture, and Health Sciences (GJBAHS)* 2: 67-71.

PAGASA, 2011. *Climate change in the Philippines*. Available at: http://tinyurl.com/yc4gf5s7.

Republic of the Philippines, 2009. Republic Act No. 9729. Retrieved from http://www.lawphil.net/statutes/repacts/ra2009/ra_9729_2009.html.

SEAMEO, 2010. *Integrating climate change issues in Southeast Asian schools: a teachers' guidebook*. Kuala Lumpur: SEAMEO RECSAM.

Shively, G.A., 2017. Infrastructure mitigates the sensitivity of child growth to local agriculture and rainfall in Nepal and Uganda. Proceedings of the National Academy of Sciences of the United States of America (PNAS) 114(5): 903-908.

Talavera, M.T.M., Tuazon, M.A.G., Narciso, M.H., Bustos, A.R., de Juras, A.R. and Padilla, L.M. 2008. Coping strategies on food insecurity among disaster-affected households in selected Philippine Provinces. *Journal of the Nutritionist-Dietitians Association of the Philippines* 22(1-2).

Thomas, V., Albert, J.R. and Perez, R.T., 2013. Climate-related disasters in Asia and the Pacific. ADB Economics Working Paper Series, No. 358. Manila, Philippines.

Tirado, M.C., Hunnes, D., Cohen, M.J. and Lartey, A., 2015. Climate change and nutrition in Africa. *Journal of Hunger & Environmental Nutrition* 10(1): 22-46.

UNICEF, 1990. *Strategy for improved nutrition of children and women in developing countries*. Policy Review Paper E/ICEF/1990/1.6. New York: JC 27/UNICEF-WHO/89.4. Available at: http://tinyurl.com/prrqo2n.

UNICEF, 2015. *Unless we act now: the impact of climate change on children*. New York: UNICEF. Available at: http://tinyurl.com/y9eg988n.

UNICEF/WHO/World Bank Group, 2017. *Levels and trends in child malnutrition. Joint child malnutrition estimates*. Available at: http://www.who.int/nutgrowthdb/jme_brochoure2017.pdf?ua=1.

Urbano, M., Deasey, M., Kusumawardani, N. and Pardosi, J. 2011. The impacts of climate change on nutrition and migration affecting children in Indonesia. Bangkok; UNICEF. Available at: http://tinyurl.com/ya7qxxzq.

Wijaya, S.A., 2014. Climate change, global warming and global inequity in developed and developing countries. Analytical perspective, issue, problem and solution. *IOP Conference Series: Earth and Environmental Science* 19: 012008.

World Bank, 2013. *Getting a grip on climate change in the Philippines: executive report*. Washington DC: The World Bank.

Yohe, G. and Tol, R.S.J., 2002. Indicators for social and economic coping capacity moving toward a working definition of adaptive capacity. *Global Environmental Change* 12: 25-40.

11. Does backyard vegetable gardening enhance perceived household food security and dietary diversity? A case study in the Philippines

J.B. Dorado[*]*, R.V. Viajar, G.P. Azaña, G.S. Caraig and M.V. Capanzana*

*Food and Nutrition Research Institute, Department of Science and Technology (FNRI-DOST), Taguig City, Metro Manila, The Philippines; *juliebdorado@yahoo.com*

Abstract

People's rights and access to foods they prefer are the core of the food sovereignty approach. In this study, the approach was used in research on the meanings of backyard vegetable gardening for people's perceptions of food security and the dietary behaviour of schoolchildren and households. The implementation and effects of the Philippine Backyard Vegetable Gardening Program (BVGP) were investigated in two provinces (Antique and Surigao del Norte) and the National Capital Region (NCR), covering four municipalities, two cities in the NCR and 10 *barangays* (villages). The findings show that gardening enhances the feeling of being food secure, but also that gardening continues to be a women's concern. Rather unexpectedly, knowledge about the importance of vegetable consumption did not differ between households with and without a garden and having a vegetable garden did not affect the dietary behaviour of schoolchildren and households. The dietary diversity scores of schoolchildren and households with and without a garden proved to be equally low. The study recommends a comprehensive review of the backyard vegetable gardening program to improve the efficacy of the program.

Keywords: dietary behaviour, food insecurity, households, schoolchildren, vegetable consumption

11.1 Introduction

Food as a basic human right implies that everyone must have access to safe, nutritious and culturally appropriate food in sufficient quantity and quality to sustain a healthy and dignified life, and that each nation should declare that such access is a constitutional right (Windfuhr and Jonsén, 2005). Food sovereignty has been defined as the right of people and communities to decide on and implement their agricultural food policies (Beuchelt and Virchow, 2012) and strategies for sustainable production and distribution of food (People's food sovereignty network, 2004 in Quaye *et al.*, 2009). To achieve food sovereignty the state has to support the development of the primary sector. An international movement of farmers, rural women and workers from 69 countries called *La Via Campesina* started to promote the idea of food sovereignty, which then became an important civil society movement. Beuchelt and Virchow (2012) emphasise that the concept of food sovereignty

DOI 10.3920/978-90-8686-864-3_11, © Wageningen Academic Publishers 2018

also applies to consumers and their rights and access to the foods they prefer, instead of having to comply with the dictates of the market. *Via Campesina* considers food sovereignty a precondition to genuine food security (Patel, 2009). In the Philippines, however, food insecurity is a national concern, addressed by the government through policies and programs on food supply and demand to guarantee people's access to sufficient and adequate food. The Backyard Vegetable Gardening Program (BVGP) of the Department of Agriculture is such a program. Especially for poorer households, keeping a backyard vegetable garden can contribute to household food security and improved dietary intake through the increased consumption of vegetables.

This study examined the implementation of the BVGP. The study was conducted among households with children of school-going age in selected cities in the National Capital Region (NCR) and municipalities in the Antique and Surigao del Norte provinces in the Philippines. The study explored the effects of backyard vegetable gardening on perceived household food security and the dietary diversity of schoolchildren and households. We also looked at program implementation in terms of the various inputs to the program, such as funds for the purchase of planting materials and fertilizers, availability of space and garden tools, and the activities of program implementers. This study addressed the following questions: (1) Does the BVGP respond to household food security needs as perceived by the households themselves? (2) Does having a backyard vegetable garden contributes to dietary improvement at the household level? (3) How is the BVGP implemented and what are the roles of program implementers?

11.2 Conceptual framework of the study

The key concepts of the study are food security, dietary improvement and implementation. In the literature, food security has been described as a situation that exists when all people, at all times, have physical, social and economic access to sufficient, safe and nutritious food that meets their dietary needs and food preferences for an active and healthy life (FAO, 2013).

In this research the concept of food security was measured subjectively, by asking the respondents whether having a backyard vegetable garden made them feel more food secure, and if so, why. Dietary improvement was measured by assessing dietary diversity, using a Household Dietary Diversity Questionnaire and, for the schoolchildren, an Individual Dietary Diversity Questionnaire. Implementation was operationalised by interviewing program implementers about the program's guidelines and their perceptions about the success of the program. The different components and their interrelationships are depicted in Figure 11.1.

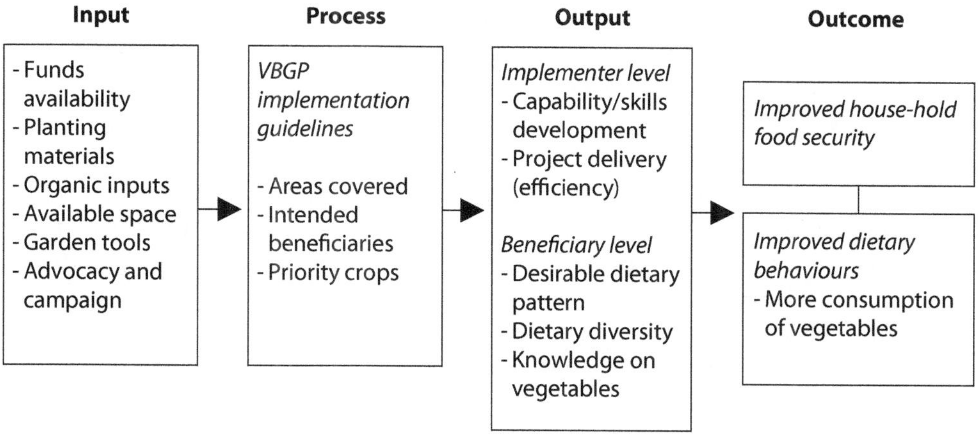

Figure 11.1. Conceptual model of the study.

11.3 Methodology

11.3.1 Data collection

The study design compared households with and without a backyard vegetable garden using two rounds of data collection, the first at the start of the study and the second after one year. Conducting two rounds of data collection was done to determine whether the households without a garden at baseline had been encouraged to start a garden after a year. The second round of data collection was also needed to determine the effects of the program on the households' perceptions about maintaining a vegetable garden and dietary behaviour.

The study areas included two Philippine provinces (Antique and Surigao del Norte) and the National Capital Region (NCR), covering four municipalities (two municipalities per province), two cities in the NCR and 10 *barangays*. *Barangay* is a term used for a village and it is the basic administrative unit in the Philippines. The study areas belong to the 10 priority one-food poor provinces (including the NCR) covered by the government's Accelerated Hunger Mitigation Program (AHMP).

Two types of participants were included in the study: households with schoolchildren and program implementers. The household sample comprised 304 households with schoolchildren aged 8-10 years old, of which 152 households had an existing vegetable garden planted with at least four vegetables at the initial data collection and 152 households that did not have such a garden. Households were selected through simple random sampling from the *barangay* lists of the households with schoolchildren in the targeted age group, distinguishing between households with and without a garden. Only one schoolchild per household was included in the research. When a selected household had two or more children in the age group concerned, only the youngest child was included in the study. The sample of 24 program implementers of the BVGP comprised local government unit (LGU) officials and staff.

Key informant interviews were conducted among 24 program implementers at the city, municipal and *barangay* levels. Six focus group discussions (FGD) were carried out, each having eight to ten participants. Pre-tested questionnaires were developed for the key informant interviews of implementers and guiding questions and topic lists for the FGDs. At the household level, mothers were interviewed face-to-face using pre-tested questionnaires. Topics included were: the household profile, health and nutrition awareness and knowledge, views on the benefits of vegetable consumption, program benefits in terms of perceptions about food (in)security, and values and work habits imbibed by the households as a consequence of the BVGP.

Food intake of the children and the households was not measured quantitatively but was determined by asking about the foods consumed for the past 24-hours for two non-consecutive days, using the dietary diversity score (DDS) questionnaire. The DDS is a simple count of food groups that the children and household members consumed in the past 24 hours. It is a proxy for nutrient intake at individual level as well as for economic access to food at household level (Kennedy *et al.*, 2011). Prior to the interview on the food intake, the interviewer determined first if the previous 24-hour period of food intake was 'normal' or 'usual' for the household without distinguishing weekdays or weekends. In case of a special occasion, another day was selected for the interview (Swindale and Bilinsky, 2006).

The household dietary diversity questionnaire comprised 12 food groups: (1) cereals; (2) roots and tubers; (3) vegetables; (4) fruits; (5) meat and poultry; (6) eggs; (7) fish and seafood; (8) legumes and nuts; (9) milk and milk products; (10) oils and fats; (11) sugar and honey; and (12) miscellaneous. It was used to obtain the household dietary diversity score (HDDS). In case of absence of the person responsible for food preparation, another adult who was present in the household and who took part in the meals during the previous day was interviewed. The interviewee was asked to recall all the dishes, snacks, or other foods consumed by household members in the home or prepared in the home. Foods purchased and consumed outside the home were excluded in the recall in the two provinces visited. However, in the National Capital Region where people often purchase cooked food (particularly viand), all foods consumed by the household members were listed.

The individual dietary diversity questionnaire consisted of nine food groups: (1) cereals, roots and tubers; (2) vitamin-A rich foods; (3) other fruits; (4) other vegetables; (5) meat, poultry and fish; (6) eggs; (7) legumes and nuts; (8) milk and milk products; and (9) fats and oils. This questionnaire was used for the schoolchildren.

11.3.2 Data analysis

The data was processed and analysed using the statistical package for the social sciences (SPSS) for Windows version 9. Descriptive statistics (i.e. frequency counts, percentages, means and standard deviations) were used to describe the profiles of the households and the implementers. T-tests between samples (i.e. independent samples) were carried out to determine the significance of the differences between the means of selected variables. A *P*-value of <0.05 was considered significant for the tests performed.

The qualitative responses to open-ended questions were incorporated into the discussion. The focus group discussions were transcribed and encoded, as was done with the results of the key informant interviews at city/municipal and *barangay* levels. Data were organised by themes and then incorporated into the manuscript.

On the food intake aspect, the study was limited to the quality (diversity) of the food consumed by the households and children, as neither the frequency of consumption nor the quantity of food consumed were measured. The dietary diversity scores were used as discrete quantitative variables and divided into two groups to distinguish diets of 'normal' and 'low' diversity. The DDS cut-offs were established by taking the mean score of the upper 33% of the children and the households with the highest dietary diversity scores.

11.4 Results and discussion

11.4.1 Profile of the respondents

About two-thirds of the households both with and without a garden (63.2 and 67.8%) have a nuclear family type. The households in both groups had on average six members. Most of the children were 6 to 10 years old. The breadwinners were predominantly labourers and unskilled workers (32.9 and 28.9%, respectively). Almost 28% of the breadwinners in the households with a garden were farmer, forestry worker or fisherman. Most respondents with and without a garden (37.5 and 48.7%, respectively) belonged to the 30-39 age group, and most had reached high school level (46.1 and 50.0%, respectively). In the majority of the cases (75.8 and 80.3%, respectively) the respondents were the mothers of the children in the study. The mean age of the children in both groups of households was eight. No significant difference was observed in the mean age of the children. More than half (50.7 and 56.6%) of the children in both groups were female. Almost all children (95.4 and 98.0%) were attending school.

The 24 program implementers who were all directly involved in the implementation of the BVGP consisted of agricultural officers, agricultural technologists, agriculture staff, *barangay* captains, *barangay* councillors on health, sanitation and environment and *barangay* nutrition scholars (BNS). They had been in the government service for 1 to 33 years.

11.4.2 Perceptions on food security, vegetable gardening and the BVGP

As Table 11.1 shows, the proportions of households feeling secure were high and did not significantly differ between households with and without a vegetable garden (99.3 and 92.1%, respectively). When probed, 34.9% of the respondents of households with a vegetable garden said to feel secure because of the availability of vegetables for consumption, which they equated with being free from sickness. Respondents were saying that with having a garden *'families are free from sickness', 'food is adequate'* and *'there is something to harvest and sell'.*

Table 11.1. Perceptions of households on food security by study group.

Characteristics	Study group			
	With garden n=152		Without garden[1] n=152	
	n	%	n	%
Security in always having vegetable for the family				
Yes	151	99.3	140	92.1
No	1	0.7	11	7.2
Don't know	-	-	1	0.7
Reasons for feeling secure in having vegetables[2]				
Family members will be free from sickness/vegetables are healthy food and nutritious/liked by the family/with garden	60	39.5	86	61.0
Always have vegetables in the garden	53	34.9	7	5.0
Sure of freshness, free from chemicals, safe with own planted vegetables	9	5.9	-	-
Mothers have clean vegetables for cooking	6	3.9	1	0.7
Always eat vegetables everyday	6	3.9	-	-
Abundant supply in the market and always buy vegetables/ always with vegetables	6	3.9	5	3.5
Budget available can afford to buy vegetables	5	3.3	12	8.5
Did not buy vegetables/no need to buy	5	3.3	1	0.7
There's money in planting of vegetables/as long as you are hardworking in planting and marketing	1	0.7	1	0.7
Buy vegetables if have money/vegetables are expensive	-	-	25	17.8
No space	-		1	0.7
No response	1	0.7	2	1.4

[1] Applies to the initial interview.
[2] Multiple response.

Interestingly, respondents from households without a garden appear to be more convinced of the importance of vegetables for health and nutrition than those from the group with a garden, which could indicate that the former have better nutrition knowledge. At the same time, 17.8% of these respondents commented that vegetables are expensive and that they buy vegetables only when they have enough money to do so. Only a rather small number of respondents in both groups (3.3 and 8.5%, respectively) said that they could afford to buy vegetables.

In the study area, vegetables were not consistently available in the local markets. Twenty-eight percent of the respondents reported times when vegetables were not available in their area. The majority (65.9%) thought that vegetables were unavailable in the local market during lean seasons, when there is no supply of vegetables coming from other areas. The households who rated the supply of vegetables in the local market as just enough comprised 53.9% of households with and 68.4% of households without a garden. Both households with and without a vegetable garden (44.7 and 40.8%, respectively) generally thought vegetables expensive, although they also said that sometimes prices were low. A contrasting picture emerges from the Cagayan de Oro study, where vegetable prices were cheap, although there seasonality was also an issue among farmers and buyers (Potutan *et al.*, 2005).

It is interesting to note that 41% of the households without a garden at the initial interview had a vegetable garden at the time of the second interview. When asked how they started it, the three most frequent answers were: the seeds provided to them (29.3%), simply their intent to plant (22.4%), and for economic reasons (17.2%). This result indicates that gardening is a 'natural' tendency among households, such that given certain inputs (seeds) they will start cultivating when there is space available.

Kortright and Wakefield (2011) found in their study on the role of home food gardening in community food security that growing food contributes to food security at all income levels. In addition, the sustainability of household's food sourcing and gardeners' health and well-being increased with keeping a backyard garden. Talukder *et al.* (2000) concluded that home gardening improves household food security and increases the diversity of foods available to the households.

At the initial interview, 63.8% of the households with a vegetable garden and 56.6% of the households without a garden affirmed that they had heard about the BVGP in their area. According to 44.5% of the households with a vegetable garden, the program is implemented by the Department of Agriculture in the Philippines. Of the households without a garden, 34.1% simply said that the program is about planting. Households in both groups, however, described the program as a livelihood support (24.4 and 20.9%, respectively). Other households had varied responses such as '*for health reasons*', '*important to plant*', and '*given seedlings*', which are indicative of being aware of the program but not knowing what it was actually about.

This study shows that the households recognized the health, nutrition and economic benefits of backyard vegetable gardening, also referring to savings earned from the garden's harvests. The households sold the extra harvest for additional earnings. Having a vegetable garden ensured a basic food source, providing for a continuous source of fresh and nutritious foods for the family. A significant finding in the study by Potutan *et al.* (2005) was the economic importance of having a garden as it allows families to save on food items they would otherwise have to buy. Additional income from the garden was used to purchase supplementary foods, such as meat, fish and cooking oil, thereby further increasing the diversity of the family's diet (see also HKI, 1998; Kiess *et al.*, 1998). In the 2008 National Nutrition Survey, most of the households used the produce from the gardens for own consumption, while some of the households both consumed and sold their produce (FNRI-DOST, 2010).

All the households with a vegetable garden at the time of the second interview recognized the benefits of backyard vegetable gardening, particularly regarding health and nutrition (39.9%), savings (26.0%) and income (22.0%). Harvested vegetables were shared with neighbours and relatives (43.8%), used for family consumption (28.8%), and both for family consumption and selling of extra harvest (26.6%). The households primarily sold their harvested vegetables for additional income (35.2%). Although the respondents agreed on the nutrition and health benefits of having a garden, in practice they sold the harvest and left only some of it for own consumption.

About 79.6% of the households with a garden reported problems in maintaining it, such as insect infestations or other pests (31.0%), the unfavourable climate or weather (16.7%), and lack of water (13.3%). A few households mentioned roaming animals of neighbours, playful children, lack of seeds, and instances of stealing their planted vegetables. For those households who were unable to plant (59.9%), some of the reasons given were unfavourable climate/weather (34.9%), busy/working (15.7%) and lack of water (15.7%). Thus, the main problems encountered in vegetable gardening were damage brought about by insects or pests, unfavourable weather and lack of water. These were similar to the constraints identified by the farmers in peri-urban areas in the Cagayan de Oro study (Potutan *et al.*, 2005).

11.4.3 Dietary diversity and vegetables consumption

Most of the foods frequently consumed by 63% to 100% of children in both groups came from four food groups: (1) cereals and starchy roots and tubers; (2) meat, poultry and fish; (3) fats and oils; and (4) vitamin A-rich fruits, tubers and vegetables. Intake of vitamin A-rich fruits, tubers and vegetables, other vegetables, eggs and fats of schoolchildren from households with a vegetable garden increased between the first and the second collection. However, intake of other fruits, meat, poultry and fish, legumes and nuts, and milk decreased during that period. Among schoolchildren in households without a garden, only intake of other fruits and other vegetables increased between the first and the second interview. The results are presented in Table 11.2.

Between 85 and 89% of the schoolchildren in households with and without a garden have low dietary diversity scores (DDS) in both rounds of data collection. The average number of food groups from which the schoolchildren consumed food was almost five out of nine. The scores did not significantly differ between the two groups and between the data collection periods (Table 11.3).

Table 11.4 shows that most of the foods frequently consumed by the households with a vegetable garden (63 to 100%) at both times of data collection were almost similar to those consumed by households without a garden (57 to 100%). Intake of eggs and fats and oils increased among households with a vegetable garden between the first and the second collection, intake of starchy roots and tubers, fruits and eggs increased among households without a vegetable garden as well.

Table 11.2. Frequently consumed food groups by schoolchildren by study group and time of data collection.

Food group	With garden				Without garden			
	1st data collection		2nd data collection		1st data collection		2nd data collection	
	n	%	n	%	n	%	n	%
1. Cereals, starchy roots and tubers	139	100.0	139	100.0	140	100.0	140	100.0
2. Vitamin A rich fruits, tubers and vegetables	88	63.3	92	66.2	89	63.2	83	58.9
3. Other fruits	38	27.0	37	26.3	33	23.6	44	31.1
4. Other vegetables	66	47.5	73	52.5	54	38.6	61	43.6
5. Meat, poultry and fish	128	92.1	128	91.7	133	95.0	131	93.6
6. Eggs	41	29.5	50	35.6	45	32.1	43	30.7
7. Legumes and nuts	26	18.3	13	9.4	24	16.8	17	11.8
8. Milk	51	36.7	36	25.9	57	40.4	52	36.8
9. Fats and oils	96	69.1	106	75.9	105	74.6	104	73.9

Table 11.3. Mean dietary diversity scores of schoolchildren by study group and time of data collection.[1]

Level	1st data collection			2nd data collection		
	With garden	Without garden	*P*-value	With garden	Without garden	*P*-value
Normal	6.9±0.57 (13.6%)	6.9±0.45 (15.0%)		6.8±0.38 (11.0%)	6.9±0.43 (11.4%)	
Low	4.5±0.98 (86.4%)	4.5±0.95 (85.0%)		4.5±0.98 (89.0%)	4.6±0.96 (88.6%)	
All	4.8±1.24	4.8±1.23	0.94	4.9±1.14	4.8±1.2	0.8
P-value[2]	with garden: 0.95			without garden: 0.73		

[1] Mean ± SD.
[2] Using independent sample t-test.

Table 11.4. Frequently consumed food groups by study group and time of data collection.

Food group	With garden				Without garden			
	1st data collection		2nd data collection		1st data collection		2nd data collection	
	n	%	n	%	n	%	n	%
1. Cereals	145	100.0	145	100.0	144	100.0	144	100.0
2. Starchy roots and tubers	21	14.5	16	10.7	14	9.4	15	10.4
3. Vegetables	129	88.6	121	83.4	112	77.8	110	76.0
4. Fruits	45	30.7	45	30.7	33	22.6	43	29.5
5. Meat and poultry	51	35.2	51	35.2	48	33.0	53	36.4
6. Eggs	47	32.4	53	36.6	49	33.7	50	34.7
7. Fish and seafood	127	87.2	123	84.8	124	85.8	120	83.3
8. Legumes and Nuts	27	18.3	18	12.4	28	19.1	16	11.1
9. Milk	56	38.3	44	30.3	64	44.1	55	38.2
10. Fats and oils	103	71.0	106	73.1	107	74.0	101	69.8
11. Sugar and honey	99	67.9	91	62.8	99	68.8	83	57.3
12. Miscellaneous	139	95.9	136	93.4	133	92.4	129	89.2

Table 11.5 presents the household dietary diversity scores in both groups.

Both groups had low dietary diversity scores, having an average DDS of six food groups at both times of data collection. There is no significant difference between households with and without a vegetable garden and between the first and second round of data collection within the groups. Dietary diversity scores were determined based on the number of food groups consumed during the 24 hours prior to the survey. The fact that the DDS of schoolchildren and households with and without a garden were low and did not significantly differ, implies that both groups consumed a less than desirable variety of foods. These results concur with the finding of the FNRI-DOST food consumption survey that the Filipino diet remains basically a rice-vegetable-fish combination (FNRI-DOST, 2010). In a study in La Trinidad, Philippines, where the food environment may be considered favourable because vegetables are abundantly grown and available, a considerable proportion of the children did not meet their recommended dietary needs, especially with regard to energy, vitamin A and vitamin C (Blijham *et al.*, 2007). The types of vegetables preferred by the children and the food served by the mothers could explain the low dietary diversity. Blijham *et al.* (2007) conclude that the children's nutritional status seemed to be primarily influenced by their food preferences and the level of parental control on their food intake.

The results in our study differ from those of a cross-sectional study in the province of Rizal on the association between home gardening and the dietary diversity of pre-schoolers (Cabalda *et al.*, 2011). Children from households with gardens had higher dietary diversity scores than children in households without a garden. In the Rizal study, having a home garden was positively associated with children's dietary diversity and with the frequency of vegetable consumption. The presence or absence of a garden was not significantly associated with food security.

Table 11.5. Mean dietary diversity scores of households by study group and time of data collection.[1]

Level	1st data collection			2nd data collection		
	With garden	Without garden	*P*-value	With garden	Without garden	*P*-value
Normal	9.14±0.7 (16.6%)	9.1±0.7 (14.6%)		9.3±0.64 (11.7%)	8.6±0.52 (18.8%)	
Low	6.3±1.2 (83.4%)	6.2±1.2 (85.4%)		6.1±1.2 (88.3%)	5.9±1.16 (81.3%)	
All	6.8±1.6	6.6±1.6	0.41	6.5±1.5	6.4±1.5	0.47
P-value[2]	with garden: 0.087			without garden: 0.106		

[1] Mean ± SD.

[2] Using independent sample t-test

The commonly consumed vegetables by the households with and without a garden, respectively, were string beans (49.3 and 59.9%); horseradish (69.7 and 71.7%) and sweet potato tops (68.4 and 73.0%). In the category of leafy vegetables, squash was commonly consumed (56.6 and 64.5%) and in the group of vegetables that contain the seeds of the vegetable this was eggplant (48.7 and 46.7%). Although the difference in the consumption of households with and without a garden is not significant, it is noteworthy that households without a garden consumed more horseradish and sweet potato tops because they are always available in the neighbourhood for free and given away by neighbours. Squash is the most commonly preferred vegetable by the children. String beans are always available in the market.

The study revealed preferences of the family members. Particularly the pre-schoolers, schoolchildren and high school students disliked specific vegetables. Almost 15% of the pre-schoolers in households with a backyard vegetable garden disliked string beans, sweet potato tops, malabar nightshade and bitter gourd, while the top three vegetables that schoolchildren did not want to eat were okra (17.0%), eggplant (13.6%) and sweet potato tops (12.5%). The children disliked the taste of these vegetables and they preferred to eat more meat and fish. Of such vegetable dishes, they only consumed the broth. Almost similar results were observed among preschool children and schoolchildren in the households without a vegetable garden. Notably, children mostly disliked green leafy vegetables. Households both with and without a garden also bought vegetables, for which they gave the following reasons: (1) they had not planted that type of vegetables; (2) these are cheap and easily available, and the respondent is used to buying these vegetables; (3) preferred by household members; and (4) for their nutritional value.

This study found that the pre-schoolers, schoolchildren and high school students do not like to eat string beans, sweet potato tops, malabar nightshade, eggplant, okra and bitter gourd because of the taste and that they preferred meat and fish. During the past three decades, Filipinos on average are increasingly eating less vegetables per day (FNRI-DOST, 2010). To explain this trend, the following reasons were identified: (1) influence of family members who do not eat vegetables; (2) dislike for the taste and preference for meat; (3) preference for fast foods and instant foods; (4) fear of chemicals and pesticides; (5) vegetables require time to prepare; and (6) lack of knowledge on the nutritional and health benefits of vegetables (FNRI-DOST, 2005). The study of Potutan *et al.* (1998) showed that although vegetables are readily accessible in the area and widely grown in backyards, small farm lots and school gardens, households with a vegetable garden still weekly bought vegetables in the market.

11.4.4 Work habits and values of households on vegetable gardening

More than half of the households with a vegetable garden had been planting vegetables for more than a decade. The respondents' spouses (49.2%) and children (31.1%) help them in maintaining their garden. During the FGDs it transpired that vegetable gardening and harvesting are activities in which husband and wife work together. Respondents with a garden thought families without a vegetable garden lazy, having no interest in planting, being dependent on their neighbours with a vegetable garden, and being pre-occupied with activities like gambling and gossip. Those with a backyard garden felt proud of having their own source of vegetables and said to feel good whenever they look at their garden.

Gardening activities can promote family cooperation, particularly between the husband and wife. One of the beneficiaries of World Vision's Morning Star Project expressed this by saying that 'the gardening venture keeps the family united'. Husband and wife tend the vegetable garden together and the children join during harvest time (Rivera, 2009). However, from our research such a picture did not emerge. Our respondents indicated that tending the vegetable garden is mainly done by women who sometimes get help from the spouse and the children. Likewise, the key informants said that the vegetable program beneficiaries were the wives of farmers and households with malnourished children, thus further confirming the image of vegetable gardening as a women's concern.

11.5 Implementation of the BVGP and the role of the program implementers

Households with a garden were asked about their perceptions on the importance of the vegetable gardening program. To all of them the program is important and implementation should be continued. Suggestions for a better implementation of the program were likewise solicited. The majority of the households identified the continuous provision of planting materials, such as seeds or seedlings and fertilizers, and the conduct of seminars and training on proper planting as scope for improvement.

Similar to the households, the program implementers recognized the nutritional and economic benefits of the BVGP. The interviews with them focused on the implementation of the program. In terms of economic benefits, the program implementers considered the program a form of assistance to households, especially in depressed areas with enough space to plant. However, in its implementation they face some challenges, such as the difficulty of encouraging communities to adopt it. According to the implementers, extra effort is needed to convince the people of its importance, despite the implementers' efforts to promote it as a good program.

The key informant interviews revealed interesting insights on the issue of actual program implementation. Asked about the guidelines for program implementation, some implementers could not recall having any, but stressed that they were doing their best to implement the program through the conduct of training and distribution of seeds or seedlings, with support from the responsible government agencies. The seeds distributed were sponge gourd, bottle gourd, okra, tomato, winged beans, sweet pepper, eggplant, squash, mustard, bitter gourd, cucumber, string beans and Chinese cabbage. Another activity undertaken by the implementers was information dissemination about the BVGP to the community and schoolchildren. They recognized the importance of the support of local government, schools, community and households in the implementation of the program.

The program implementers agreed on the existence of administrative guidelines for implementing the vegetable gardening program. Although individually they could not completely recall these, they could give the following summary: (1) participating households should have a planting area of 5 square metres and the community concerned of 50 square metres; and (2) priority participants are farmer households, poor families, and households with malnourished children (in view of the hunger mitigation efforts of the government). Participants are identified based on a list of the city/municipal nutrition action officer. As recalled by the program implementers, the guidelines also include the conduct of lectures by the municipal agriculture staff on the kind of fertilizers to use, about refraining from buying recycled materials, and on the kind of plants that are suitable for the type of soil in the planting area. Part of the guidelines is about the need to convince people to plant, to give seeds or seedlings to those who are interested to plant, to help people negotiate with owners of fallow land to use it for planting, and to monitor the program. Although from their harvests the households should produce seeds for the next planting season, in reality each planting season the clients ask for seeds from the agriculture office. The program implementers submit monthly reports to the province and region including the name of recipients, quantity of seeds received and given, and harvest and production in kilograms. Some of the implementers commented that the guidelines were not clear.

Program implementers who did not acknowledge the implementation guidelines just went on with the routine tasks of giving seeds or seedlings to clients who come to their office (walk-in clients) and household beneficiaries listed by the *barangay*. They also gave seeds to the schools in their area that maintain vegetable gardens. The recipient households and schools are instructed on planting the vegetables through seminars that are funded by the *barangay* and/or the city. The implementers also do monitoring to check whether the clients planted the seeds received.

The program implementers' perceptions on the status of the implementation of BVGP in their locality appeared ambivalent. Some implementers consider the program successful because clients were given seeds to continuously plant and joined community vegetable garden contests that demonstrated the volunteerism of helping each other. In the Philippines, there is a traditional practice of *bayanihan* which denotes volunteerism or cooperative work without expecting any payment. They also said that the backyard vegetable garden served as an additional source of income to the households and that, in fact, some clients were bringing their harvested vegetables to other municipalities. The people were helping each other to sell their harvested vegetables.

However, to other implementers (41.67%), the program is not successful. They said that vegetable gardening is labour intensive and that most of the beneficiaries could not invest more time in gardening. Even if they had the seeds, they were sometimes unable to plant because of more important tasks like childcare. Other seed beneficiaries were more interested in activities like playing local card games (*tong-its)* with neighbours and friends. If they had this option, mothers preferred to engage in daily wage jobs with cash payments at the end of the day. The program implementers also hoped that households in depressed areas with limited space could plant vegetables in cans and plastic containers, particularly in urban areas. Experiences like heavy rains and typhoons that destroyed vegetable gardens affected program implementation negatively. The implementers said that the program is more successful in school gardens because the teachers were focused on its implementation, unlike the mothers who seemed unaware of the importance of having a vegetable garden in their backyard or did not have enough time for gardening. In the study of Conway (2015), from in-depth interviews it transpired that lack of time was the common barrier for the respondents to growing edible plants.

Backyard gardening is considered as a strategy to support the basic needs of the family by planting vegetables, even in small vacant lots and both in rural and urban settings. However, in some areas, vegetable gardening is not a priority program, but, instead, solid waste management and environmental beautification are priorities. In the program participation survey of the 2008 National Nutrition Survey (NNS), it was found that only 51.8% of households participated in the government's food production program that promoted integrated backyard gardening. High participation was noted in Region IV-B in Southern Tagalog, in Region 6 in Visayas and in the Caraga Region in Mindanao, while there was low participation in the NCR (FNRI-DOST, 2010).

11.6 Conclusion and recommendations

This research brings into focus basic issues and concerns related to vegetable gardening, consumption of and perceptions about vegetables, and the implementation of the backyard vegetable garden program. These issues were investigated in relation to dietary behaviour of the schoolchildren and households, and perceptions on food security.

The program on vegetable gardening appeared as a regular component of the municipal agriculture office's activities particularly in terms of seed distribution to the *barangays* and clients. The delivery system of vegetable gardening as a program changes depending on the priorities and agenda of the current government administration. If home food production is a priority, investments through resource allocations are put in place, the backyard vegetable garden as a case in point. Implementation of national programs like this is often well covered by the media. Thus, information and updates on program activities, conditions and criteria, coverage and performance are captured and reported in public. This research explored the realities beyond the public exposure by interviewing the (intended) participants of the program and the people who are assigned the task of running the program.

At the time of the study, the 'enhanced' vegetable gardening program had just concluded its last phase of implementation. In this research, we asked about the program's implementation guidelines. In all study areas, however, the interviewed program implementers were divided about the presence of such guidelines and those who acknowledged the guidelines could not completely recall these. There is a need to formulate appropriate guidelines on vegetable gardening or improve existing guidelines. Based on this study, such guidelines should include items like checks on access to garden inputs like seeds, seedlings and fertilizers, available space to cultivate, mastering planting techniques, and access to water. The respondents themselves also identified the need for innovative technologies in vegetable gardening. Furthermore, monitoring of the vegetable gardens should be pursued as an important program component. This is worth doing because of the positive view of both program implementers and households on the benefits of vegetable gardening, regardless of study site.

This research was conducted in a natural condition in the *barangays* where just the regular BVGP activities were carried out. The findings show that primarily mothers are involved in tending the garden, with some help from the spouse and children. In the program implementers' view, the gardening program is intended for housewives of farmers and households with malnourished children, indicating that they see gardening as a women's concern. Households with and without a garden did not differ in knowledge about the importance of vegetables consumption. Unexpectedly, having a vegetable garden did not affect the dietary behaviour of children and households. Schoolchildren and households with and without a garden had equally low dietary diversity scores, which could be due to the children's general lack of preference for vegetables. Hence, there is a need to strengthen nutrition education to increase the mothers' and caregivers' knowledge and awareness on the importance and functions of vegetables for bodily health.

It is recommended that the backyard vegetable gardening program be put under more comprehensive scrutiny, taking into account the annual resources allocated for it. As revealed in this study, at the conceptual level the people do recognize the benefits of gardening. However, the actual implementation of the gardening activities at the local level does not yet yield the desired effects. So, the program should be revisited to enhance its efficacy. Issues that could be important in this respect include the way in which seeds are distributed, the timing of seeds distribution, the type of seeds distributed and the strategies for client-households to sustain their vegetable garden.

Acknowledgment

The authors would like to thank the Neys-Van Hoogstraten Foundation, the Netherlands, for the financial support to conduct this research project. Our deepest appreciation goes to the Department of Agriculture and the Bureau of Plant Industry (BPI-DA) for their administrative support. Grateful recognition is extended to the *barangay* officials, community workers, study households and the mothers and schoolchildren for providing data and valuable insights.

References

Beuchelt, T.D. and Virchow, D., 2012. Food Sovereignty or the human right to adequate food: which concept serves better as international development policy for global hunger and poverty reduction? *Agriculture and Human Values* 29: 259-273.

Blijham, N., de Kan L. and Niehof, A., 2007. Determinants and adequacy of food consumption of children in La Trinidad, Philippines. *International Journal of Consumer Studies* 31(3): 195-203.

Cabalda, A.B., Rayco-Solon, P., Solon J.A. and Solon, F.S., 2011. Home gardening is associated with Filipino preschool children's dietary diversity. *Journal of the American Dietetic Association* 111(5):711-715.

Conway, T., 2015. Motivations and barriers to home-based edible gardening. (Working Paper) Mississauga, Toronto: University of Toronto, Department of Geography. Available at: http://tinyurl.com/yb4da6rg.

FNRI-DOST, 2005. FNRI-DOST Annual Report 2005. Manila: Food and Nutrition Research Institute.

FNRI-DOST, 2010. Philippine Nutrition Facts and Figures 2008. Manila: Food and Nutrition Research Institute.

FAO, 2013. *The state of food insecurity in the world: the multiple dimensions of food insecurity*. Rome: Food and Agriculture Organization of the United Nations.

HKI, 1998. Gardening and nutrition education surveillance project (NGNESP) monitoring report, Round 12, October 1997 to January 1998. Dhaka: Helen Keller International.

Kennedy, G., Ballard, T. and Dop, M.C., 2011. *Guidelines for measuring household and individual dietary diversity*. Rome: Food and Agriculture Organization of the United Nations.

Kiess, L., Bloem, M.W., De Pee, S., Hye, A., Khan, T., Talukder, A. and Huq, N., 1998. Bangladesh: Xerophthalmia free. The results of an effective vitamin A capsule program and homestead gardening. (Abstract). In: *American Public Health Association 126th Annual Meeting Report*. Washington DC: American Public Health Association, p.361.

Kortright, R. and Wakefield, S., 2011. edible backyards: a qualitative study of household food growing and its contributions to food security. *Agriculture and Human Values* 28(1): 39-53.

Patel, R., 2009. Grassroots voices: food sovereignty. *The Journal of Peasant Studies* 36(3): 663-706.

Potutan, G.E., Janubas, L.G., Arnado, J.M., Holmer, R.J, and Schnitzler, W.H., 1998. Peri-urban vegetable production consumption and marketing in Cagayan de Oro, Philippines. (Conference paper) *Peri-Urban Vegetable Production in the Asia Pacific Region for the 21st Century*. Bangkok, 28 September -1 October 1998: KURDI and AVRDC.

Potutan, G. E., Janubas, L.G., Arnado, J.M., Holmer, R.J. and Schnitzler, W.H., 2005. Urban agriculture in Cagayan de Oro: a favourable response of city government and NGOs. In: Bakker, N., Dubbeling, M., Gundel, S., Sabel-Koschella, U. and De Zeeuw, H. (eds) *Growing cities, growing food: urban agriculture on the policy agenda. A reader on urban agriculture.* Feldafing, Germany: Deutsche Stiftung für Internationale Entwicklung, Zentralstelle für Ernährung und Landwirtschaft, pp. 413-428.

Quaye, W., Adofo, K., Madode, Y.E. and Abizari, A.R., 2009. Exploratory and multidisciplinary survey of the cowpea network in Tolon-Kumbungu district of Ghana: a food sovereignty perspective. *African Journal of Agricultural Research* 4(4): 311-320.

Rivera, K., 2009. Philippines: backyard gardening – ticket out of hunger and malnutrition. Available at: http://reliefweb.int/node/304451.

Swindale, A. and Bilinsky, P., 2006. *Household dietary diversity score (HDDS) for measurement of household food access: indicator guide.* Washington DC: Academy for Educational Development, Food and Nutrition Technical Assistance Project.

Talukder, A., Kiess, L., Huq, N., De Pee, S., Darton-Hill, I. and Bloem, M.W., 2000. Increasing the production and consumption of vitamin A-rich fruits and vegetables: Lessons learned in taking the Bangladesh homestead gardening programme to a national scale. *Food and Nutrition Bulletin* 21(2):165-172.

Windfuhr, M. and Jonsén J., 2005. *Food sovereignty: towards democracy in localized food systems.* United Kingdom, ITDG Publishing: FIAN International. Available at: http://www.ukabc.org/foodsovpaper.htm.

12. Developing homestead food production in Muslim Mindanao, Philippines: does mode of implementation matter?

M.F.D. Reario[1], P.J. Pacheco[1], E. Lafuente[1] and A. Talukder[2]*

*[1]Helen Keller International, the Philippines; [2]Helen Keller International, Asia Pacific Region, Cambodia; *dreario@hki.org*

Abstract

In the Philippines, the Autonomous Region in Muslim Mindanao (ARMM) suffers the most from food insecurity and has the least varied food intake. This study compared the implementation of sustainable homestead food production (HFP) by a local government unit (LGU) and a non-governmental organization (NGO), and it assessed the impacts of the HFP intervention on the consumption of micronutrient-rich foods and livelihood among 300 mothers with children under five. A baseline and an endline survey were conducted in November 2011 and December 2012, in Lanao del Sur province, in the municipalities of Bumbaran (LGU) and Wao (NGO). Overall, the two models yielded similar benefits from the HFP intervention. HFP improved food and nutrition security practices in both NGO and LGU households by increasing food production and the diversity of micronutrient-rich foods. The households' economic capacity improved by the income generated from home gardens and poultry sales. Moreover, women's empowerment was reflected in more control over resources and decision-making power on allocating the income received from HFP. The results indicate that where food availability is affected by economic crisis, armed conflict as well as climate change, HFP can benefit poor households, whether managed by the local government or a NGO.

Keywords: homestead food production intervention, Philippines

12.1 Introduction

Micronutrient malnutrition, a chronic phenomenon that largely results from an inadequate intake of foods rich in micronutrients like vitamin A, iron, and zinc as well as other essential vitamins and minerals, is prevalent in the Philippines, specifically among women and children in the Autonomous Region in Muslim Mindanao (ARMM). Poor and marginal households lack adequate access to foods rich in micronutrients. General malnutrition and deficiencies in vitamin A, iron and iodine have been confirmed as public health problems and are reflected in anthropometric and biochemical data.

Rising food prices and economic crises in Asia had a considerable impact on the nutritional status of the poor, particularly that of women and young children. Evidence from financial crises in Asia demonstrated a reduction in non-staple food consumption, which was associated with increased prevalence of anaemia in women and children (Block *et al.*, 2004). In 2009, the Global Hunger Index, an indicator reflecting negative household effects of the high food prices and economic recession, identified Bangladesh, Cambodia and Nepal as having 'alarming' levels of hunger, and Philippines was ranked as 'serious' (IFPRI, 2009).

The Philippine 2013 National Nutrition Survey (NNS) showed that among children of 0-5 years old 21.5% are underweight and 33.4% are under height, and that anaemia prevalence is 39.4% among children aged 6 to 12 months. Anaemia prevalence among pregnant and lactating women proved to be 25.2 and 16.6%, respectively. At the national level, the most consumed food groups were the main staples (99.7%), followed by meat and fish (99.1%), vegetables (98.6%), condiments (96.1%) and oil (91.2%), while the least consumed food groups are milk (51.8%) and pulses (38.9%). ARMM had the lowest intake of all food groups except for sugar and main staples. Of all regions, ARMM suffered most from food insecurity and had the least varied food intake, consuming only five to eight food groups. It was followed by Eastern Visayas and other provinces in Mindanao (FNRI-DOST, 2013). The Accelerated Hunger Mitigation Program (AHMP), a package of strategies and interventions incorporated in the Medium-Term Philippines Development Plan (MTPDP) 2004-2010, aims at addressing poverty and hunger in a holistic manner.

The Mindanao region including ARMM has suffered from over four decades of armed conflict, resulting in internal displacements and overall deterioration of living standards. The people of this region are among the country's poorest. The most common type of livelihood is agriculture. As there are limited crops to harvest, food becomes scarce to the extent that no one is willing to sell. This leads to a spike in malnutrition among children in these armed conflict areas.

The education of the Muslim mothers and children has clearly suffered from the long history of conflict in the southern Philippines. Armed conflict has forced many families to flee from their homes and evacuate to another place. Such displacements aggravate difficulties in getting education for children in conflict-affected areas. Families' worries about safety and daily needs relegate schooling to the background. For those who managed to continue education, the conditions are challenging. There is a poor learning environment where school facilities, if at all available, are often below standard and teachers cannot focus on their teaching activities due to fear for their own safety and that of their own families.

The National Nutrition Council (NNC), the government policy-making body on nutrition, oversees the AHMP's implementation in the provinces. The food production program called *Gulayan ng Masa* (Vegetables for the masses) was started in 2006 to promote integrated vegetable backyard gardening with animal raising. On average, more than half of the households interviewed by NNS have participated in the program The NNC also implemented the Nutrition Education program that aims at the adoption of positive nutrition and nutrition-related practices for the consumption of nutritious and safe foods. However, only 13% of households in the country participated in this program (FNRI-DOST, 2008). From 2011 to 2016, the *Agrikulturang Pilipino* or Agri-Pinoy is the over-all strategic framework of the Department of Agriculture (DA) that guides the various services and programs to help address food security, poverty, and sustainable growth. In a survey in 2013, it was found that in all regions only 12% of the households took part in the food production program (FNRI-DOST, 2013).

Agricultural interventions to improve household food availability and dietary diversity are considered a sustainable solution, as these are expected to resolve household food insecurity and malnutrition through access to diverse foods and consumption of micronutrient-rich food. These programs can also lead to reduced household poverty (Bloem *et al.*, 2001), improved nutritional status of household members (De Pee *et al.*, 2000) and potentially empower women beneficiaries (Bushamuka *et al.*, 2005; De Pee *et al.*, 2008).

Helen Keller International (HKI) has been implementing homestead food production (HFP) programs in several countries in Asia for the past two decades. HFP started as a pilot home gardening project in Bangladesh. It focused on improving vitamin A status by encouraging households to produce and consume vitamin A-rich fruits and vegetables. However, during the 1990s research produced evidence that bioavailability of micronutrients such as vitamin A from plant foods was lower than had been originally thought. Therefore, HKI revised the model and added small animal husbandry and nutrition education. This integrated model is currently one of the major HKI interventions in Bangladesh, Cambodia, Nepal, Vietnam and the Philippines.

HKI Philippines piloted the HFP programme from July 2004 until September 2007 in Wao municipality, Lanao del Sur, an ARMM province found to score much lower on nutritional and health parameters than the national average. The results demonstrated that excess food production provided opportunities to market the products. The improved agricultural and food processing technologies enhanced the farmers' income and the community and women were empowered through higher incomes and the women's control over household resources. The evaluation of the pilot project suggest an increased level of food production at the household level between the two models. The project was a success in terms of food availability, consumption and awareness on the importance of micronutrient-rich foods. This provided evidence that the HFP program with nutrition education increased the quantity, quality (diversity) and frequency of household consumption of micronutrient-rich foods.

Given this evidence, the question arose what the most appropriate model or approach is for establishing sustainable HFP, i.e. collaborating with local government units (LGUs) or with non-governmental organizations (NGOs). The present study compared two approaches of implementing sustainable HFP: Model 1, managed by HKI partnership with the LGU, and Model 2, managed in partnership with a local NGO. We compared the two models for impacts on income generation and on dietary diversity and vitamin A-rich food intake as indicators of micronutrient intake. The research assessed the impact of household food security on the accessibility and consumption of micronutrient-rich foods and livelihood, to find the most appropriate model for the sustainability of the food security intervention through HFP. The research was targeted at households with children under five and their mothers.

The model works by providing technical assistance, training, agricultural supplies and management support to either partner, NGO or LGU, targeting primarily women farmers from poor households. In establishing the HFP model, HKI first provides initial inputs, such as seeds, seedlings, saplings and chickens to the participating households and then establishes village model farms (VMFs) in the target communities. After the external assistance by HKI is withdrawn, the VMFs serve as a continuous community-based resource to the participating households for their year-round food production. The NGO and LGU also support participation in the program by facilitating access health and agriculture services as well as to markets to sell the excess produce for additional household income.

In comparing the performances of the respective models, it should be noted that for the LGU concerned it was the first time to be involved in the implementation of a HFP project. The local NGO had been organised prior to the HFP implementation and had participated in a previous HFP project, hence it could be more of a hands-on implementer than the LGU.

To summarise, this study reports about the HFP programme as a way to achieve food and nutrition security and investigates whether the mode of implementation makes a difference. The main determinant of food insecurity is the vulnerability of people, which in turn is induced by poverty. Poverty makes people unable to feed themselves, i.e. to produce sufficient and adequate food or to raise sufficient income to buy food. Poverty eradication is thus a key factor in the achievement of household food security. HFP must be recognized as a major source of food and livelihood for the poor in developing countries like the Philippines. This is also an issue of food sovereignty, a concept that underlines the right of nations and their people to define their own food production systems (production, distribution, and consumption).

12.2 Conceptual framework

Food security is described by FAO (2013) as all people having access to sufficient, safe and nutritious food that meets their dietary needs and food preferences for an active and healthy life. Food security is anchored at the household level, since households are resource-managing, family-based groups that are geared towards meeting the primary and daily needs of their members (Niehof, 2011). Nutrition security is a sub-set of food security. It relates to the quality of individual food intake given a certain level of household food availability and access.

Food and nutrition security are key components of food well-being. Food sovereignty is another component. It has been defined as the right of people and communities to decide and implement their agricultural food policies and strategies for sustainable food production and distribution (Quaye *et al.*, 2009). Figure 12.1 highlights the principles of food sovereignty.

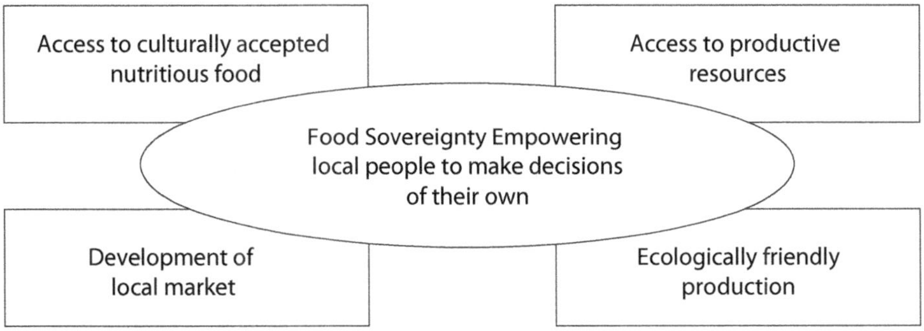

Source: Quaye *et al.*, 2009

Figure 12.1. The conceptual framework of food sovereignty (Quaye et al., 2009).

The following principles of food sovereignty can be distinguished:
- Focus is on food as a basic human right, not just a commodity for agri-business.
- Policies that undermine and threaten the livelihoods of food providers – small-scale farmers, fisher folk, indigenous peoples – are rejected.
- Emphasis on the local level by bringing together food providers and consumers through local markets, rejecting inequitable international trade and unaccountable corporations.
- Emphasis on local control by food providers over land, water, seeds and livestock, and rejecting privatising natural resources through commercial contracts and property rights.
- Support for the building of local knowledge and skills that conserve, develop, and manage local food systems.
- Using the contribution of ecosystems is used to its full extent with low external input, and seeking to improve resilience and adaptation, especially in the light of climate change.

12.3 Methods

12.3.1 Study design and study area

Two cross-sectional surveys (baseline and endline) were conducted to assess how the HFP implementation was managed by the LGU model, which involved 100% Muslim household beneficiaries versus the NGO model, in which had 80% of the households were Roman Catholic. The baseline survey was conducted in November 2011 and the endline survey in December 2012. Changes in the outcome variables between baseline and endline were compared between the two models. Two municipalities were selected in the province of Lanao del Sur, the municipality of Bumbaran (LGU model) and the municipality of Wao (NGO model).

The identification of the four beneficiary villages was based on the prevalence of undernutrition as depicted in the occurrence of underweight children in the child weighing records or the *Operation Timbang* (OPT) of the municipality. The OPT is the annual weighing and height measurement of all children under 6 years old (pre-schoolers) to identify and locate malnourished children. Other selection criteria were the availability of space (land) for household vegetable and poultry production and an interest to participate in the project. All respondents were informed about the purpose of the survey and their verbal and written consent was obtained beforehand. They were assured that the data would be dealt with confidentially.

12.3.2 Respondents, survey tool and data collection

Respondents for the two surveys were 300 mothers or the head of households with children under five years of age from the villages of the two municipalities. They were randomly selected based on the master list of the Rural Health Midwife (RHM) assigned in the study area and had been a recipient of the HFP project.

A pre-coded and structured survey questionnaire was used to collect data on household characteristics, economic and livelihood status, vegetable and animal poultry production, household food consumption, vitamin A knowledge and women's empowerment. The survey tool was adapted from a previous HFP project conducted by HKI, Philippines, i.e. the HFPP Baseline Survey (HKI/Philippines, 2005). The questionnaire was translated into the local language and was further improved during the enumerators training.

For each model, a survey team of five members, a supervisor and four enumerators, conducted the data collection. The enumerators were trained prior to the baseline and endline surveys. During the practicum, the survey tool was pretested for consistency, correct translation, and sequence of the questions. Data collection was conducted simultaneously in the two municipalities (models). HKI field supervisors supervised the data collection and conducted the data quality check to ensure valid and consistent data.

12.3.3 Data processing and analysis

Prior to data processing, the accomplished questionnaires were checked for completeness. The quantitative data from baseline and endline surveys were analysed by generating statistics using Epi Info 7.0 and Stata. The two models were then compared for impact assessment.

12.4 Results

12.4.1 Characteristics of the study population

Table 12.1 presents the characteristics of the two groups of households at baseline and endline.

As shown in Table 12.1, Roman Catholicism is the predominant religion in the NGO model, Islam (100%) in the LGU model. The households in the NGO group include settlers from other parts of the country that are mainly Roman Catholic. The NGO concerned (RIC) is religion-neutral; it is not a Roman Catholic NGO. The average household size is slightly higher in the LGU model (7 baseline and 8 endline) than in the NGO model (5 baseline and 6 endline). The majority of the household heads was married (97% in the NGO model and 96% in the LGU model). Most of the households have one to two income earners (85 and 91% in the NGO model and 83 and 86% in the LGU model, at baseline and endline, respectively). The household monthly income pattern reveals differences between the NGO- and LGU households and also shifts between the baseline and endline measurement. Educational levels were higher in the NGO group.

Table 12.2 shows that agriculture is the main source of income for most households at baseline and endline in both municipalities (NGO and LGU model) and that it gained importance as a source of income at endline, at the expense of entrepreneurial activities as a source of income. Main crops produced are rice and maize. The vegetable home garden and poultry products have increased income from agriculture for both groups between baseline and endline. Because the households focused more on their efforts in agriculture, entrepreneurial activity declined.

The respondents were asked about their source of drinking water and the presence of a toilet facility in the house. For the two groups, the main source of drinking water was piped water. Most households in the NGO group use toilet facilities that are not water-sealed like a latrine/pit or drop/overhang (57% at baseline and 53% at endline), followed by flush toilet (38 and 40%), while few defecate in the bush, open field, river or pond (4 and 6%). Most households in the LGU group used a flush toilet (40% at baseline and 52% at endline),

Table 12.1. Characteristics of NGO- and LGU-households at baseline and endline.[1]

Information	Baseline		Endline	
	NGO (n=146)	LGU (n=145)	NGO (n=142)	LGU (n=135)
Education				
No education	1.4%	15.9%	0.7%	5.9%
Did not finish elementary	24.7%	29.7%	28.9%	37.8%
Elementary graduate	18.5%	21.4%	14.8%	22.2%
Did not finish high school	27.4%	17.2%	28.9%	14.1%
High school graduate	15.8%	9.7%	14.8%	12.6%
Some college	6.2%	3.4%	7.7%	4.4%
College graduate	3.4%	-	0.7%	0.7%
Vocational graduate	1.4%	-	3.5%	0.7%
Arabic	-	-	-	1.5%
Civil status	n=150	n=150	n=146	n=142
Single	2.7%	-	2.7%	-
Married	97.3%	96.7%	97.3%	95.1%
Widow	-	1.3%	-	2.1%
Separated	-	2.0%	-	2.8%
Religion				
Roman catholic	80.1%	0.7%	78.2%	-
Islam	-	99.3%	-	100%
Other religion	19.9%	-	21.8%	-
Household size				
3-5	58.7%	23.3%	61.0%	31.7%
6-8	31.3%	34.0%	32.2%	44.4%
≥9	10.0%	42.7%	6.8%	23.9%
Monthly income				
<1000	34.0%	10.0%	1.4%	2.1%
1000-5,000	41.3%	62.0%	68.5%	81.0%
5,001-10,000	22.0%	25.3%	14.4%	12.0%
>10,000	2.7%	0.7%	15.8%	4.9%
Don't know	-	3.0%	-	-

[1] NGO and LGU refer to mode of implementation: NGO-managed or LGU(local government unit)-managed. Baseline and endline refer to the survey results before the intervention and one year after.

Table 12.2. Source of income of NGO and LGU households at baseline and endline.[1]

Source of income	Baseline		Endline	
	NGO (n=150)	LGU (n=150)	NGO (n=146)	LGU (n=141)
Agriculture	56.7%	57.3%	71.9%	71.6%
Service	-	-	26.0%	23.4%
Entrepreneurial	40.0%	37.3%	0.7%	2.8%
Professional	3.3%	4.0%	1.4%	1.4%

[1] NGO and LGU refer to mode of implementation: NGO-managed or LGU(local government unit)-managed. Baseline and endline refer to the survey results before the intervention and one year after.

closely followed by those with no toilet facility (40 and 28%) and a smaller group who made use of facilities that are not water-sealed (17 and 15%).

12.4.2 Home gardening

Type of gardens

In the Philippines, households traditionally produce vegetables and fruits in home gardens but these are not consistently available and consumed throughout the year. During the lean season, the total yield of annual and seasonal vegetables is lower. In the research, gardens were identified as traditional, improved or developed. Traditional gardens are gardens with no fixed plots, sparsely planted with indigenous vegetables like sweet potato (*camote*), horse radish (*malunggay*) or bitter gourd (*ampalaya*) and planted only in certain months of the year. Improved gardens have fixed plots and a combination of indigenous and improved vegetables but planted only in certain months of the year. Developed gardens are the same as improved gardens but are maintained throughout the year.

At baseline, the majority (71%) of the NGO households and more than one third (36%) of the LGU households had a traditional garden. Almost two thirds (62%) of LGU households and 19.3% of the NGO households did not have gardens.

Table 12.3 shows a shift towards an improved and – to a lesser extent – a developed garden for both groups, most prominent in the LGU group. This shift can be attributed to the project inputs, i.e. capacity building through the establishment of the Barangay Nursery and Poultry Farm (BNPF), nutrition education, facilitating household income generating activities, and the free provision of seeds and seedlings. Furthermore, involvement of women in decision-making played a major role in the improved practices, and the LGU monitoring is a relevant factor as well.

Table 12.3. Type of garden of NGO and LGU models, baseline and endline.[1]

Type of Garden	Baseline		Endline	
	NGO (n=150)	LGU (n=150)	NGO (n=146)	LGU (n=142)
Traditional	71.3%	36.0%	24.0%	13.4%
Improved	8.7%	2.0%	67.1%	79.6%
Developed	0.7%	-	6.8%	4.2%
No garden	19.3%	62.0%	2.1%	2.8%

[1] NGO and LGU refer to mode of implementation: NGO-managed or LGU(local government unit)-managed. Baseline and endline refer to the survey results before the intervention and one year after.

Decision-making on gardening results

Table 12.4 shows that it is mostly wives who receive the money from selling vegetables and decide on its spending. Their agricultural skills were enhanced by the establishment of the BNPF, a consistent source of seeds, seedlings, and chicken, as well as training and demonstrations. The BNPFs are managed by trained operators and run as micro-enterprises that primarily function as resource centres in the *barangay*, through which inputs and technical assistance are channelled to the farmer households (cf. HKI/Philippines, 2008).

Table 12.4. Decision-making on vegetable sales in NGO and LGU households at baseline and endline.[1]

Decision-making	Baseline		Endline	
	NGO (n=16)	LGU (n=5)	NGO (n=35)	LGU (n=45)
Who receives the money from selling vegetables?				
Wife	87.5%	100.0%	91.4%	97.8%
Husband	6.3%	-	5.7%	-
Other (father-in-law, child, mother/ mother-in-law)	6.3%	-	5.7%	2.2%
Who decides to spend the money?				
Wife	75.0%	100.0%	94.3%	95.6%
Husband	18.8%	-	5.7%	4.4%
Other (father-in-law, child)	6.3%	-	2.9%	-

[1] NGO and LGU refer to mode of implementation: NGO-managed or LGU(local government unit)-managed. Baseline and endline refer to the survey results before the intervention and one year after.

HKI's HFP model encourages both the improved and developed gardens, but promotes and assists households to ultimately establish the developed type of garden together with animal husbandry.

Although the NGO households could benefit from a more experienced implementer than the LGU households because of the NGO's active involvement in the earlier HFP Project, the NGO has a limited support in terms of resources. The NGO implementers closely coordinated with the *barangay* officials for assistance in transportation and monitoring activities. They always conducted feedback meetings for every activity and laid out future plans during meetings. For the LGU, manpower and financial support are available from the government agricultural sector, which played an integral role in the improved LGU performance.

Irrigation and source of seeds

At baseline, 54% of the NGO households and 49% of the LGU households had no irrigation for their home gardens. The sources of irrigation were the community system (23% in NGO and 39% in LGU) which was piped into either their yard or plot, a public tap, or water from a protected well or spring (5% NGO, and 15% LGU). At endline, the majority of households in both models irrigated their home gardens by using the community water system (52% NGO and 78% LGU) or a protected well or spring (34% NGO and 11% LGU). Only a few households (4% NGO and 5% LGU) claimed their home gardens were not irrigated.

Regarding the source of vegetable and fruit seeds, the baseline survey results indicate that the large majority of households (83% NGO model and 88% LGU model) sourced vegetable seeds and seedlings in the market. Other households from the NGO model claimed they used seeds and seedlings from their own garden (13%) and from neighbours (9%). In the LGU model, vegetable seeds were provided by the local government (8%). At endline, 89% of the NGO households and 80% of the LGU households derived seeds and seedlings from various NGOs. Still, 9% of the NGO households used seeds from their own garden, while 16% of the LGU households obtained seeds from the local government. At baseline, the saplings were either from own garden (26% NGO and 44% LGU), neighbours (26% NGO and 14% LGU), relatives (18% NGO and 8% LGU) or already existed (17% NGO and 11% LGU). At endline this pattern was similar.

Vegetable and fruit varieties cultivated

At baseline, 81% of the NGO households and more than one third (38%) of the LGU households planted vegetables. At endline, nearly all households in both municipalities (98 and 97%, respectively) did so, which implies a substantial increase of LGU households growing vegetables.

Households cultivated a variety of vegetables at both survey times. At baseline, for the NGO households this ranged from 1 to 13 varieties and for the LGU model from 1 to 10 varieties, which is slightly less. At endline, the range had increased to 16 for the NGO households and 17 for the LGU households. The proportions of households growing fruit trees were almost the same at baseline and endline, but higher for NGO households (83 and 84%) than for LGU households (24 and 25%). The NGO households demonstrated a slightly higher median number of varieties of trees in fruit-bearing stage than the LGU households.

Quantities produced, and use of production

Figure 12.2 illustrates the median quantities of vegetables (in kilograms) produced among households with home gardens in the month preceding the surveys. At baseline, households in both models produced on average 2 kilos of vegetables. At endline, production had increased in both models, for LGU households to 20 kilos and for NGO households to 12 kilos.

As can be seen in Figure 12.3, for almost all households (NGO 98 and 99%, LGU 93 and 99%, at baseline and endline, respectively) the vegetables produced were primarily for household consumption. Only in a few cases the home garden produce was sold.

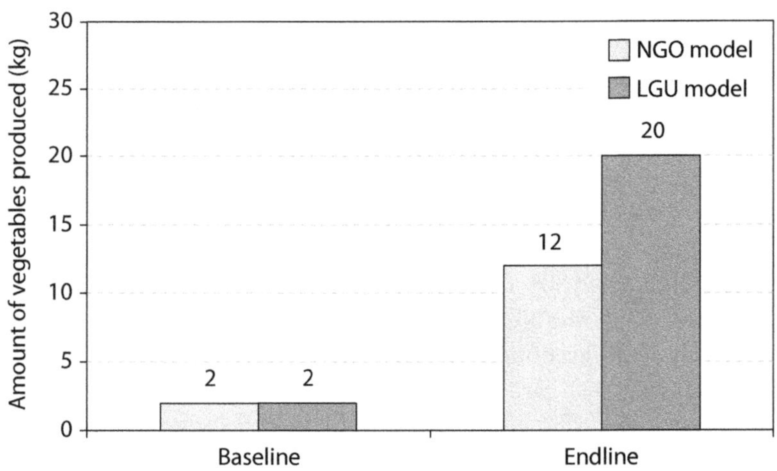

Figure 12.2. Median quantity of vegetables produced during the month prior to the surveys.

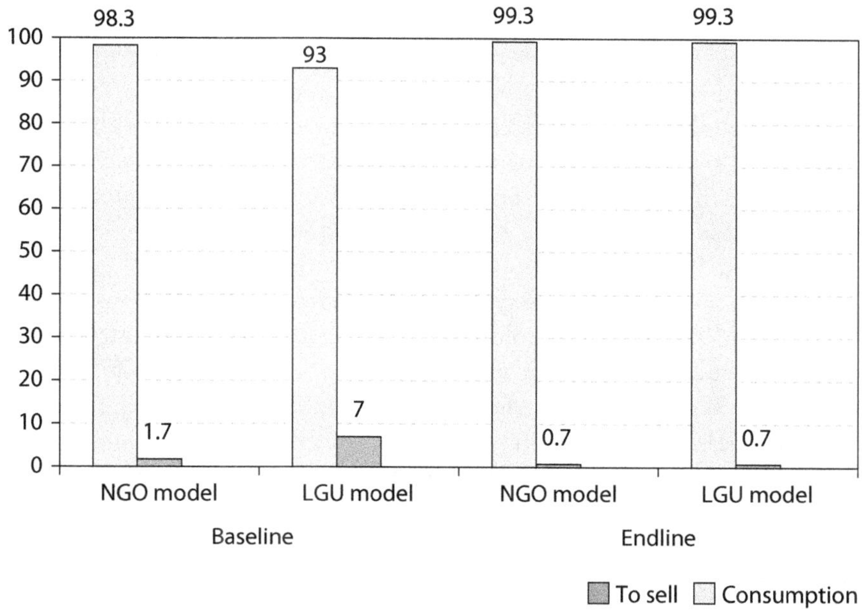

Figure 12.3. Main use of vegetables from home garden.

12.4.3 Poultry production

Households practiced poultry and duck raising. At baseline, a higher proportion of NGO households (64%) than LGU households (39%) raised poultry. At endline, the proportions were higher in both models (NGO 87% and LGU 63%). At baseline, few households had ducks (3%) and this remained so. Table 12.5 shows the median numbers of chicken possessed by NGO and LGU households on the day of interview, with that of the NGO households increasing at endline.

Table 12.5. Number of chickens owned at baseline and endline.[1]

Number of chickens	Baseline		Endline	
	NGO	LGU	NGO	LGU
Minimum, Maximum	1 – 30	0 – 25	1 – 100	0 – 24
Median	7	3	9	3

[1] NGO and LGU refer to mode of implementation: NGO-managed or LGU(local government unit)-managed. Baseline and endline refer to the survey results before the intervention and one year after.

More than half of NGO households (56 and 53% at baseline and endline, respectively) had the practice of letting chickens roam in open field. Others confined the chickens in the house at night (18 and 37%, at baseline and endline, respectively) or in cages (17% at baseline, 32% at endline). LGU households had the practice of keeping chickens in the house at night (34 and 44% at baseline and endline, respectively) or keeping them in cages (21 and 42% at baseline and endline, respectively). At baseline, some respondents claimed chickens were kept in a poultry house while none did so at endline.

Most respondents said to feed the chickens with feeds from their own household (NGO households 82 and 59%, LGU households 66 and 71%, at baseline and endline, respectively), while others said the animals fed in an open field (30 and 43%, and 34 and 38%, in the same order). Only some households bought feeds from the market (NGO households 8 and 11% and LGU households12-8%, at baseline and endline, respectively). At endline, other sources of poultry feeds included cornfields, stores, and provided by HKI and friends.

As Figure 12.4 shows, the majority of households (77.1-93.7%) in both models and at both times raised chickens primarily for home consumption. Relatively few households in both models (NGO: 10 and 6%; LGU: 12 and 18%, baseline and endline, respectively) raised chickens for additional income. Other purposes included raising chickens for cockfight or for egg production.

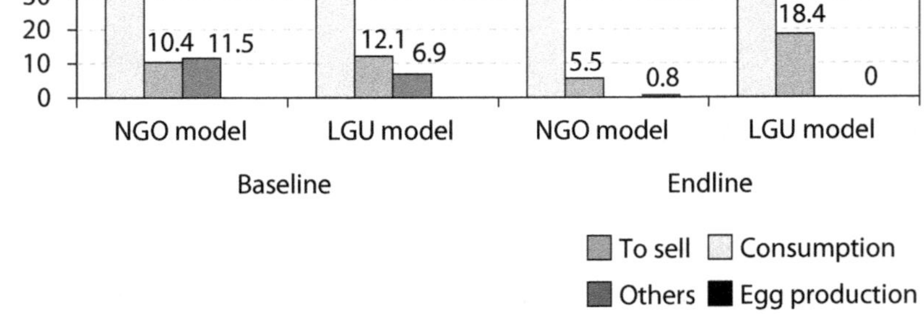

Figure 12.4. Main purpose of raising poultry.

12.4.4 Technical assistance

At endline, the majority of the households in both models (NGO: 89.7%, LGU: 91.5%) declared the NGO as the source of technical assistance on HFP (Table 12.6). Few households (5.5 and 4.2% for NGO and LGU households, respectively) did not receive technical assistance.

Table 12.6. Source of technical assistance on HFP activities, endline, November 2012.[1]

Source of technical assistance	NGO model (n=146)	LGU model (n=142)
No assistance	5.5%	4.2%
GO (government organization)	2.7%	2.1%
NGO	89.7%	91.5%
Others (such as *barangay* captain, mother or neighbour)	2.1%	2.1%

[1] NGO and LGU refer to mode of implementation: NGO-managed or LGU(local government unit)-managed.

12.4.5 Income from home gardening and poultry production and its utilisation

One of the intended benefits of the HFP program is an increase in household income from the sale of surplus produce from HFP. This is critical for poor families who are mainly subsistence farmers. There was a remarkable improvement of income from vegetable garden sales a month prior to interview among the NGO households from baseline (7%) to endline (30%). Among LGU households, there was a negligible decline (33 and 31%, respectively). Notably, at endline the proportion of households receiving income from garden sales is for both models at more or less the same level (30-31%). Poultry is another source of income generation. Results showed that income from poultry sales increased for both NGO households (from 67% to 89%) and LGU (from 75% to 100%) households between baseline and endline.

Table 12.7 shows how households in both models utilise the income generated from sales of home garden products and poultry. In general, such income was used for purchasing foods, education, housing, medicines and other things, with a higher proportion of NGO households (93.8 and 88.6%, baseline and endline, respectively) spending the income from the home gardens on food. A few NGO households (2.9%) saved the additional income.

The LGU households used the income from home gardens more for food and less for housing and clothing when comparing baseline and endline. For NGO households education expenditures increased between baseline and endline. Similarly, additional income from poultry was also utilised for foods, education, housing, and clothing.

Table 12.7. Income utilisation from home gardening, NGO and LGU, baseline and endline.[1]

Items[2]	NGO		LGU	
	Baseline (n=16)	Endline(n=35)	Baseline(n=5)	Endline(n=45)
Food	93.8%	88.6%	60.0%	80.0%
Education	6.3%	22.9%	20.0%	24.4%
Housing	12.5%	17.1%	40.0%	15.6%
Medicine	-	-	-	4.4%
Saved	-	2.9%	-	-
Clothing	-	-	20.0%	-
Other (soap, toothpaste, fare, seeds)	6.3%	17.1%	20.0%	4.4%

[1] NGO and LGU refer to mode of implementation: NGO-managed or LGU(local government unit)-managed. Baseline and endline refer to the survey results before the intervention and one year after.
[2] Multiple answers possible.

Figure 12.5 reveals the type of foods purchased from the garden sales at baseline and endline. The category of other food is interesting. It includes canned goods, noodles and bread.

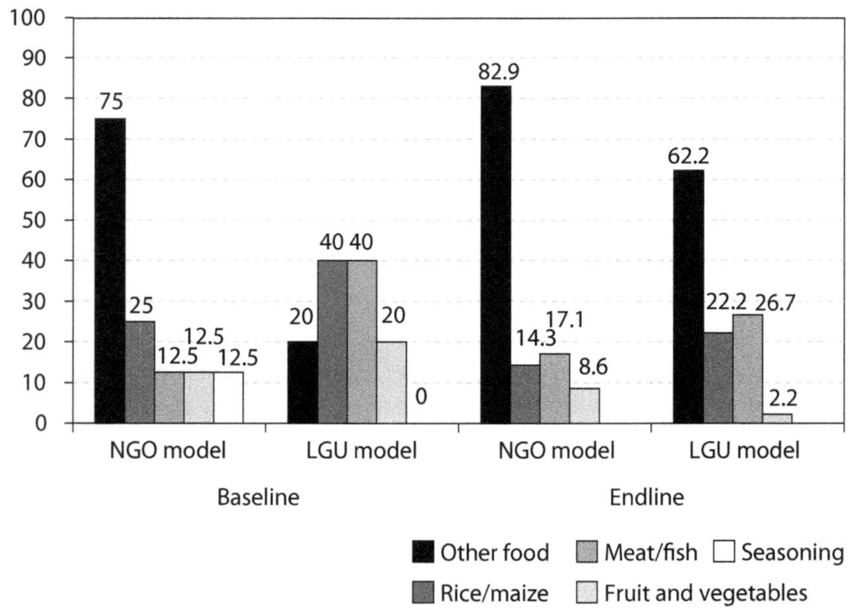

Figure 12.5. Food items purchased with income earned from home gardening.

Especially the NGO households spent the money from sales on this category of food, at both times. For LGU households spending on other food increased considerably between baseline and endline. Similar purchases were observed for the income from poultry production in both models at baseline and endline, i.e. mainly rice or maize and other foods (canned goods, noodles, bread).

12.4.6 Household food consumption

One of the main goals of the HFPP is to increase micronutrient-rich food availability and diversity in the households, and particularly to increase consumption of micronutrient-rich food that are essential for mothers and children under five years of age. Table 12.8 presents the answers to the question about the food items consumed in the last seven days prior to the survey.

The findings suggest that the HFP program improved overall household access to a wide variety of foods, although not necessarily all micronutrient-rich foods. Overall, in the majority of both the NGO and LGU households dark green leafy vegetables were consumed almost daily, slightly more at endline than at baseline. Consumption of other foods such as yellow vegetables, chicken and other poultry products, eggs and egg products also increased for both groups. However, so did the consumption of canned foods, which from a nutrition perspective is less desirable.

Table 12.8. Household food consumption, NGO and LGU, baseline and endline.[1]

Food item[2]	NGO		LGU	
	Baseline (n=150)	Endline (n=146)	Baseline (n=150)	Endline (n=142)
Dark green leafy vegetables	96.7%	97.3%	81.3%	95.1%
Yellow vegetables	47.3%	59.6%	59.3%	78.9%
Chicken and other poultry products	38.7%	51.4%	14.0%	19.7%
Eggs and egg products	55.3%	71.2%	38.0%	53.5%
Canned goods	55.3%	61.6%	40.7%	45.8%

[1] NGO and LGU refer to mode of implementation: NGO-managed or LGU(local government unit)-managed. Baseline and endline refer to the survey results before the intervention and one year after.
[2] Multiple answers possible.

12.4.7 Food shortages

Respondents were also asked how they dealt with food shortages. Table 12.9 presents the answers.

Table 12.9. Food shortage scenarios of households with food shortage during 30 days prior to the survey, baseline and endline.[1]

Food shortage scenarios	NGO		LGU	
	Baseline (n=150)	Endline (n=146)	Baseline (n=150)	Endline (n=142)
Household member(s) were not able to eat the kinds of foods they like to eat	88.7%	71.9%	94.6%	85.2%
Sometimes (three to ten times)	27.1%	47.1%	27%	32.2%
Often (more than ten times)	61.7%	38.5%	63.8%	47.9%
Household member(s) ate only a few foods	81.3%	74%	93.3%	85.9%
Sometimes (three to ten times)	35.2%	38%	29.3%	54.1%
Often (more than ten times)	44.3%	34.3%	60%	28.7%
Household member(s) ate some foods that they really did not want to eat	58.7%	38.4%	84%	75.4%
Sometimes (three to ten times)	37.5%	48.2%	43.7%	52.3%
Often (more than ten times)	25%	21.4%	26.2%	17.8%
Household member(s) ate a smaller meal	66.0%	58.2%	91.3%	81.7%
Sometimes (three to ten times)	52.5%	48.2%	36.5%	54.3%
Often (more than ten times)	27.3%	18.8%	52.6%	24.1%
Household member(s) ate fewer meals a day	36.0%	21.2%	71.3%	50.0%
Sometimes (three to ten times)	42.6%	48.4%	39.3%	40.8%
Often (more than ten times)	20.4%	22.6%	37.4%	28.2%
No food to eat of any kind in the house	46.7%	30.8%	68.7%	43.7%
Sometimes (three to ten times)	35.7%	40%	35.9%	45.2%
Often (more than ten times)	25.7%	15.6%	35%	16.1%
Household member(s) go to sleep hungry	32.0%	15.8%	71.3%	38.0%
Sometimes (three to ten times)	31.3%	34.8%	46.7%	29.6%
Often (more than ten times)	10.4%	8.7%	14%	11.1%

[1] NGO and LGU refer to mode of implementation: NGO-managed or LGU(local government unit)-managed. Baseline and endline refer to the survey results before the intervention and one year after.

At baseline, the majority of both NGO households (88%) and LGU households (97%) reported that within the past 30 days they were worried that the family would not have enough food. At endline, the proportions were slightly lower: 60 and 82%, respectively. Among the LGU respondents at baseline 57% said to worry 'often', at endline 'only sometimes' (38%). For the NGO households there was not much difference between the two times. Table 12.9 shows that the group of respondents that answers that there were occasions when there is just no food of any kind in the house to eat is quite large, in both NGO and LGU households, though declining between baseline and endline. For both NGO and LGU households the number of respondents who report that household member(s) may have to go to sleep hungry declines by almost 50% between baseline and endline. The scenario of the reduction of number of meals per day also prevails less at endline than at baseline. The other scenarios are not as dramatic as the previous ones, but do occur more frequently.

12.4.8 Breastfeeding and complementary feeding

The WHO guidelines for infant and young child feeding recommend exclusive breastfeeding for infants up to six months old, to initiate breastfeeding within one hour at birth, to feed colostrum, and to introduce adequate and safe complementary foods at about six months. Nearly all mothers breastfed their child in both the NGO and LGU sites at baseline and endline. Breastfeeding initiation within one hour after birth was higher for NGO mothers than for LGU mothers at baseline and endline (47.2 and 69.0%; 27.1 and 54.5%, respectively). Overall, giving of colostrum to the reference child was higher at endline. Both NGO and LGU mothers gave prelacteals such as honey, *ampalaya* (bitter gourd), water, and sugar water to the reference child at both times, the proportion being higher for LGU mothers. Table 12.10 summarises the mothers' knowledge and practice on when to start complementary feeding.

Four months is the mean duration of exclusive breastfeeding for both NGO and LGU mothers. As Table 12.10 shows, knowledge and practice of introducing others foods in addition to breastmilk varies in the two models and between baseline and endline. Both knowledge and practice about postponing giving complementary food to an age of at least six months improved for NGO as well as LGU mothers between baseline and endline. The findings indicate the need for nutrition education to mothers on the timely introduction of complementary foods to their children. This could be done in mothers' classes, together with informing the mothers on topics such as child immunisation, deworming, the function of weighing, and Vitamin A supplementation.

Table 12.10. Mothers' knowledge and practice on complementary feeding, baseline and endline, NGO and LGU.

Age of introduction of (semi-)solid or soft foods	NGO		LGU	
	Baseline (n=54)	Endline (n=146)	Baseline (n=55)	Endline (n=142)
Knowledge				
0-5 months	20.4%	15.1%	40.0%	21.8%
6 months	44.4%	55.5%	40.0%	57.7%
>6 months	35.2%	29.4%	20.0%	20.4%
Practice				
0-5 months	31.4%	25.3%	47.2%	40.1%
6 months	44.4%	54.1%	32.7%	44.3%
>6 months	22.2%	19.9%	20.0%	14.8%

[1] NGO and LGU refer to mode of implementation: NGO-managed or LGU(local government unit)-managed. Baseline and endline refer to the survey results before the intervention and one year after.

12.4.9 Health and nutrition education

In both municipalities, the mothers were asked about the functions of Vitamin A and the possible consequences of Vitamin A deficiency. Between baseline and endline, this topic was addressed in mothers' classes and during counselling sessions. Table 12.11 shows an improvement of knowledge on all aspects of the problem – although on some more than others – of both NGO and LGU mothers, especially among the latter. Among LGU mothers, there is also a substantial decline of the percentage of mothers who 'don't know'.

12.4.10 Shared responsibility

As could be expected, Mindanao household organisation is gendered, with men and women having different tasks and responsibilities. In Mindanao, the husband is considered the head of the household by 75 and 86% of the NGO households (at baseline and endline, respectively) and by 66 and 81% (baseline and endline) of the LGU households. There is a slight increase in acknowledging the husband as household head. There is also an increase between baseline and endline in considering the wife as the main caretaker of the home garden (NGO households: 67 to 87%; LGU households: 84% to 91%). At the same time, at baseline in 88% of the NGO households the wife received the income from vegetables sales (at endline 91%), and in LGU households this was even 100% at baseline (and 98% at endline). Decision making on spending the money reveals a similar same pattern. The wife is the decision maker in 75 and 94% (baseline and endline) of the NGO households, and in 100 and 96% (baseline and endline) of the LGU households.

Table 12.11. Mothers' knowledge on Vitamin A, NGO and LGU, baseline and endline.

Knowledge	NGO		LGU	
	Baseline (n=144)	Endline (n=131)	Baseline (n=137)	Endline (n=97)
Functions of vitamin A				
Promotes growth and development	53.5%	79.4%	27.0%	83.5%
Good for the eyes	18.1%	39.7%	8.0%	37.1%
Boosts immune system	18.8%	31.3%	5.8%	16.5%
Others (good for the skin, hair)	6.9%	15.3%	3.6%	19.8%
Don't know	28.5%	7.6%	60.6%	5.2%
Consequences of vitamin A deficiencies				
Weak body	52.8%	77.1%	28.5%	70.1%
Retarded growth	29.9%	26.7%	14.6%	47.7%
Prone to infection	19.4%	33.6%	5.1%	26.8%
Night blindness	6.9%	7.6%	2.2%	12.4%
Others (dizziness, pale)	3.5%	9.2%	6.6%	9.3%
Don't know	15.3%	7.6%	51.1%	6.2%

[1] NGO and LGU refer to mode of implementation: NGO-managed or LGU(local government unit)-managed. Baseline and endline refer to the survey results before the intervention and one year after.

The picture is different and more diverse for poultry rearing, where the husband is more often identified as the main caretaker. In NGO households, the percentages were 64 and 67% (at baseline and endline, respectively). In LGU households, this percentage was only 41% at baseline, whereas at endline in 48% of the cases the wife was identified as the main caretaker. In all cases, however, the wives were the primary managers of income generated from the sales of poultry and its products. Table 12.12 further shows the more prominent role of wives in several aspects of HFP management (as measured at endline, November 2012).

Table 12.12. Person responsible for homestead food production (HFP) management, NGO and LGU at endline.

	Endline	
	NGO	LGU
Person responsible for the management of HFP, in general?	(n=146)	(n=142)
Wife	67.8%	88.7%
Husband	24.0%	6.3%
Both husband and wife	6.2%	-
Other (mother/ mother-in-law, child, son-in-law)	2.1%	4.9%
Who decides in general what to grow, when to sell the production?	(n=144)	(n=140)
Wife	79.2%	88.6%
Husband	17.4%	10.0%
Other (mother/mother-in-law, relative, brother)	3.5%	1.4%
Who receives the income from different activities of HFP (gardening, poultry, etc.)?	(n=90)	(n=82)
Wife	92.2%	92.7%
Husband	1.1%	4.9%
Other (child, mother/mother-in-law)	6.7%	2.4%

[1] NGO and LGU refer to mode of implementation: NGO-managed or LGU(local government unit)-managed. Baseline and endline refer to the survey results before the intervention and one year after.

12.5 Discussion and conclusion

The two models presented HFP results reflect enhanced sovereignty in using resources for food at the levels of households and communities. The household beneficiaries used their backyard for producing a variety of micronutrient-rich vegetables as well as for raising poultry. The results demonstrate that HFPP improved the food and nutrition security practices in both types of households by increasing food production and the diversity of micronutrient-rich foods. Moreover, the economic capacity of the households improved by generating income from home garden and poultry sales. Nutrition education was intensified, which shows in the enhanced knowledge of the respondents about the importance of adequate intake of Vitamin A.

Involvement of women, particularly the mothers, in the two models provided them with an opportunity for empowerment by giving them more control over resources and decision-making power on allocating the income generated from the HFP activities. The findings show that the surplus garden produce and poultry products are used for important expenditures, including food purchases, thereby increasing the diversity of the family's diet. The increased expenditure on canned foods, however, is a matter of concern.

The innovation in the LGU model is the passing of a policy that will increase production and overall agricultural productivity of small farmers with the provision of seeds, fertilizers, technology and credit, to ensure food self-sufficiency and promote sustainable agriculture. The *barangay* nursery and poultry farm BNPF is a consistent source of seeds, seedlings and animals, as well as the resource centre for the training of households. Partnership with the health and agriculture offices enabled the community health and nutrition workers to conduct nutrition classes, monitor the process of project implementation, and disseminate information on services available at the health facility.

When conflicts arise, food sovereignty and household food security are affected. In ARMM, due to the conflict, people have experienced losing their home, their land and a sense of community. The high levels of hunger, food insecurity and poverty in ARMM have often been attributed to the conflicts in the region. Despite the government's claims that food production has improved through the years, the region continues to be confronted with problems in feeding its people. This is why an intervention such as HFP is so important.

Overall, the two models yielded similar benefits in operating the HFP, regardless of factors affecting food sovereignty (religion, education, armed conflict) and the different strategies in managing the project. The NGO had participated in the previous HFP project. For the LGU it was a new experience. However, the LGU managed the project very well.

Although the NGO households could benefit from a more experienced implementer of the HFP program than the LGU households, the latter households could benefit from the LGU's access to government manpower and financial support in the agricultural sector. This played an important role in the LGU performance. It can be concluded that in conditions where food availability is affected because of economic crisis, armed conflict as well as climate change, programs like HFP benefit poor households, whether managed by the local government or a NGO.

Given the positive effect of the LGU model on food production, the local government could identify fallow land in the community to use for cultivation, together with providing capacity building for new agricultural technologies and livelihood generation. The NGO model with an organised cooperative strengthened the households' livelihood by generating additional income. This could be scaled up in the LGU model by organising a women's group to manage and market the produce of both LGU and NGO households.

Considering the almost similar benefits that could be observed for both models, it can also be concluded that it is the commitment of those who are directly involved in the implementation rather than mode of implementation that really matters. In terms of sustainability, however, the LGU model has more potential because of the manpower and financial support from the agricultural sector that can be made available. To realize this potential, proper monitoring and supervision are crucial.

Acknowledgement

Our acknowledgements go to the Neys-Van Hoogstraten Foundation.

References

Block, S.A., Kiess, L., Webb, P., Kosen, S., Moench-Pfanner, R., Bloem, M.W. and Timmer, C.P., 2004. Macro shocks and micro outcomes: child nutrition during Indonesia's crisis. *Economics and Human Biology* 21(1): 21-44.

Bloem, M.W., Moench-Pfanner, R. and Kiess, L., 2001. Combating micronutrient deficiencies – an important component of poverty reduction. *Biomedical Environmental Science* 14: 92-97.

Bushamuka, V.N., De Pee, S., Talukder, A., Kiess, L., Panagides, D., Taher, A. and Bloem, M.W., 2005. Impact of a homestead gardening program on household food security and empowerment of women in Bangladesh. *Food and Nutrition Bulletin* 26: 17-25.

De Pee, S., Bloem, M.W. and Kiess, L., 2000. Evaluating food-based programmes for their reduction of vitamin A deficiency and its consequences. *Food Nutrition Bulletin* 21: 232-238.

De Pee, S., Talukder, A. and Bloem, M.W., 2008. Homestead food production for improving nutritional status and health. In: Semba, R.D. and Bloem, M.W. (eds) *Nutrition and health in developing countries.* [2nd edition] Totowa, NJ: Humana Press, pp. 753-779.

FAO, 2013. The state of food insecurity in the world: the multiple dimensions of food security. Rome: Food and Agriculture Organization of the United Nations. Available at: http://www.fao.org/docrep/018/i3434e/i3434e.pdf.

FNRI-DOST, 2008. *Philippine nutrition facts and figures 2008.* Metro Manila, Philippines: Food and Nutrition Research Institute – Department of Science and Technology.

FFNRI-DOST, 2013. *Philippine nutrition facts and figures 2013.* Metro Manila, Philippines: Food and Nutrition Research Institute – Department of Science and Technology.

HKI/Philippines, 2005. *Baseline survey for the homestead food production in selected municipalities of Lanao del Sur.* Manila, Philippines: Helen Keller International.

HKI/Philippines, 2008. Homestead food production – an effective integrated approach to improving food and nutrition security among vulnerable households in the Philippines' Autonomous Region in Muslim Mindanao. *Philippines: Nutrition Bulletin* 3(1): 1-8.

IFPRI, 2009. Global hunger index: the challenge of hunger: focus on financial crises and gender inequality. Available at: http://tinyurl.com/ya73j529.

Niehof, A., 2011. Conceptualizing the household as an object of study. *International Journal of Consumer Studies* 35(5): 488-497.

Quaye, W., Adofo, K., Madode, Y.E. and Abizari, A.R., 2009. Exploratory and multidisciplinary survey of the cowpea network in Tolon-Kumbungu district of Ghana: a food sovereignty perspective. *African Journal of Agricultural Research* 4(4): 311-321.

13. How non-targeted households benefited from implementing a homestead food production program in Son La, Vietnam

K. Shaw[1]*, P.K. Ngoc*[2]*, H. Lan*[2]*, G. Mundy*[3*] *and Z. Talukder*[4]

[1]*Helen Keller International (internship from Mailman School of Public Health, Columbia University, New York City, USA;* [2]*Helen Keller International, Vietnam;* [3]*Helen Keller International, Asia Pacific Regional Office, Cambodia;* [4]*Helen Keller International, Bangladesh; *gmundy@hki.org*

Abstract

There is a growing body of work documenting the reach, coverage and outcomes of nutrition sensitive agriculture programs and their contribution to the nutritional status of young children. These programs operate through the provision of agricultural inputs, training in effective agricultural technologies and practice, nutrition education, along with efforts to address inequities between men and women that contribute to poor nutrition outcomes. Little is known about the potential effects that these programs have on the wider communities in which they are implemented. We conducted a survey of 250 project households and 152 non-project households from the same villages, to assess the coverage of an enhanced homestead food production (EHFP) project in Son La province, Vietnam. We found that 77% of non-project households had been exposed to at least one project intervention, with 25% engaging with at least five. The most common mechanisms of exposure were attendance at education sessions and cooking demonstrations, and receipt of nutrition counselling by community health workers. We conclude that program coverage goes substantially beyond the direct program beneficiaries and that future monitoring, evaluation and research strategies for EHFP programs should widen their scope to assess impact and outcomes at the community level.

Keywords: agriculture and nutrition intervention, spillover effects

13.1 Introduction

13.1.1 Malnutrition in Vietnam

Although one in five children are stunted in Vietnam, this figure masks significant disparities by wealth, region, and ethnicity. Stunting in the lowest wealth quintile is 41%, as compared to 6% in the highest. Additionally, the proportion of the population below the national poverty line is 17%, but 50% of the ethnic minorities are poor (UNICEF, 2014). Roughly one in three children in the Northern Midland and mountain areas and the Central Highlands are stunted, and 41% of ethnic minority children under five are stunted, 14%

DOI 10.3920/978-90-8686-864-3_13,

severely so. Postnatal factors appear to be important determinants of stunting in Vietnam. Infant and young child feeding practices include delayed initiation of breastfeeding, very low rates of exclusive breastfeeding, delayed introduction of complementary foods, and insufficient frequency of complementary feeding. Further, the prevalence of underweight among women of reproductive age (18%) still exceeds the prevalence of overweight in Vietnam and is highest among women under 30 and in the North Central, Coastal and the Red River Delta areas (UNICEF, 2014).

13.1.2 Helen Keller International's enhanced homestead food production model

Over the last 25 years, Helen Keller International (HKI) has developed an integrated nutrition-specific and nutrition-sensitive program model that uses agriculture as the delivery platform through which disadvantaged households receive a package of services to improve nutritional status targeted at young children and pregnant and lactating women. This program, referred to as the enhanced homestead food production (EHFP), was designed for implementation in rural areas, which, in most cases, are food or nutrition insecure (Haselow *et al.*, 2016).

The main objective of the program is to increase and ensure year-round availability and intake of micronutrient-rich foods in poor households, particularly by women and children. The EHFP model works through two pathways linking food production with food consumption and human nutrition: 1. subsistence-oriented production for the household's own consumption, and 2. income-oriented production for sale in markets. The building of household capacity to produce nutritious food is enhanced through behaviour change communications that promote essential nutrition and hygiene behaviours, and through the empowerment of women to be agents of change.

Although there is considerable anecdotal evidence and grey literature supporting HKI's approach as a means of improving dietary diversity, income and nutrition outcomes, there have only been four peer-reviewed publications that have assessed HKI's intervention models for linkages between any agricultural inputs and any nutrition outcomes to date. Three early studies looked at HKI's homestead gardening programs and found homestead gardening to be associated with increased production and consumption of micronutrient-rich fruits and vegetables (Bushamuka *et al.*, 2005; Jones *et al.*, 2005; Talukder *et al.*, 2000). Evidence emerged from these studies showing that HFP can also improve income, especially when income earned from farming activities remains under the control of women and their household-level decision-making power increases.

HKI's model further evolved through incorporating a greater focus on production and consumption of small-animal sourced foods. Four more recent studies have looked at this full EHFP model (Olney *et al.*, 2009, 2013, 2015; Talukder *et al.*, 2010). The most recent study was the first cluster-randomized controlled trial on the topic, and found that a two-year HKI EHFP intervention in Burkina Faso reduced anaemia, wasting and diarrhea in very young children (Olney *et al.*, 2009). Thus, although the evidence for how EHFP is linked to nutrition is growing, much work remains to be done to evaluate these programs globally and build evidence on whether and how EHFP can improve health and nutrition outcomes (Talukder *et al.*, 2000).

Based on the evidence generated to date, EHFP has evolved into an integrated package that includes agriculture, nutrition, water/hygiene/sanitation, linkages to health care, women's empowerment, income generation and advocacy for government and non-government stakeholders to adopt policies and initiatives that promote improved nutrition for mothers and young children (Haselow *et al.*, 2016).

Spillover and community level effects of enhanced homestead food production programs

While there is peer reviewed and grey literature evaluating the effects of EHFP programming on households that directly receive support from HKI programs, until this point data have not been collected or used to evaluate the effects of programming on the communities in which these households are situated. During implementation of the EHFP project in Son La province, Vietnam, implementers observed that many households close to the direct beneficiary households had begun to adopt various components of the EHFP programming. This raised questions about both the extent to which practices may have been adopted by households that had not been receiving direct support through the project, as well as about the community mechanisms through which the adoption of these practices could have been brought about.

These community level changes – also known as 'spillover' – have largely gone unmeasured and unevaluated both in HKI programming and by other organizations implementing similar agriculture and nutrition programs. For the purposes of this study, spillover is defined as the reach and coverage of program activities beyond the immediate project beneficiaries. HKI is in the process of developing methodologies and measures to standardize how community-level spillover is captured. The research reported on here represents the first step towards a better understanding of the wider community level transformations brought about by targeted EHFP programs.

The objectives of this study are limited to understanding the extent of community level spillover, and the mechanisms through which this has occurred. This can be phrased as the research question: What is the additional coverage of the project – beyond the intervention households – and how is this occurring? The study does not look at the consequences of spillover for household behaviour or nutrition outcomes. To do this would require a design that tracked community level effects from baseline to endline, together with a control population. While this is planned for future studies, this study investigated the spillover that emerged during the final phases of the project as a first attempt at understanding its extent and mechanisms.

The enhanced homestead food production project in Son La, Vietnam

Helen Keller International Vietnam conducted a pilot project known as the '*Improved household food and security and nutrition through enhanced homestead food production in Son La Province, Vietnam*' to address the nutrition and food security challenges in the area. The EHFP broadly includes agricultural production activities and training, as well as nutrition, sanitation and hygiene education using behaviour change communication approaches. It aims at increased production and consumption of micronutrient rich foods by establishing homestead food production support for target beneficiaries. The ultimate goal of EHFP in Vietnam is to improve the nutrition and micronutrient status of mothers and children aged less than two years in Tan Lang Commune, Phu Yen District, Son La Province.

Vietnam has 53 minority ethnic groups that together represent 16% of the total population. Studies and national data identify most of these minority groups as being poorer and suffering worse health outcomes than the majority Kinh ethnic group (Målqvist *et al.*, 2013). The EHFP program in Son La gives particular focus to ethnic minority households, particularly those from the Dao and Muong ethnic groups, who are commonly found to have poor health outcomes compared with the Kinh majority (Målqvist *et al.*, 2013).

EHEP linked households to health services, provided household financial management training, and facilitated health and nutrition behaviour change. The nutrition sensitive agricultural activities included six seasonal distributions of vegetable seeds and seedlings to Village Model Farms (VMFs) and project households, the distribution of 2,684 chicks to VMFs and households (complete with vaccinations and disinfection solution); the distribution of fifteen pigs to VMFs, and the circulation of bio-compost education and tools to VMFs and households. These activities began in 115 households with children under the age of two in 10 communes located in the Son La province in 2013, later expanding to 352 households in 12 communes in 2014, reaching a total of 415 households in mid-2016.

At the program's start, VMFs were created to assist in the adoption of EHFP practices. These farms served as sites for training and demonstration of improved agriculture and livestock techniques, with each providing technical assistance to 15-20 beneficiary households in their commune. While the VMF was established as a resource centre for households enrolled in HKI's EHFP programming, the sites were open resources that any community member could access. A total of 20 Village Model Farmers were trained to serve as a community resource centre, promoting the garden-to-plate approach, which emphasizes that households need not only know how to cultivate nutrient rich crops, but also how to prepare nutritious balanced meals.

13.2 Methodology

We conducted a one-time, cross-sectional survey of women in households receiving enhanced homestead food production program inputs from HKI, together households not receiving these inputs. Both intervention and non-intervention households were from the same villages. In total, 205 project households, out of a total 416, and 152 non-project households were sampled across the 12 villages. We used simple random sampling from a list of households provided by HKI program staff and Vietnam census data.

Data were collected during September 2016 by enumerators trained by HKI and the Vietnam National Institute of Nutrition (NIN). All households included in the study had children under five years of age and the mother of the household was the respondent for the interview. In the event that a mother was not present, grandmothers were accepted substitutes. Interviews were conducted with recruited project and non-project mothers or grandmothers using a standardized consent script.

In the questionnaire, ethnicity was recorded via self-report and wealth quintiles were constructed from information on household assets provided by respondents. The extent of household isolation was measured via a dichotomous variable that recorded whether villages were located along a main road. The respondent's highest educational attainment was self-reported, as was household participation in EHFP program activities.

Data were analysed unweighted as households were not stratified and there was no clustering of sample units; all households were selected using simple random sampling, with all households in the program having equal probability of selection. Households not enrolled in HKI programming were also randomly sampled; hence, all households with a mother of reproductive age and a child under five had an equal probability of selection.

13.3 Results

Table 13.1 presents the demographic and household characteristics of the households that were participating in the project and receiving direct support, and those that were not participating in the project.

The mean ages of women across these two samples was 27 in project households and 32 in non-project households. Approximately 38% of women in project households and 36% of women in non-project households had not completed secondary school, and 36% of women from project households completed secondary school or higher, whilst only 20% of non-project household women did the same. Forty-five percent of the project households and 43% of non-project households were members of the Muong ethnic group and about 21% of project households and 19% of non-project households were members of the Dao ethnic group. Nearly 44% of participant household women fell into either the lowest or

Table 13.1. Population characteristics of project and non-project households.

	Participant	Non-participant
Ethnic group		
Muong	44.9% (92)	43.4% (66)
Kinh	34.6% (71)	37.5% (57)
Dao	20.5% (42)	19.1% (29)
Proximity to road		
Near main road	62.9% (129)	61.2% (93)
Not near main road	37.1% (76)	38.8% (59)
Mother education level		
Not finished secondary school	26% (53)	43% (65)
Not finished high school	37.7% (77)	36.4% (55)
Finished high school or higher	36.3% (74)	20.5% (31)
Wealth quintile		
Lowest	25.4% (52)	13.2% (20)
Second	18.5% (38)	21.7% (33)
Third	21.5% (44)	18.4% (28)
Fourth	17.6% (36)	23% (35)
Highest	17.1% (35)	23.7% (36)

second lowest wealth quintiles while only 35% of non-project household women were in these quintiles. A similar proportion of participant and non-project households were living on or close to a main road (62.9% of participants and 61.2% of non-participants). Only the educational attainment of the mothers in the project group was slightly higher than those in the non-project group. These data show that the mothers in the project households and those in the non-project households had sufficiently similar profiles.

As Table 13.2 shows, we found a high degree of exposure to EHFP project interventions in both the project and non-project households. We found that 94.6% of project households and 77.5% of non-project households reported participating in at least one intervention in the past one year, and one quarter of non-project households engaged in at least three activities in the same period.

Table 13.3 summarizes the activities that project and non-project households reported engagement with. We found that 57.1% of project households and 20.4% of non-project households reported having received nutrition counselling, breastfeeding guidance and complementary feeding guidance in the previous year. Sixty percent of project and one quarter (25.7%) of non-project households reported attending a project health and nutrition

Table 13.2. Project and non-project household enhanced home food production activity engagement.

	Participant	Non-participant
Exposed to at least 1 intervention	94.6% (192)	77.5% (127)
Exposed to just 1 intervention	13.7% (28)	18.4% (28)
Exposed to 2-4 interventions	39.5% (81)	44.1% (67)
Exposed at least 5 interventions	43.4% (89)	25.7% (39)

'festival day'. Almost half of both project (48.8%) and non-project (48.7%) households had attended an education session on gender equality and domestic violence. Almost three-quarters of project households (73.6%) reported having attended a cooking demonstration in the past year, and just over one quarter of non-project households (28.4%) reported that they had also attended a cooking demonstration.

Just under one-third (31.2%) of project households reported having received support from project trained agriculture extension workers, compared with 6.6% of non-project households. We found that a slightly higher percentage of non-project households (51.3%) compared with project households (45.4%) reported having listened to a project led community radio talk show, discussing nutrition and health issues in the previous year.

Table 13.3. Types of program activities that households report engagement with.

	Participant	Non-participant
Attended cooking demonstration	73.6% (145)	28.4% (40)
Received messages from HKI or commune health staff about hygiene	30.7% (63)	23.7% (36)
Received support from agricultural extension worker	31.2% (64)	6.6% (10)
Received nutrition/ feeding/ breastfeeding counselling from commune health staff	57.1% (117)	20.4% (31)
Attendance at training on finance and economic management	13.4% (26)	4.9% (7)
Attendance at sessions on gender equity and domestic violence	48.8% (100)	48.7% (74)
Listened to community radio program on nutrition	45.4% (93)	51.3% (78)
Received any support from agriculture extension workers	31.2% (64)	6.6% (10)
Attended health and nutrition festival day	60.0% (123)	25.7% (39)

Almost one third of project households (30.7%) and one quarter of non-project households (23.7%) reported having received environment and hygiene counselling.

Whilst it is perhaps not surprising that such a large number of non-project households had listened to the community radio program, it is noteworthy that half of them report having attended a session on gender equity and domestic violence, and over a quarter had attended a cooking demonstration – both activities that are specifically targeted at project households. Project households are directly invited to these activities by program staff and these activities are not widely advertised within villages, suggesting that social networks between project and non-project households are a strong influence on the behaviour of the latter.

Furthermore, 20.4% of non-project households reporting having received nutrition and/or breastfeeding counselling. Commune health staff may have delivered nutrition counselling either at their clinic or during household visits. This suggests that capacity building of commune health staff is having wider community benefits than just among project households. The relatively low percentage of non-project households (6.6%) that report support from agriculture extension workers suggests that these staff are primarily focusing on project households.

Table 13.4 summarizes the extent to which households reported engagement with village model farms, and motivations for this engagement, across both project and non-project households. We found that just over one third of project households (36%) and one-fifth (19.1%) of non-project households reported having visited a VMF in the past one year. Among project households that reported having visited a VMF in the past year, the most commonly given motivations were for advice on growing vegetables (91.4%), advice on garden bedding (84.2%), and support with animal husbandry (81.1%). Among non-project households, the most commonly cited reasons were advice on growing vegetables (82.4%), assistance with garden bedding (64.5%), and advice on animal health and veterinary practice (63.3%).

These data suggest that VMFs are acting as a wider community resource, and are not just a resource for households participating in the project. Interestingly, the rank order of motivations for visiting a VMF was the same across both project and non-project households, indicating that the wider community needs for support are broadly similar to those of project households.

We found that 65.9% of project households and 17.1% of non-project households reported receiving either seeds, chicks, pigs, or seedlings in the past year. Among project households that reported having received an input of some kind, the most commonly cited inputs received were seeds (93.6%) and seedlings (92.1%). Just over a quarter reported having received chickens, and 1.2% reported receiving pigs from a VMF. Among non-project households that reported having received an input of some kind from a VMF over the past

Table 13.4. Project and non-program household engagement with village model farms (VMFs).

	Participant	Non-participant
Visited VMF	36% (74)	19.10% (29)
Reason visited VMF		
Garden bedding assistance	84.2% (64)	64.5% (20)
Vegetable fence assistance	66.2% (47)	35.5% (11)
Animal husbandry	81.1% (60)	54.8% (17)
Animal health and veterinary practice	64.7% (44)	63.3% (19)
Bio-compost	57.1% (44)	34.4% (11)
Growing vegetables	91.4% (74)	82.4% (28)
Vegetable diversification	53.7% (36)	36.7% (11)
Education on fruits and vegetables rich in micronutrients	43.1% (28)	30% (9)
Other	0% (0)	11.1% (2)
Received an input from VMF	65.9% (135)	17.1% (26)
Input received		
Seeds	93.6% (131)	75% (24)
Chicks	26.6% (29)	10.7% (3)
Pigs	1.2% (3)	0% (0)
Seedlings	92.1% (128)	41.9% (14)
Other	2.6% (1)	0% (0)

year, the most commonly cited inputs received were seeds (75.0%) and seedlings (41.9%). About one tenth of non-project households reported having received chickens, and none reported receiving pigs.

These data indicate that, despite the project model being one where VMF inputs are primarily intended for project households, a large proportion of non-project households also benefit from the availability of seeds, seedlings and chickens at their local VMF. Given that the VMF is not widely advertised as a source of seeds and poultry, these findings are suggestive of a strong, informal social network that influences community behaviour and gives a larger number of households than reported in project monitoring access to program benefits.

13.4 Discussion and conclusions

Typically, EHFP evaluations focus on households that are direct recipients and beneficiaries of the intervention. However, the findings presented here suggest that coverage of interventions goes some way beyond these direct beneficiaries. Although we were not able to look at the consequences of this coverage in terms of behavioural and health outcomes, it does raise important questions about how EHFP programs are evaluated. We found that over three-quarters of households that were not direct project beneficiaries had been reached by and/or had participated in at least one program intervention, with almost half reached by two to four interventions, and a quarter exposed to at least five interventions. This is considerably higher than anticipated, and, for some activities, such as attending community sessions on gender equality and domestic violence, not greatly different from what we observed among participating households.

We found multiple mechanisms through which non-project households were reached. At one end of the spectrum, there is what can be considered as being a passive form of participation. This would include listening to community radio broadcasts on nutrition and benefiting from the increased capacity of services that households were perhaps already using before the project, such as nutrition counselling by commune health workers. At the other end of the spectrum, some households demonstrated more active forms of engagement, which included attending cooking demonstration sessions and workshops on gender equality, and visiting VMFs. All of these activities require some degree of motivation by household members to participate. We also found that motivations for visiting a VMF are almost identical between project and non-project households, suggesting that support and service needs are not so different within communities, regardless of whether or not a household is targeted to participate in the program.

A factor in explaining the high levels of active participation in non-project households may be the relatively small size of the villages in Son La where the EHFP program was implemented. Villages typically consist of 100 households, located in relatively close proximity to each other. Future implementations of the EHFP program in areas with larger and more geographically dispersed villages might expand to study whether exposure to program activities is higher among non-direct beneficiary households that are closer to VMFs or participating households. We did not take proximity as a variable into account in this study, but it would add to an understanding of the mechanisms through which EHFP interventions go beyond direct beneficiary households.

This study did not fully explore informal 'reach', whereby non-participating households are exposed to knowledge and practice via their social networks, of which direct beneficiary households may be a part. In this sense, there may be wider or deeper community reach than has been captured in the findings presented here. Future studies should seek to capture both the formal and informal mechanisms through which community level exposure takes place.

The findings also highlight the complexity of assessing or estimating the per-household program costs. Typically, per-household costs are determined by a simple calculation of the total project costs divided by the total number of direct beneficiaries. The findings presented in this paper suggest that such estimates should consider wider actual community reach, and that 'costs per household reached' are in fact lower if this is taken into account.

Finally, research plans for future EHFP programs should be designed in such a way that an assessment can be made of the extent to which participation in program activities by non-project households affects household behaviours. Such research should include research questions the answers to which can reveal how information about program goals and activities is disseminated in the community, across participating and non-participating households, and what role kinship and other social networks play in this.

Acknowledgement

We would like to thank the Vietnam National Institute of Nutrition (NIN) for their support in developing and delivering the survey, and the Neys Van Hoogstraten Foundation for providing the funding for this work.

References

Bushamuka V.N., De Pee, S., Talukder, A., Kiess, L., Panagides, D., Taher, A. and Bloem, M., 2005. Impact of a homestead gardening program on household food security and empowerment of women in Bangladesh. *Food and Nutrition Bulletin* 26(1):17-25.

Haselow, N.J., Stormer, A. and Pries, A., 2016. Evidence-based evolution of an integrated nutrition-focused agriculture approach to address the underlying determinants of stunting. *Maternal and Child Nutrition* 12(S1): 155-168.

Jones, K.M., Specio, S.E., Shrestha P., Brown, K.H. and Lindsay, H., 2005. Nutrition knowledge and practices, and consumption of vitamin A – rich plants by rural Nepali participants and nonparticipants in a kitchen-garden program. *Food and Nutrition Bulletin* 26(2):198-208.

Målqvist, M., Hoa, D.T.P., Liem, N.T., Thorson, A. and Thomsen, S., 2013. Ethnic minority health in Vietnam: a review exposing horizontal inequity. *Global Health Action* 6(1). http://doi.org/10.3402/gha.v6i0.19803.

Olney, D.K., Talukder, A., Iannotti, L.L., Ruel, M.T. and Quinn, V., 2009. Assessing impact and impact pathways of a homestead food production program on household and child nutrition in Cambodia. *Food and Nutrition Bulletin*. 30(4):355-369.

Olney, D.K., Vicheka, S., Kro, M., Chakriya, C., Kroeun, H., Sok Hoing, L., Talukder, A., Quinn, V., Lannotti, L., Becker, E. and Roopnaraine, T., 2013. Using program impact pathways to understand and improve program delivery, utilization, and potential for impact of Helen Keller International's Homestead Food Production Program in Cambodia. *Food and Nutrition Bulletin*. 34(2):169-184.

Olney, D.K., Ruel, M.T., Peedehombga, A. and Dillon, A., 2015. A 2-year integrated agriculture and nutrition and health behavior change communication program targeted to women in Burkina Faso reduces anemia, wasting and diarrhea in children 3-12.9 months of age at baseline: a cluster-randomized controlled trial. *The Journal of Nutrition* 145(6): 1317-1324.

Talukder A., Kiess, L., Huq, N., De Pee, S., Darnton-Hill, S.I. and Bloem, M.W., 2000. Increasing the production and consumption of vitamin A-rich fruits and vegetables: lessons learned in taking the Bangladesh homestead gardening programme to a national scale. *Food and Nutrition Bulletin* 21(2):165-172.

Talukder A., Haselow, A.K., Osei, E. Villate, E., Reario, D., Kroeun, H., Sok Hoing, L., Uddin, A., Dhunge, S. and Quinn, V., 2010. Homestead food production model contributes to improved household food security and nutrition status of young children and women in poor populations. *Field Action Science Reports [Online]*. Special Issue, Urban Agriculture. Institut Veolia. Aubervilliers, France. https://factsreports.revues.org/404.

UNICEF, 2014. *Multiple indicator cluster survey*. UNICEF, Vietnam. Available at: https://www.unicef.org/statistics/index_24302.html.

14. Conclusion: diversity and dimensions of food wellbeing

A. Niehof[1], H. Gartaula[2] and M. Quetulio-Navarra[3]*

*[1]Professor Emeritus, Wageningen University, Wageningen School of Social Sciences (WASS), the Netherlands; [2]International Program Coordinator, Manitoba Council for International Cooperation (MCIC), Winnipeg, Canada; [3]Presidential Commission for the Urban Poor, Quezon City, Metro Manila, the Philippines; *anke.niehof@wur.nl / niehofanke@gmail.com*

The kaleidoscope of food wellbeing

The studies presented in the book together form a kaleidoscope of images and aspects of food wellbeing. The research for the studies was conducted in different environments, varying from rural villages, areas where rice-farming systems prevail, upland or more remote locations, the habitat of indigenous peoples, and an urban area, in the countries of the Philippines, Indonesia, Nepal and Vietnam. The focus shifts from food security to food consumption of families, children and women in particular, and from food to nutrition. In the analyses, food sovereignty – one of the two dimensions of food wellbeing (Chapter 1; Gartaula *et al.*, 2018) – plays a role as well, albeit in some studies more than in others. Together, these studies testify to the diversity of ways in which people try to achieve food wellbeing. They may succeed in doing so in their own eyes and according to their own emic perceptions, or as measured by standard etic measurements, or both.

To answer the question of what we learned from the studies about food wellbeing, we shall first highlight key findings of the research presented in the book. The following section gives an overview of the studies, which follows the order of the chapters.

What do the studies tell us?

The study by Gartaula and Niehof (2018; Chapter 2) is about the effects of male out-migration on household food security and agricultural practice in Nepal. The findings indicate a positive impact of migration on household food security in the study area, thereby confirming the general picture in the literature that remittances from migration result in improved access to food among migrant households (cf. Zezza *et al.*, 2011). However, in the study area labour out-migration also generated cultural change and altered local food security dynamics. Lack of agricultural labour and considerations of convenience and social status have led to a preference for purchased food and agricultural work has acquired an image of working hard and gaining little. For the younger generation, land has become an asset rather than a resource for food production. If these trends persist, subsistence agriculture can no longer provide a safety net for the food provision of rural households. Since the *terai* area where out-migration is high also is the country's granary, the authors conclude that these developments may threaten Nepal's food sovereignty in the long run.

A. Niehof *et al.* (eds) **Diversity and change in food wellbeing – Cases from Southeast Asia and Nepal**
DOI 10.3920/978-90-8686-864-3_14, © Wageningen Academic Publishers 2018

Gayao *et al.* (2018; Chapter 3) investigated the knowledge on root and tuber crops and the role of these crops in the food system among indigenous groups in the Cordilleras, northern Philippines. They identified twenty kinds of roots and tubers, cultivated as well as wild species. The crops are used for family consumption, sold, bartered, used as animal feed and as gift. Knowledge on roots and tubers is learned from family, kin, and neighbours, but also through schools, church and local health and nutrition workers. The latter were the most mentioned single source of knowledge. Loss of knowledge and decline of planting and consuming roots and tubers have occurred among the generation born after the 1980s. The differences found in retention or loss of traditional knowledge by ethnic group, age group and gender, indicate that understanding the transfer of traditional knowledge on roots and tubers requires a disaggregated approach (cf. Zent, 2010). The authors conclude that building on the traditional knowledge of root and tuber crops improves food security, food diversity and wellbeing, creates livelihood opportunities, and strengthens environment-friendly farming practices.

Licyayo (2018; Chapter 4) conducted her research on wild edible mushrooms in the Cordilleras as well. She identified thirty species from 23 genera of wild edible fungi. Most of these were found in the Benguet and Mt. Province areas, where the coniferous forests are conducive to the growth of edible fungi. Especially the male members of the family collect the wild edible mushrooms in the forest, which they combine with gathering firewood and managing swidden farms. Women prepare the mushrooms for food and market the surplus. Large households were found to collect more mushrooms, because of their higher food demand and having more family members to engage in the activity. In the Cordillera provinces, wild edible mushrooms are primarily collected for family consumption. Additional income comes second. So far, the commercial demand for edible mushrooms is low and only little income derived from sales. However, there is a potential market, especially since local restaurants that cater for tourists increasingly put mushrooms on the card. Seasonality, lack of knowledge, and poor packaging and marketing are constraints in the utilisation of wild edible mushrooms.

Although also focusing on an indigenous group, i.e. the Tengger in East Java, Indonesia, the study of *Hastuti* (2018; Chapter 5) moves from food security to food care and child nutrition. When measured by dietary intake in relation to recommended levels, about two-thirds of the families were moderately to severely food insecure. However, most villagers consider themselves food secure. The discrepancy between objective and subjective assessment of food security is greatest for the mothers, whose nutrient intake also compares unfavourably with that of their children. This could be due to mothers prioritising their child(ren) and husband when serving food. The Tengger children's energy and nutrients adequacy levels were better than those of the children of the Baduy, another indigenous group in Indonesia (Khomsan *et al.*, 2009). Tengger mothers were also higher educated than Baduy mothers. The Tengger people mainly engage in farming (horticulture). Perhaps the fact that they live in the Mount Bromo area, a national park, explains why most households (79.2%) were not poor; tourism generates employment and a market for agricultural produce. Also in this regard, the situation is quite different from that of the rather isolated Baduy. For difficult times, such as during the eruptions of Mount Bromo in 2011, there is a community-level local social safety net in the form of rice collections and redistribution, from which poor households can benefit. The conclusion is that in terms of food wellbeing and children's nutritional status, the Tengger families are doing better than expected, even though there is scope for improvement.

Quetulio-Navarra *et al.* (2018; Chapter 6) conducted research in Bantarpanjang Translok, a resettlement site in Cilacap district, Central Java, Indonesia, which was built for people whose houses in the mountains had been destroyed by the 2001 landslides. The study investigated the role of institutions and social capital for the food security of 58 resettled households and the wellbeing of 95 children under five in these households. A cohort approach combined retrospective data on the first year of residency in the site and several years later with cross-sectional data on the situation in 2012. Surprisingly, household food security did not deteriorate in the first year after resettlement and several years after. One explanatory factor is that the relatively short distance to the area of origin enabled women to keep vegetable gardens there; another one is the support of the community leaders. However, most important is that the residents were able to build new social capital, composed of friends and neighbours. Continuing the cultural practice of *slametan* (communal meals) strengthened social capital building as well. Since household food security is a necessary but not sufficient condition for the wellbeing of under-fives, other factors than food are relevant as well. The findings show that the institutional context proved to be more favourable for under-fives after resettlement than before. The physical layout of the site allowed for easy supervision of children and health facilities and schools were closer to the home in the new site. The overall picture is more positive than could be expected based on previous findings about the negative effects of involuntary displacement on the social capital and wellbeing of resettled households (e.g. Cernea and McDowell, 2000).

The study by Purwestri *et al.* (2018; Chapter 7) also focussed on the wellbeing of children under five, in this case in the Central Javanese district of Demak. Demak ranks high in rice productivity, but also in chronic child malnutrition. Review studies (e.g. Masset *et al.*, 2012) have indeed found that among farmer families high agricultural productivity does not automatically translate into good nutritional status of all family members. Hence, this study investigated the relationship between food accessibility and utilisation of households and the prevalence of stunting among under-fives. The research was conducted in three sub-districts in Demak where the contrast was most pronounced, among 335 children aged 6-60 months from farmer families. The study did not find an association between household food availability and children's malnutrition. However, a monotonous rice-based diet combined with high consumption of widely available and affordable snacks seemed to contribute to stunting; non-stunted children had a lower snack intake and were provided with a significantly better home-based diet quality than stunted children. The study also shows that for the youngest children breastfeeding protects against stunting by narrowing the gap between actual and required daily nutrient intake.

Madjdian (2018; Chapter 8) conducted research on women's vulnerability to food and nutrition insecurity in the remote and poor district of Humla, Nepal. In South Asia, food scarcity and malnutrition are persistent problems, particularly among women and children (DeRose *et al.*, 2000; De Schutter, 2013). Intra-household food allocation that disfavours women combined with deficient household food availability may cause nutritional deprivation of women. The study focuses on the influence of household, kinship and culture in a situation of social change by assessing intra-household food allocation and women's food and nutrition (in)security in two ethnically different communities in the same region. In-depth interviews were done and BMI values were collected among 30 women of 15 Nyinba (Buddhist) and 15 Hindu (7 Chhetri and 8 Dalit) families. In both groups women's low position in the household hierarchy was reflected in the food serving order but this was more pronounced in the Hindu households, where food allocation clearly disfavoured women and adolescent girls in terms of quantity and quality. Hindu households, especially those of the Dalit, suffered more from food scarcity than the Buddhist households did. During pregnancy, immediate post-partum and lactation, the Nyinba women could eat more nutritious foods. Hindu women said they were denied those, in addition to being seen as unclean, which for Dalit women could entail confinement in animal sheds. While food scarcity and poverty play a major role in women's nutrition, the findings reveal the strong influence of ethnicity and family system on women's vulnerability to malnutrition. Social change seemed to benefit especially the Buddhist women. Education, income-earning opportunities, and the ongoing shift towards nuclear-family living arrangements gained them more power in their households. To the Hindu women, however, this did not seem to apply.

The research by Ekawidyani *et al.* (2018; Chapter 9) was conducted in the city of Bogor, Indonesia. The study assessed the impacts of food habits and physical activity on the nutritional status and physical fitness of 224 middle-class children in the 5th grade of primary school by comparing eating habits, physical activity and performance between normal and overweight children. Overweight in schoolchildren has been associated with low physical activity and consumption of fast foods. In the contemporary urban context, fast foods are easily available around the school and affordable with pocket money. Low physical activity relates to children watching television or playing on-line or video games rather than playing outdoors (cf. Kimbro *et al.*, 2011). This study found higher proportions of over-nutrition (overweight and obese) in the study group than at national level. Overweight proved to be strongly associated with low physical fitness, high amount of pocket money, the mother's education and low protein adequacy. Among these, fitness status was the strongest predictor of overweight. From a gender perspective, the finding that children of higher educated mothers were more at risk of overweight is rather alarming. These mothers are more often working, which means that they have less time for food care and can afford to give their children ample pocket money.

Talavera *et al.* (2018; Chapter 10) investigated climate change impacts on the nutritional status of young children from farming households in Infanta (Quezon) and Los Baños (Laguna) in the Philippines. There is growing evidence in the literature on linkages between climate change and nutritional status (e.g. Lloyd *et al.*, 2011). The fact that the country has a high prevalence of natural disasters that become more frequent and intense, adds relevance to the study. The methodological problem to prove causality between the longer-term effects of climate change and children's nutritional at the time of the survey was mitigated by selecting two study areas with a different climate type and using a mix of methods. Climate data on temperature and rainfall were obtained from the Philippine climate agency PAGASA, and children's nutritional and health status was assessed by survey interviews and monthly weight and height measurements for six months. Although malnutrition could not be directly attributed to climate change, the study found a significant association between rainfall, location and undernutrition. In Infanta, with a climate characterised by high rainfall, the odds of stunting were higher than in Los Baños. A similar result was found in a panel study among Ethiopian children (Hagos *et al.*, 2014). Climate change aggravates the malnutrition problem, particularly given the increasing incidence of typhoons and flooding. By being exposed to a local climate in an area with given geographic features, farming households are vulnerable to the effects of climate change. They need technical support to enable them to adapt, so that they can be self-reliant in meeting their food and nutrition needs.

The chapter by Dorado *et al.* (2018; Chapter 11) is the first of three chapters that present studies about the effects of food interventions, in this case the Backyard Vegetable Gardening Program (BVGP) of the Philippine Department of Agriculture. The study investigated whether the BVGP contributed positively to household food security and dietary diversity. The latter was measured in households with 8-10 year-old children. Issues of program implementation were addressed as well. Rather unexpectedly, households with and without a vegetable garden (152 in each group) equally had a positive attitude to and knowledge about the importance of vegetables consumption. Having a vegetable garden also did not make a difference for their dietary behaviour. Both groups had equally low dietary diversity scores. Although program implementation could be improved, the study reveals more fundamental issues. One is children's general distaste of (especially green leafy) vegetables, which mothers apparently could not do much about (see also Blijham *et al.*, 2007). The second is that backyard vegetable gardening is mostly left to women and that mothers often have no time for it, which would explain the contrast between positive attitude and poor practice. It also noteworthy that harvests from the garden are given to neighbours and friends or sold to augment household income rather than used for family consumption.

In the study by Reario *et al.* (2018; Chapter 12), the intervention is sustainable homestead food production (HFP), a model developed by Helen Keller International. The researchers investigated the results of household participation in the HFP program and whether mode of implementation of the program made a difference. For the research, two municipalities were selected in Lanao del Sur province in the Autonomous Region in Muslim Mindanao, a region burdened by armed conflict and high levels of poverty and food insecurity. In the first municipality a local government unit (LGU) managed the project, in the second a non-governmental organisation (NGO). To assess impacts, a baseline and – one year later – an endline survey were conducted. Regardless of mode of implementation, the intervention yielded positive results in terms of increased food diversity, income, nutrition knowledge and women's empowerment. The NGO households could benefit from a more experienced implementer of the HFP program than the LGU households, but the latter could benefit from the LGU's access to government manpower and financial support in the agricultural sector. The authors suggest that it could be the commitment of those directly involved in the implementation rather than mode of implementation that really matters.

The study by Shaw *et al.* (2018; Chapter 13) follows the implementation of the enhanced home food production program in Son La, Vietnam, which is targeted at rural food or nutrition insecure areas (cf. Haselow *et al.*, 2016). However, instead of just assessing project impacts on the project households, the study aimed at identifying possible coverage of the intervention beyond these households. This 'spillover effect' was investigated by comparing outcome variables for 205 project and 152 non-project households in 12 villages. Both groups had similar background characteristics. It was found that 94.6% of project households and 77.5% of non-project households participated in at least one project intervention, and one quarter of non-project households engaged in at least three activities. More non-project households than project households had listened to a project led radio talk show on nutrition and health issues. Half of the non-project households attended a session on gender equity and domestic violence, and over a quarter a cooking demonstration, both project activities. One-fifth of non-project households reported having visited a village model farm (36% of project households), mostly for advice on growing vegetables. The participation of non-projected households in project activities proved to be much higher than anticipated. The small size of the villages and close proximity of the houses in the project area could explain the spillover effect. Exposure to project information and activities through social networks presumably played a role as well, but this was not explored in the study.

Food wellbeing: dimensions and constraints

Food wellbeing is the lead concept of the book. There can be no overall wellbeing without food wellbeing and the degree of food wellbeing largely determines overall wellbeing. Through their diversity, the studies in the book document the different dimensions of food wellbeing and identify the enabling and constraining factors that play a role in the processes involved. Food wellbeing is very much a layered concept. It links insights on food security, food sovereignty and social wellbeing, in this way generating an understanding of the complex nature of food and security in a context of social and environmental changes and people's subjective perceptions about those changes (Gartaula *et al.*, 2017).

FAO (2013) describes food security as a condition whereby all people having access to sufficient, safe and nutritious food that meets their dietary needs and food preferences for an active and healthy life. Food security is seen as anchored at the household level, because households are resource-managing (family-based) groups that are geared towards meeting the primary needs of their members (Niehof, 2011). Nutrition security is a sub-set of household food security and relates to individual food consumption and utilisation. Even in food secure households, not all members may get the nutritious food they need, women and young children in particular. The food sovereignty approach highlights the role and right of people to define their own food and agricultural system to achieve self-sufficiency and sustainable food security. Though originally defined as the right to food self-reliance of nations, the approach was extended to all actors involved in the food chain, including family farms and consumers (Agarwal, 2014)

The Flip-Coin Model of Food Wellbeing (Figure 1.1, Gartaula *et al.*, 2018) pictures food wellbeing as a two-sided coin with food and nutrition security on the one side and food sovereignty on the other. The value of the coin is determined by the degree to which people accomplish food and nutrition security and food sovereignty. To achieve this, the actors involved need access to and control over the necessary resources. As shown in the model (Figure 1.1), we see food and nutrition security and food sovereignty as having three dimensions. First, there is the objective dimension that relates to etic measurements of food availability, access, distribution and consumption, nutritional status, and the relevant factors in the social and physical (including climatic) environment. Second is the subjective dimension of people's experiences, perceptions and own emic explanations. Third is the relational or social dimension, which has an impact on the former through social relations (influenced by gender, age, class, and ethnicity), social inequality and institutions. Table 14.1 applies the model to the review of the studies in the foregoing section.

The food sovereignty approach is a rights-based approach. There is a long history of defining and formalising the right to food at the international level, which – to a greater or lesser degree – has been incorporated in national agenda's. Van der Meulen (2010: 99) points at the emergence of 'a human right to acquire access to food that is sufficient in quality to satisfy one's dietary needs, that is sufficient in safety not to cause adverse effects and that is culturally acceptable.' He adds that, 'this right can be exercised by applying within the prevailing legal environment all available economic and legal means.' Food sovereignty also has the connotation of food self-reliance. Although substantially addressing the same subject, in food and nutrition security approaches the emphasis is on the resources and agency people need to exercise the right to food and the problems they experience in trying to do so.

Table 14.1. Constraining and enabling factors in food wellbeing according to dimension.[1]

Dimensions	Factors affecting the components of food wellbeing[2]		
	House food security	Nutritional status	Food sovereignty
Objective dimension	Remittances: + (C2). Roots and tubers (R&T): + (C3). Commercial value of mushrooms: -; potential + (C4). Many households not food secure, not clear why: - (C5). Social capital, cultural traditions: + (C6). Rice production, household income: + (C7). Poverty: - (C8). Climate change: - (C10). Homestead food production (HFP) model: + (C12). HFP benefits beyond project beneficiaries: + (C13).	More purchased foods: ± (C2). Dietary diversity from R&T: + (C3). Dietary diversity from mushrooms: + (C4). Diets and snacks consumption of children under five: - (C7). Ethnicity (Hindu) women's nutrition and role of family system: - (C8). Snacks consumption, low physical activity, higher educated/working mothers causing overweight of schoolchildren: - (C9). Climate change: - (C10). Vegetables in children's diets: - (C11). HFP income: ± (C12). HFP benefits beyond project households: + (C13).	Decline of agricultural production: - (C2) Loss of knowledge on R&T: - (C3). Threats to mushrooms habitat: - (C4). Value of horticultural produce because of tourism: + (C5). Food preferences of under-fives: - (C7). Women's lack of control over own food intake: - (C8). Quality of urban food environment: - (C9). Climate change: - (C10). HFP: + (C12). HFP benefits enhance food environment in the whole area: + (C13).
Subjective dimension	More income: + (C2) Households feel secure; community safety net: + (C5). Resettlement setting: + (C6). Fears about food crop damages: - (C10). Pride in having a vegetable garden: + (C11).	Lifestyle changes: ± (C2) Perceived nutritional value of R&T: - (C3). Awareness about food needs of under-fives: - (C7). Women aware of their food deprivation: - (C8). Urban middle-class lifestyles: - (C9).	Land as asset not for food: - (C2). Mushrooms are a treat for the family: + (C4). Access to vegetable gardens: + (C6). Impact of climate change variables on environment: - (C10). Having a vegetable garden: + (C11).

Table 14.1. Continued.

Dimensions	Factors affecting the components of food wellbeing[2]		
	House food security	Nutritional status	Food sovereignty
Relational imension	Women's empowerment: + (C2). Younger generation less involved in R&T: - (C3) Men collect, women prepare and sell mushrooms: ±; family size: + (C4). Solidarity among neighbours: + (C6). Gendered power inequalities: - (C8). Sharing vegetables with friends and neighbours: ± (C11). Social networks and proximity: + (C13).	Women prioritise husband and children, have lower nutrition indicators: - (C5). Role of institutions and community leaders for under-fives: + (C6). Culture/ethnicity: women and girls lowest in food serving order: - (C8). Food care time of working mothers: - (C9). Community transfers of nutrition knowledge: + (C13).	Feminisation of agriculture: ± (C2). Women's role as gardeners: + (C6). Women prepare food but lack control: - (C8). Women manage vegetable gardens: ± (C11). Impact of women's empowerment on the communities: + (C12). Community relations: + (C13).

[1] Referring to the studies reviewed in the foregoing section by using chapter numbers (Cx).
[2] Enabling factors: +; constraining factors: -; neutral or mixed: ±.

When scrutinising the evidence in Table 14.1, together the studies bring out all elements in the descriptions of food sovereignty and food and nutrition security. Gender is a factor that features in several guises. Gender inequality may deprive women of their right to nutritious food, which in the case of the Hindu women in Nepal was objectively observed and subjectively felt by the women themselves. Women's role as caregiver and their moral responsibility for feeding the family may cause mothers to neglect their own nutrition needs on the one hand, but enables them to exercise their agency by tending vegetable gardens and benefit from community food and nutrition interventions on the other. A noteworthy finding is that in the urban context women's higher education and gainful employment, positive from a women's empowerment perspective, lead to more children being overweight. This all attests to the importance of gender and underscores the relevance of Agarwal's (2014) critique on the conceptualisation of food sovereignty as being gender blind.

The evidence summarised in Table 14.1 reveals several paradoxes. To begin with, the Nepal study on impacts of outmigration shows that remittances make households more food secure but that cultural change and agricultural decline may jeopardize Nepal's food sovereignty in the end. The discrepancy between food and nutrition inadequacy as measured by objective standards and subjective feelings of wellbeing and food security among the Tengger families is another paradox. Also paradoxical is that children from food secure households in a rice surplus area can still be undernourished in terms of nutrients, because of snacks consumption and inadequate diets. Another unexpected finding is that involuntary displacement did not have the anticipated negative effect on the food security and children's wellbeing in the resettled households. The explanation is the positive role of institutions and rebuilt social capital in the new setting. Social relations and proximity of neighbours could also explain why in the Vietnamese villages the non-project households could benefit from project activities almost as much as the project households did. It could be that a possible bias towards formal structures in the food sovereignty approach would cause underexposure of the role of informal social networks.

A last point to draw attention to is that, as stated in descriptions of food sovereignty and food security, food should be is culturally acceptable and in accordance food preferences. Taking cultural acceptability and preferences into account is important, because people should not be forced by need to eat foods that are culturally unacceptable or profoundly disliked. An example of the latter would be the tulip bulbs people in the Netherlands ate for lack of an alternative during the hunger winter of 1944-45. However, in the studies there are examples of nutritious foods not eaten because these are found distasteful (children not eating green leafy vegetables) and of foods that are popular and culturally acceptable but lack nutritional value. In two studies, the deficient nutritional status of children is related to the latter, i.e. habitual snacks consumption. Especially, but not exclusively, in urban areas snacks increasingly dominate the easily accessible and affordable food market. This has a negative impact on nutrition security but does not violate sovereign consumer rights.

To conclude, the issues raised by the results of the studies demonstrated the relevance of hands-on research about food and nutrition among households and children in all kinds of social settings. Processes of social change and climate change add urgency to these issues. The challenge is to bridge the gap between practices at the level of households and local communities on the one hand and policies and formal structures on the other. Research such as reported in this book could inform policymakers and improve food and nutrition interventions.

References

Agarwal, B., 2014. Food sovereignty, food security and democratic choice: critical contradictions, difficult conciliations. *The Journal of Peasant Studies* 41(6): 1247-1268.

Blijham, N., De Kan L. and Niehof, A., 2007. *Determinants and adequacy of food consumption of children in La Trinidad, Philippines. International Journal of Consumer Studies, 31(3): 195-203.*

Cernea, M.M. and McDowell, C., 2000. *Risks and reconstruction: experiences of resettlers and refugees.* Washington, DC: World Bank.

Corazon M. Licyayo, D., 2018. Gathering practices and actual use of wild edible mushrooms among ethnic groups in the Cordilleras, Philippines. Chapter 4. In: Niehof, A., Gartaula, H.N. and Quetulio-Navarra, M. (eds) *Diversity and change in food wellbeing: cases from Southeast Asia and Nepal.* Wageningen Academic Publishers, Wageningen, the Netherlands, pp. 71-86.

De Schutter, O., 2013. *Gender equality and food security. women's empowerment as a tool against hunger.* Mandaluyong City, Philippines: Asian Development Bank.

DeRose, L.F., Das, M. and Millman, S.R., 2000. Does female disadvantage mean lower access to food? *Population and Development Review,* 26(3): 517-547.

Dorado, J.B., Viajar, R.V., Azaña, G.P., Caraig G.S. and Capanzana, M.V., 2018. Does backyard vegetable gardening enhance perceived household food security and dietary diversity? A case study in the Philippines. Chapter 11. In: Niehof, A., Gartaula, H.N. and Quetulio-Navarra, M. (eds) *Diversity and change in food wellbeing: cases from Southeast Asia and Nepal.* Wageningen Academic Publishers, Wageningen, the Netherlands, pp. 225-242.

Ekawidyani, K.R., Karimah, I., Setiawan, B. and Khomsan, A., 2018. Parents' characteristics, food habits and physical activity of overweight schoolchildren in Bogor City, Indonesia. Chapter 9. In: Niehof, A., Gartaula, H.N. and Quetulio-Navarra, M. (eds) *Diversity and change in food wellbeing: cases from Southeast Asia and Nepal.* Wageningen Academic Publishers, Wageningen, the Netherlands, pp. 177-194.

FAO, 2013. *The state of food security in the world.* Rome: Food and Agriculture Organization.

Gartaula, H.N. and Niehof, A., 2018. Changing value of food and perception of food security in the context of male out-migration in Nepal. Chapter 2. In: Niehof, A., Gartaula, H.N. and Quetulio-Navarra, M. (eds) *Diversity and change in food wellbeing: cases from Southeast Asia and Nepal.* Wageningen Academic Publishers, Wageningen, the Netherlands, pp. 27-42.

Gartaula, H., Patel, K., Johnson, D., Devkota, R., Khadka, K. and Chaudhary, P. 2017. From food security to food wellbeing: examining food security through the lens of food wellbeing in Nepal's rapidly changing agrarian landscape. *Agriculture and Human Values,* 34(3): 573-589.

Gartaula, H.N., Niehof, A. and Quetulio Navarra, M., 2018. Introduction. Chapter 1. In: Niehof, A., Gartaula, H.N. and Quetulio-Navarra, M. (eds) *Diversity and change in food wellbeing: cases from Southeast Asia and Nepal.* Wageningen Academic Publishers, Wageningen, the Netherlands, pp. 15-26.

Gayao, B.T., Meldoz, D.T. and Backian, G.S., 2018. Indigenous knowledge and household food security: the role of root and tuber crops among indigenous peoples in the Northern Philippines. Chapter 3. In: Niehof, A., Gartaula, H.N. and Quetulio-Navarra, M. (eds) *Diversity and change in food wellbeing: cases from Southeast Asia and Nepal.* Wageningen Academic Publishers, Wageningen, the Netherlands, pp. 43-70.

Hagos, S., Lunde, T., Mariam, D.H., Woldehanna, T. and Lindtjør, B., 2014. Climate change, crop production and child under nutrition in Ethiopia; a longitudinal panel study. *BMC Public Health*, 14:884.

Haselow, N., Stormer, A. and Pries, A., 2016. Evidence-based evolution of an integrated nutrition-focused agriculture approach to address the underlying determinants of stunting. *Maternal and Child Nutrition,* 12(S1): 155-168.

Hastuti, D., 2018. Household food security, food care, child nutrition, and wellbeing of ethnic Tengger families in East Java, Indonesia. Chapter 5. In: Niehof, A., Gartaula, H.N. and Quetulio-Navarra, M. (eds) *Diversity and change in food wellbeing: cases from Southeast Asia and Nepal.* Wageningen Academic Publishers, Wageningen, the Netherlands, pp. 87-113.

Khomsan, A., Anwar, F., Sukandar, D., Riyadi, H., Mudjajanto, E.S. and Wigna, W., 2009. Socio-cultural aspects of nutrition and food systems of the Baduy. Bogor: Faculty of Human Ecology, Bogor Agricultural University. [Unpublished report]

Kimbro, R.T., Brooks-Gunn, J. and McLanahan, S., 2011. Young children in urban areas: links among neighborhood characteristics, weight status, outdoor play, and television-watching. *Social Science & Medicine*, 72(5): 668-676.

Lloyd, S.J., Kovats, R.S. and Chalabi, Z, 2011. Climate change, crop yields, and under-nutrition: development of a model to quantify the impact of climate scenarios on child under-nutrition. *Environmental Health Perspectives* 119(12): 1817-1823.

Madjdian, D.S., 2018. Gender, intra-household food allocation and social change in two Himalayan communities in Nepal. Chapter 8. In: Niehof, A., Gartaula, H.N. and Quetulio-Navarra, M. (eds) *Diversity and change in food wellbeing: cases from Southeast Asia and Nepal.* Wageningen Academic Publishers, Wageningen, the Netherlands, pp. 153-175.

Masset, E., Haddad, L., Cornelius, A. and Isaza-Castro, J., 2012. Effectiveness of agricultural interventions that aim to improve nutritional status of children: Systematic review. *British Medical Journal* 344: d8222. https://doi.org/10.1136/bmj.d8222.

Niehof, A., 2011. Conceptualizing the household as an object of study. *International Journal of Consumer Studies,* 35(5): 488-497.

Purwestri, R.C., Barati, Z., Wirawan, N.N., Fahmi, I., Lauvai J. and Scherbaum, V., 2018. What explains stunting among children living in a rice surplus area in Central Java, Indonesia? Chapter 7. In: Niehof, A., Gartaula, H.N. and Quetulio-Navarra, M. (eds) *Diversity and change in food wellbeing: cases from Southeast Asia and Nepal.* Wageningen Academic Publishers, Wageningen, the Netherlands, pp. 173-152.

Quetulio-Navarra, M., Frunt, E. and Niehof, A., 2018. The role of social capital and institutions in food security and wellbeing of children under five for resettled households in Central Java, Indonesia, Indonesia. Chapter 6. In: Niehof, A., Gartaula, H.N. and Quetulio-Navarra, M. (eds) *Diversity and change in food wellbeing: cases from Southeast Asia and Nepal.* Wageningen Academic Publishers, Wageningen, the Netherlands, pp. 115-136.

Reario, M.F.D., Pacheco, P.J., Lafuente, E. and Talukder, A., 2018. Developing homestead food production in Muslim Mindanao, Philippines: Does mode of implementation matter? Chapter 12. In: Niehof, A., Gartaula, H.N. and Quetulio-Navarra, M. (eds) *Diversity and change in food wellbeing: cases from Southeast Asia and Nepal.* Wageningen Academic Publishers, Wageningen, the Netherlands, pp. 243-266.

Shaw, K., Ngoc, P.K., Lan, H., Mundy, G. and Talukder, Z., 2018. How non-targeted households benefited from implementing a homestead food production program in Son La, Vietnam. Chapter 13. In: Niehof, A., Gartaula, H.N. and Quetulio-Navarra, M. (eds) *Diversity and change in food wellbeing: cases from Southeast Asia and Nepal.* Wageningen Academic Publishers, Wageningen, the Netherlands, pp. 267-278.

Talavera, M.T.M., Pangan, M.R.L., Colladilla, J.O., Sarol, J.N., Agdeppa, I.A. and Lorenzo, F.M.E., 2018. Climate change and nutritional status of preschool children living in Infanta, Quezon, and Los Baños, Laguna, Philippines. Chapter 10. In: Niehof, A., Gartaula, H.N. and Quetulio-Navarra, M. (eds) *Diversity and change in food wellbeing: cases from Southeast Asia and Nepal.* Wageningen Academic Publishers, Wageningen, the Netherlands, pp. 195-224.

Van der Meulen, B., 2010. The freedom to feed oneself. In: Hospes, O. and Hadiprayitno, I. (eds) *Governing food security: law, politics and the right to food.* Wageningen: Wageningen Academic Publishers.

Zent, S. 2010. *Resilience and vitality of traditional econological knowledge.* Proceedings of the 1st National Conference on Biodiversity, Gender and Indigenous Knowledge. October 6-8, 2010. Agricultural Training Institute, Benguet State University, La Trinidad, Benguet, Philippines.

Zezza, A., Carletto, C., Davis, B. and Winters, P., 2011. Assessing the impact of migration on food and nutrition security. *Food Policy* 36(1): 1-6.

About the authors

I.A. (Imelda) Agdeppa did her MSc and PhD in nutrition at the Community Nutrition Study Program, a joint program of the SEAMEO Regional Center for Food and the Universitas Indonesia in Jakarta. She is senior researcher at the Philippine Food and Nutrition Research Institute, Department of Science and Technology (FNRI-DOST). Email: iangelesagdeppa@yahoo.com.ph

G.P. (Glenda) Azaña is a nutritionist-dietitian from the University of the Philippines at Los Baños. As a senior researcher at the Food and Nutrition Research Institute of the Department of Science and Technology (FNRI-DOST), she is involved in efficacy and effectiveness studies and in the advocacy of nutrition interventions. Email: glendaazana@yahoo.com

G. S. (Grace) Backian obtained her MSc in plant pathology at Benguet State University. She is a senior researcher at the Northern Philippines Root Crops Research and Training Center, Benguet State University. Currently she is the Yam Program Leader. Research interests include crop management and seed production. Email: gracebsu_nprcrtc@yahoo.com

Z. (Ziba) Barati, MSc in environmental protection and agricultural food production at the University of Hohenheim, Germany, is a PhD student at the Institute for Agricultural Engineering in the Tropics and Subtropics, University of Hohenheim. Email: Barati@uni-hohenheim.de.

M.V. (Mario) Capanzana is currently the director of the Food and Nutrition Research Institute, Department of Science and Technology (FNRI-DOST) in the Philippines. He works in research and technology development in food fortification, functional foods, food quality and safety and food product development. He has several patents to his name. Email: mar_v_c@yahoo.com

G.S. (Georgina) Caraig is a senior researcher at the Food and Nutrition Research Institute, Department of Science and Technology, University of the Philippines Los Baños. Her research is about testing nutrition intervention models for households and communities and assessment of food-related policies. Email: gcaraig@yahoo.com

J.O. (Jesusita) Colladilla obtained her MSc and PhD in agriculture from the University of the Philippines at Los Baños. Currently, she is a faculty member of the College of Agriculture, Southern Luzon State University. Email: jocoladilla@gmail.com

J.B. (Julieta) Dorado is a supervising researcher at the Food and Nutrition Research Institute, Department of Science and Technology (FNRI-DOST) in the Philippines, having a background in sociology, social development and communication. Her work focuses on food and nutrition intervention assessment and policies. Email: juliebdorado@yahoo.com

Dwi Hastuti obtained her MSc from Bogor Agricultural University. She is currently working at the Division of Child Development, Department of Family and Consumer Sciences, College of Human Ecology, Bogor Agricultural University, Indonesia. E-mail: tutimartianto@yahoo.com

K.R. (Karina) Ekawidyani, a medical doctor, did her MSC in nutrition at the Community Nutrition Study Program, a joint program of the SEAMEO Regional Center for Food and the Universitas Indonesia in Jakarta. She is a lecturer at the Department of Community Nutrition, Bogor Agricultural University, Indonesia. Her research interests include mother-and-child health and nutrition, and community nutrition. Email: karina_rahmadia@yahoo.co.id

I. (Ilmia) Fahmi obtained her MSc in nutrition at the Community Nutrition Study Program, a joint program of the SEAMEO Regional Center for Food and the Universitas Indonesia in Jakarta. She currently works at the Nutrition Department, Faculty of Medicine, Universitas Brawijaya, Malang. Email: ilmia.fahmi@gmail.com

E. (Eden) Frunt did his MSc in health and society from Wageningen University in the Netherlands. His master's research was about the wellbeing of under-fives in a context of involuntary resettlement in Cimanggu, Indonesia. Since then, he has worked in the fields of occupational health, information security and quality management. He currently works at Bouw & Van de Meerendonk, a Dutch consultancy agency for certification in the areas of health and environment. Email: frunteden@gmail.com

H.N. (Hom) Gartaula has a PhD in social sciences from Wageningen University. Currently, he is international program coordinator at Manitoba Council for International Cooperation and he teaches at Menno Simons College, Canadian Mennonite University, Winnipeg. His work focuses on household food and livelihood security, climate change and human wellbeing, and agricultural and rural development. Email: hom.gartaula@gmail.com

B.T. (Betty) Gayao is a senior researcher at the Northern Philippines Root Crops Research and Training Center, Benguet State University, where she heads the Social Science and Training Division. She holds an MSc in rural development. Research interests include indigenous knowledge and economic and social aspects of root crops production. Email: btgayao@yahoo.com

L.N. (Lan) Hoang is a health education and nutrition specialist who graduated from Flinders University, Australia, and the University of South Carolina, USA. Her work focuses on infant and young child feeding, health education, ethnic minorities partnership, food security, nutrition counselling, and nutrition policy advocacy in Vietnam. Email: hoangngoclan@yahoo.com

I. (Ima) Karimah obtained her MSc in community nutrition at Bogor Agricultural University, Indonesia. She currently works as a lecturer in the Nutrition Department of the Health Polytechnic Tasikmalaya, Indonesia. Email: ima.chic55@gmail.com

A. (Ali) Khomsan obtained his PhD in family and consumer sciences education from Iowa State University, USA. He is a full professor at the Department of Community Nutrition, Faculty of Human Ecology, Bogor Agricultural University, Indonesia. His research focuses on community nutrition. Email: erlangga259@yahoo.com

E. (Eleanore) Lafuente is a nutritionist-dietitian with postgraduate diploma on Food Science and Nutrition from the International Agricultural Centre, Wageningen, the Netherlands. She has more than 30 years of working experience in community nutrition programs in developing countries. Email: elafuente@hki.org

J. (Judith) Lauvai is a post-doctoral researcher at the Institute of Biological Chemistry and Nutritional Science at the University of Hohenheim, Germany. Her work focuses on influence of dietary components on adipogenesis, dietary intake of pregnant and breastfeeding women and young child feeding. Email: j.lauvai@uni-hohenheim.de

D.C.M. (Dinah) Licyayo is a professor at Ifugao State University, Philippines. Her research interests are in the field of agriculture and mycology, indigenous food enterprise development, and household food security. She is the current Dean of the College of Agriculture and Home Science. Email: dmlicyayo@yahoo.com

F.M.E. (Fely) Lorenzo, PhD in public health from the University of California at Berkeley, is a registered nurse, and is currently is professor of health policy and management at the University of the Philippines Manila. Research and policy interests are care pathways, benefit packages and valuation of climate change adaptation and health. Email: felorenzo@up.edu.ph

D. (Donya) Madjdian is a PhD student at Wageningen University, the Netherlands, studying the interplay of social, economic and nutritional trajectories of adolescent girls in Nepal. In her research, she uses interdisciplinary and participatory approaches. Email: donya.madjdian@wur.nl

D.T. (Dalen) Meldoz holds a MSc in rural development and is senior researcher at the Northern Philippines Root Crops Research and Training Center, Benguet State University, where she is the Program Leader of Minor Root Crops RDE. Her research interests include social science and extension aspects of root and tuber crops production. Email: d.meldoz@yahoo.com

G. (Gary) Mundy is regional advisor for monitoring and evaluation at Helen Keller International, Asia and Pacific Region. He has worked in the field of public health for over 15 years, across Asia and Africa, to support programs with high quality monitoring of implementation and impact evaluation. Email: gmundy@hki.org

P.K. (Pham) Ngoc holds an MBA and is the country director of Helen Keller International Vietnam. She and has been working in the public health sector, especially in nutrition and eye health for over 14 years. Her main interest is in project management, providing guidance and direction in the areas of assessment, planning and general implementation. Email: pngoc@hki.org

A. (Anke) Niehof, PhD in anthropology from Leiden University, is professor emeritus of Sociology of Consumption and Households at Wageningen University, the Netherlands. She publishes on household food and livelihood security, informal care, gender issues, and social impacts of HIV and AIDS in sub-Saharan Africa. Email: anke.niehof@wur.nl or niehofanke@gmail.com

P.J. (Pamela) Pacheco holds a BA in development studies. She has 10 years of work experience in public health nutrition and research and is currently working as a monitoring and evaluation officer at Helen Keller International Philippines. Email: pjpacheco@hki.org

M.R.L. (Maria) Pangan, MSc and PhD in nutrition from the University of the Philippines Diliman, is a retired professor from the Department of Nutrition of the University of the Philippines Manila. She heads a consultancy agency on health and nutrition. Email: mrlpangan@gmail.com

R.C. (Ratna) Purwestri is a nutrition specialist with a special interest in women and children and food security and nutrition issues in less privileged communities in Indonesia and Africa. She works as a postdoctoral researcher at the Institute of Biological Chemistry and Nutrition, University of Hohenheim, Germany. Email: rc.purwestri@uni-hohenheim.de

M. (Melissa) Quetulio-Navarra, MA in international studies from Kobe University, Japan, and a PhD from Wageningen University, the Netherlands. She heads the Relocation and Resettlement Monitoring and Development Division of the Presidential Commission for the Urban Poor, in the Philippines. She publishes on involuntary resettlement in Southeast Asia, social capital, and use of the calendar method and third-party help in surveys. Email: melisnavarra@gmail.com

M.F.D. (Dolly) Reario, a nutritionist-dietitian with postgraduate degree in public health, is the Senior Programs Manager of Helen Keller International Philippines. She has more than 20 years of experience in nutrition programs management and in monitoring and evaluating maternal and child health and nutrition interventions. Email: dreario@hki.org

J.N. (Jesus) Sarol obtained a PhD in epidemiology from the University of California in Los Angeles, USA. He is now retired as a full professor after 22 years of teaching at the Department of Epidemiology and Biostatistics, College of Public Health, University of the Philippines Manila. Email: jsaroljr@gmail.com

V. (Veronika) Scherbaum, MSc in mother and child health from the University of London and PhD from the University of Hohenheim, Germany, is a senior researcher at the Institute of Biological Chemistry and Nutrition, and at the Food Security Center of the University of Hohenheim. Research topics include strategies to prevent/treat iron deficiency anemia using locally grown foods and the development of improved recipes for prevention of maternal and child malnutrition. Email: veronika.scherbaum@uni-hohenheim.de

B. (Budi) Setiawan obtained his PhD in human nutrition from the University of Nebraska-Lincoln, USA. As a lecturer at the Department of Community Nutrition of Bogor Agricultural University, Indonesia, he is responsible for teaching, research, and community services in nutrition sciences and gender and child studies. Email: bsetiawan.ipb@gmail.com

K. (Kaitlin) Shaw is a clinical research data manager at the Department of Surgery, Pancreas Center, Columbia University, New York City. Her research focuses on nutrition, genetics, and pancreatic cancer. She worked with Helen Keller International on the efficacy of an Enhanced Homestead Food Production project in Northern Vietnam. Email: kes2196@columbia.edu

M.T. (Theresa) Talavera, postgraduate diploma in public health from Otago University, New Zealand and MSc and PhD in public health at the University of the Philippines Manila, is associate professor at the Institute of Human Nutrition and Food, University of the Philippines Los Baños. Research interests include infant and young child feeding, nutrition labelling, and climate change and nutritional status. Email: mmtalavera@up.edu.ph

A. (Zaman) Talukder is the country director of Helen Keller International (HKI) in Bangladesh and Regional Food Security Advisor for the HKI Asia-Pacific Region. He has been working with HKI for more than 25 years in the field of food security and nutrition, and has been responsible for the development and implementation of the HKI homestead food production program in Asian and African countries. Email: ztalukder@hki.org

R.V. (Rowena) Viajar is a researcher at FNRI-DOST in the Philippines, where she works on nutritional assessment and nutrition interventions, monitoring and evaluation, and policy issues. Email: wenavelasco@yahoo.com

N. (Nia) Wirawan obtained her MSC in nutrition at the Community Nutrition Study Program, a joint program of the SEAMEO Regional Center for Food and the Universitas Indonesia, Jakarta, and is now a PhD student at the same university. She is also a faculty member at the Faculty of Medicine, Universitas Brawijaya, Malang, Indonesia. Email: nia_wirawan.fk@ub.ac.id

Neys - van Hoogstraten Foundation

More than two decades ago, the late Dr Karel Neijs bequeathed his estate to establish a foundation to support research on household food security and nutrition in Indonesia and other Asian countries, especially in Southeast Asia. The Neys-Van Hoogstraten Foundation (NHF) has been funding over 200 research projects in the region since it became operational in 2004. The following may be helpful for researchers considering applying for a research grant.

Who can apply?

Persons affiliated with research institutes, universities (including PhD students), or NGOs in the region. NHF also has a fellowship programme to support the research of master students. PhDs and master students can be affiliated with a university outside the region.

Topic range

Eligible fields of research include food production and food security at household level, fragile food environments, food and social and economic change, food con¬sumption and nutrition of vulnerable groups, and (new focus since 2016) urban poverty and nutrition, upscaling of traditional food crops. The background of the researcher may vary from social to agronomic or nutrition sciences. Research proposals from a multidisciplinary team are welcome. The website (see address below) provides more detailed information.

Where?

Southeast Asia, Nepal, Bangladesh.

What can be funded?

Costs pertaining to research data collection and analysis and dissemination of results, up to a maximum of € 25,000 (for the fellowship programme € 5,000). NHF does not fund salaries of applicants but can compensate the costs of hiring persons needed for the research (e.g. enumerators, research assistant).

Contact details and website address

External Secretariat Neys-Van Hoogstraten Foundation
c/o FEMCONSULT, Consultants on Gender and Development
Nassaulaan 5, 2514 JS The Hague
The Netherlands
Tel: +31 (0)70 3655744
E-maill: info@neys-vanhoogstraten.nl
Website: www.neys-vanhoogstraten.nl

www.ingramcontent.com/pod-product-compliance
Lightning Source LLC
Chambersburg PA
CBHW061202240926
51846CB00006B/1090
9789086863167